SPRINGER
LAB MANUAL

Demetrios Kyriacou

Modern Electroorganic Chemistry

With 16 Figures

Springer-Verlag
Berlin Heidelberg New York London Paris
Tokyo Hong Kong Barcelona Budapest

Professor DEMETRIOS KYRIACOU

University of Massachusetts
Department of Chemistry
One University Avenue
Lowell, MA 01854, USA

ISBN-13:978-3-642-78679-2   e-ISBN-13:978-3-642-78677-8
DOI: 10.1007/978-3-642-78677-8

Library of Congress Cataloging-in-Publication Data. Kyriacou, Demetrios K. Modern electroorganic chemistry/Demetrios Kyriacou. p. cm. – (Springer laboratory) Includes bibliographical references. ISBN-1-3:978-3-642-78679-2 1.Organic electrochemistry. I. Title. II. Series. QD273.K98 1994 547.1′37 – dc20 94-13316

Typesetting: Best-set Typesetter Ltd., Hong Kong
51/3130-5 4 3 2 1 – Printed on acid-free paper

Preface

The last two decades (1970–1990) have probably seen more reviews and books about electroorganic chemistry than any other such short period in the history of this special branch of organic chemistry. The continuous progress in organic electrochemistry certainly requires frequent exposition and systematization of the new knowledge in the form of readily accessible up-to-date textbooks that could serve as guides and references for the expert and the neophyte alike.

In preparing this book as a retired industrial chemist, I had primarily in mind the widest possible community of industrial researchers and practioners in the chemical and closely related biochemical, pharmaceutical and agrochemical fields. Organic electrochemical methodology can contribute very significantly to these fields, technologically and economically, and especially as regards the efforts toward a clean manufacturing technology. I would also hope that advanced students in the aforementioned fields would benefit from using the book in their studies as a textbook and a reference.

The book consists of five chapters. The first chapter is a synoptic introduction to organic electrochemistry. The second and third chapters deal separately with anodic and cathodic reactions of synthetic interest. The reactions are conveniently classified on the basis of functional groups in the organic molecules. Selected representative preparative examples are described in detail for illustrative purposes. The fourth and fifth chapters are concerned with indirect electrochemical reactions and some interesting special topics.

I would like to express my thanks to the Department of Chemistry of the University of Massachusetts Lowell for providing the necessary facilities for this work. Also, I thank Eleni, my wife, for her devoted efforts in the preparation of the manuscript. I thank professor E.G.E. Jahngen for valuable discussions.

Lowell, Massachusetts
May 1994

DEMETRIOS KYRIACOU

Contents

1	**Introduction**	1
1.1	Electroorganic Chemistry. Historical Perspective	1
1.2	Pertinent Fundamental Concepts and Principles in Organic Electrochemistry	2
1.2.1	The Overall Cell Reaction	2
1.2.2	The Generalized Electroorganic Reaction	3
1.2.2.1	Electrochemical Couplings and Cross Couplings	6
1.2.2.2	The Electron Transfer Step. Basic Qualitative Description	6
1.2.3	Thermodynamic and Kinetic Aspects	7
1.2.3.1	Nernstian Equation. Electrode Potential	7
1.2.3.1a	Free Energy of Cell Reaction and Cell Potential	9
1.2.3.2	Kinetic Aspects	10
1.2.4	Preparative Electrolysis under Activation (Kinetic) or Mass-Transfer Controlled Conditions	12
1.2.5	The Electrical Double Layer and its Significance in Electrosynthesis	14
References		17
2	**Anodic Reactions**	18
2.1	Fundamental Nature	18
2.2	Hydrocarbons	19
2.2.1	Saturated Hydrocarbons	19
2.2.2	Unsaturated Hydrocarbons	22
2.2.3	Functionalizations of Hydrocarbons	25
2.3	Hydroxy Compounds	28
2.4	Oxidations at Oxide Covered Electrodes	30
2.5	Phenolic Compounds	36
2.6	Carbonyl Compounds	43
2.7	Carboxylic Acids	47
2.7.1	General Perspective	47
2.7.2	Oxidations in Fluorosulfuric Acid	52
2.7.3	Kolbe Synthesis in SPE Cells	52
2.7.4	Photokolbe Synthesis	54
2.8	Amines and Amides	54

2.8.1   Amines, in General   54
2.8.2   Amides   57
2.9   Ethers and Esters   60
2.10   Organic Halides   63
2.10.1 Alkyl Halides   63
2.10.2 Aromatic Halides   64
2.11   N-Heterocyclics   65
2.12   Organosulfur Compounds   74
2.12.1 Sulfides   74
2.12.2 Thiolates and Dithianes   75
2.12.3 Sulfonium Salts   77
2.13   Anodic Cyanations   79
2.14   Acetoxylations and Acetamidations   83
2.14.1 Acetoxylations   83
2.14.2 Acetamidations – Electrochemical Ritter-Type
        Reactions   86
2.15   Anodic Halogenations   88
2.16   Hydroxylations   91
2.17   Alkoxylations   92
2.18   Pyridinations   100
2.19   Nitrations   101
References   102

3   Cathodic Organic Reactions   110

3.1   Phenomenological Nature   110
3.2   Carbon-Hydrogen Bond Formations.
        Unsaturated Systems   111
3.3   Carbonyl Compounds   113
3.3.1   General Reduction Mechanism   113
3.3.2   Reactions via the Carbonyl Electrophore   114
3.4   Carboxylic Acids and Esters   119
3.4.1   Preparation of Metal Carboxylates
        and Metal Alkoxides   120
3.4.2   Reduction of Esters   121
3.5   Organohalides   122
3.5.1   General Electroreduction Mechanism   122
3.5.2   Products from Aromatic and Aliphatic Halides   122
3.5.3   Various Reductive Dehalogenations   127
3.5.4   Electrohydrogenolysis of Polyhalogenated
        Compounds   128
3.5.5   Electrochemically Induced Nucleophilic $S_{RN}1$
        Substitutions   131
3.5.6   Organic Fluorides   133

3.6   Nitrocompounds   133
3.6.1   *N*-Heterocyclizations   134
3.6.2   Reduction on Transition Metal Electrodes   138
3.7   *N*-Heterocyclics and Other Heteroatom Structures   138
3.7.1   Pyridines, Purines, and Similar Compounds   138
3.7.2   Pyridinium and Similar Salts   141
3.7.3   *N*-Heterocyclics with Carbonyl Groups   142
3.8   Organosulfur Compounds   145
3.9   Some Special Cleavages of C—O Bonds
          in Hydroxy Compounds   148
3.9.1   Cleavage of Ethers   150
3.10   Electrocarboxylations   150
3.10.1   Electroreduction of Carbon Dioxide   150
3.10.2 CO_2 as Electrophile   151
3.10.3   Photoelectrochemical Reduction of Carbon Dioxide   153
3.11   Nitriles and Oximes   154
3.12   Various Cathodic Couplings......................   155
3.12.1   Acylations   155
3.12.2   Organophosphorus Compounds   156
3.12.3   Arene-Metal Complexes   157
3.12.4   Cathodic Reactions (Couplings) of Organometallics   158
References   160

4   **Indirect Electroorganic Reactions (Indirect Electrolysis)** ...   166

4.1   Introduction   166
4.1.1   Basic Mechanisms   166
4.2   Oxidative Indirect Reactions   169
4.2.1   Metallic Ions and Oxyanions Mediators   169
4.2.2   Paired Oxidative Processes.......................   172
4.2.3   Halide Ions as Mediators   174
4.2.4   Nitrate Ion as Mediator   178
4.2.5   Organic Mediators for Oxidations...................   178
4.3   Indirect Reductive Reactions   182
4.3.1   Metallic Ions a Catalysts   182
4.3.2   Reductions with Amalgams   185
4.3.3   Reductions with Diborane Formed in Situ..............   186
4.3.4   Organic Mediators   187
References   188

5   **Some Special Topics Relevant to Electrosynthesis**   192

5.1   Electrolysis in Two-Phase Solvent Systems   192
5.2   Solid Polymer Electrolyte – Electrolysis...............   194
5.3   Reactions via Electrode Dissolution   196

5.4   Electropolymerizations. Conductive Polymers 198
5.4.1   General View 198
5.4.2   Conductive Polymers................................. 201
5.5   Applications of Electrochemical Methodology
        to the Synthesis of Natural Products 203
5.6   Electroreductive Cyclizations for Natural Product
        Synthesis (Little-Baizer ERC Reaction)................ 206
5.7   Electrogenerative Systems 208
5.8   Electrogenerated Bases.............................. 212
5.9   Electroreduction of Enones (α,β-Unsaturated Ketones) ... 214
5.10   Oxygenations. The Superoxide Ion.................... 215
References 219

Subject Index 225

1 Introduction

1.1 Electroorganic Chemistry. Historical Perspective

Electroorganic chemistry is a multidisciplinary science overlapping the vast fields of organic chemistry, biochemistry, physical chemistry and electrochemistry. As a branch of organic chemistry its principal practical aim is to enhance synthetic organic methodology in all its aspects.

Although as a pure science it has advanced to a very respectable status, as a technology, unfortunately, it has not, despite the fact that it is an old technology. Industrially, it is limited to the manufacture of some high-cost chemicals and pharmaceutical intermediates and perhaps a small number of specialty products.

Actually thousands of organic electrochemical transformations have been achieved in the laboratory but only relatively few of them advanced behind the laboratory, apparently because of economic and technical barriers. The continuing progress in the fields of chemical engineering and materials science should provide the stimulus for the development of electroorganic technology on a much broader scale than has been possible heretofore.

The first electroorganic synthesis was performed by Michael Faraday (1834). It was the anodic decarboxylation of acetic acid in aqueous medium, and the formation of ethane via creation of a new carbon-carbon bond:

$$2\ CH_3COO^- \xrightarrow[Pt]{-2\,e} CH_3CH_3 + 2\ CO_2$$

Henry Kolbe (1849) electrolyzed fatty acids and half-esters of dicarboxylic acids and established the practical basis of electroorganic synthesis. Near the end of the 19th century various electrolytic industrial processes emerged, mostly reductive, stimulated by the pioneering works of Kolbe, Haber, Fichter, Tafel and other notable contemporaries who expanded the foundation of organic electrochemical technology. Great progress in the understanding of fundamental mechanisms of electrochemical reactions has been achieved during the last few decades in parallel with the spectacular advances of electroanalytical and spectroanalytical methodologies [1] and the commercial availability of the relevant instruments.

1.2 Pertinent Fundamental Concepts and Principles in Organic Electrochemistry

1.2.1 The Overall Cell Reaction

Electrochemical reactions are performed in reactors known as electrolytic cells or electrolyzers, and are commonly referred to as cell reactions [2]. All overall cell reactions are *necessarily* the result of two *half-cell* reactions, the cathodic and the anodic. These reactions take place simul-

Half-cell reactions taneously and involve transfer of electrons at the electrode/solution interface, as is symbolically indicated by expressions 1.1, 1.2. The sum of the two half-reactions is the overall cell reaction:

$$A + e \longrightarrow C \qquad \text{cathodic} \qquad (1.1)$$

$$B - e \longrightarrow D \qquad \text{anodic} \qquad (1.2)$$

$$A + B \longrightarrow C + D \qquad \text{overall cell reaction} \qquad (1.3)$$

Overall cell reactions can proceed either spontaneously or by imposing on the cell a certain external electrical potential or a *driving* force [2]. Because of inhibiting thermodynamic or kinetic factors most organic cell reactions are not spontaneous. Therefore they must be *driven* by an imposed potential.

Electrodes In organic electrosynthesis we are mainly interested in one of the half-cell reactions. This reaction occurs at the *working* electrode, which may be either the anode or the cathode. The other half-cell reaction takes place at the *auxiliary* or *counter* electrode and is usually of no *direct* synthetic interest. However, there are cases in which both the anodic and the cathodic reactions are synthetically useful (paired electrosynthesis).

Basic apparatus In principle, the electrochemical method of synthesis, as practiced in the organic laboratory, is a relatively simple method. The basic laboratory apparatus is shown schematically in Fig. 1.1. It consists of an ordinary glass beaker, two electrodes, the anode and the cathode, a reference electrode (if needed) and a dc-power source. A voltmeter and an ammeter are inserted in the circuit for voltage and current measurements. Today's potentiostats incorporate the basic electrical components into one compact assembly. The electrochemical method of synthesis might be described very briefly as involving the following fundamental physical and chemical processes:

Physical and chemical processes Motion of electrons in the external cell circuit (electronic current) and of ions in solution (ionic current) while electronations (reductions) and deelectronations (oxidations) of chemical species take place at the cathode

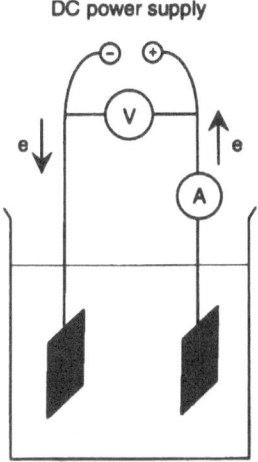
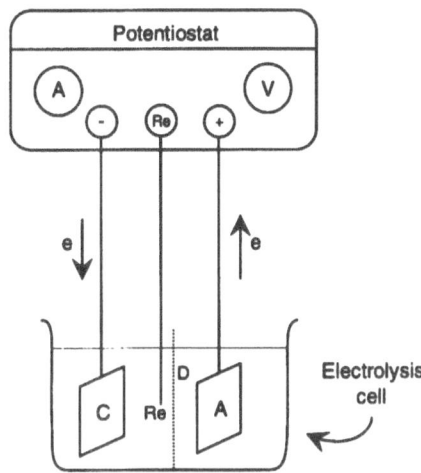

Fig. 1.1. Schematic of basic electrolysis apparatus, showing principle components Ⓐ, ammeter, Ⓥ, voltmeter, Ⓡₑ, reference electrode, ⊖, cathode, ⊕, anode, D, diaphragm, if needed

and the anode, respectively. These primary electrodic redox reactions lead eventually to the isolated chemical products.

Preparative electrolyses are generally performed under either controlled potential or constant current conditions, in one-compartment or two-compartment cells depending on the particular electrosynthesis. Three types of solutions are commonly used: a. aqueous; b. aqueous-organic (homogeneous or heterogeneous mixtures) and c. Organic solutions made with protic or aprotic solvents. The organic substrates must be soluble or at least partially soluble in the solvent system. The solvents must be able to dissolve sufficient amounts of supporting electrolyte to become ionically conductive. The electrolyte may be a neutral salt, an acid or a base and must be ionized in the electrolysis medium. The ions of the electrolyte may or may not participate in the electrochemical reactions. It is now possible in some special electrosyntheses to avoid the need of a soluble electrolyte by employing cells with a solid polymer electrolyte (see Chapter 5).

Electrolysis conditions

Solvents and electrolytes

1.2.2 The Generalized Electroorganic Reaction

The fundamental events in electroorganic reactions are shown schematically in Fig. 1.2. The organic substrate, R, is considered to undergo an electron-transfer reaction (electroreaction) as a primary step in the overall electro-chemical reaction. This primary electron transfer (ET) reaction is defined

Direct and Indirect ET reactions

Anodic reactions

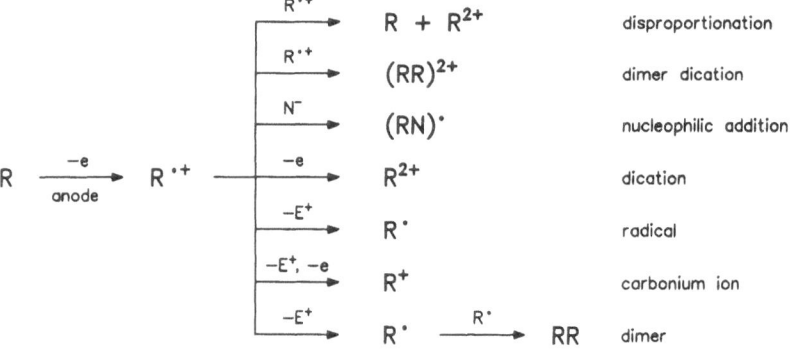

Cathodic reactions

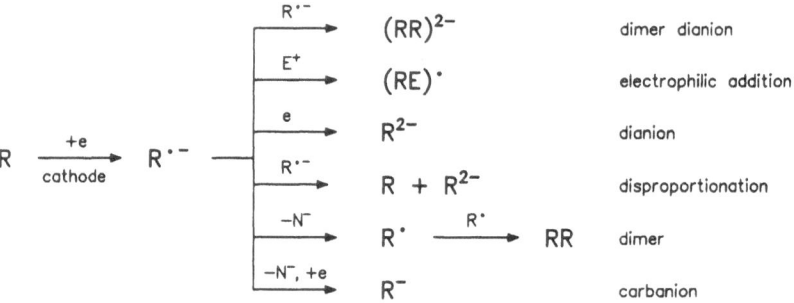

Indirect or mediated reactions

$$M \xrightarrow[\text{cathode}]{+e} M^{\cdot-} \xrightarrow{R} R^{\cdot-} + M$$

$$M \xrightarrow[\text{anode}]{-e} M^{\cdot+} \xrightarrow{R} R^{\cdot+} + M$$

Fig. 1.2. Phenomenological classification of the generalized electroorganic reaction's fundamental events

as a *direct* reaction if the electron exchange takes place directly between the electrode and the organic substrate. If the organic substrate undergoes the primary electron exchange via a mediator the reaction is *indirect*. Direct ET reactions are typical heterogeneous reactions. Indirect reactions may be either homogeneous or heterogeneous, depending on whether the mediator is dissolved in the solution or immobilized physically or chemically on the electrode surface.

From a purely synthetic perspective organic electrode reactions are perceived as heterogeneous chemical processes involving transfer of electrons in the overall mechanistic reaction scheme.

Symmetrical Couplings

Anodic addition

Cathodic addition

Anodic elimination

$$2 \text{ RE} - 2 \text{ e} \longrightarrow \text{RR} + 2 \text{ E}^+$$

Cathodic elimination

$$2 \text{ RNu} + 2 \text{ e} \longrightarrow \text{RR} + 2 \text{ Nu}^-$$

Unsymmetrical Couplings

Anodic

$$\text{R} - \text{e} + \text{XY} \longrightarrow [\text{RY}]^\cdot + \text{X}^+$$

Cathodic

$$\text{R} + \text{e} + \text{XY} \longrightarrow [\text{RY}]^\cdot + \text{Y}^-$$

Scheme 1.1

1.2.2.1 Electrochemical Couplings and Cross Couplings

Electrochemical couplings and cross couplings constitute perhaps the broadest category of organic electrosynthetic reactions. Therefore, it appears that a formal synopsis of this type of reaction would be appropriate at this point. Most electrochemical couplings involve formations of C—C, C—O, C—N, C—S, N—N, and S—S bonds. These reactions are classified as *symmetrical* and *unsymmetrical*. They are accomplished either anodically (oxidatively) or cathodically (reductively) as is shown in Scheme 1.1. The symbols Nu^- and E^+ stand for nucleophile and electrophile, respectively.

1.2.2.2 The Electron Transfer Step. Basic Qualitative Description

The experimentally observed high rates (currents) of electrodic reactions could not be adequately understood on the basis of the classical theory of rate processes. Modern views explain the high rates of electrodic reactions by considering the wave nature of the electron which enables *electron tunneling* through the energy barrier rather than passage of the electron over the barrier [2–4]. The applied potential or the electronic energy supplied to the electrode in contact with the solution brings about formation of equal electronic energy levels on both sides of the electrified interface, and thus the probability for radiationless electron transfer increases. This enhanced probability manifests itself in the observed high rates of electrodic reactions. Thermal reaction rates are expressed as functions of $e^{-E/RT}$ and electrodic reaction rates as functions of $e^{-\alpha\eta F/RT}$ where the symbols E and $\alpha\eta F$ denote thermal and electronic activation energies [6a]. Figure 1.3 is intended to represent a simple profile of possible energy barriers along the path of an electrodic reaction.

Molecular orbitals   When an organic molecule attains the transition complex state at the cathode the electron leaves the electrode and is accepted into the lowest in energy unoccupied molecular orbital (LUMO) of the molecule. The molecule is electronated. The opposite occurs when the molecule is de-electronated or oxidized at the anode. The electron leaves from the highest occupied molecular orbital (HOMO) and is accepted by the anode. In both cases available electronic energy levels must be present on both sides of the electrode-solution interface for the ET reaction to take place [2]. This availability of energy levels is determined by the prevailing electrode potential.

Electron transfers   The electroactivity of organic molecules depends heavily on their structure. For example, conjugated double bonds are, as a rule, more electroactive than nonconjugated double bonds. Conjugation lowers the LUMO and raises the HOMO energy levels, so that both cathodic and anodic ET reactions are facilitated. Functional group electroactivities are significantly

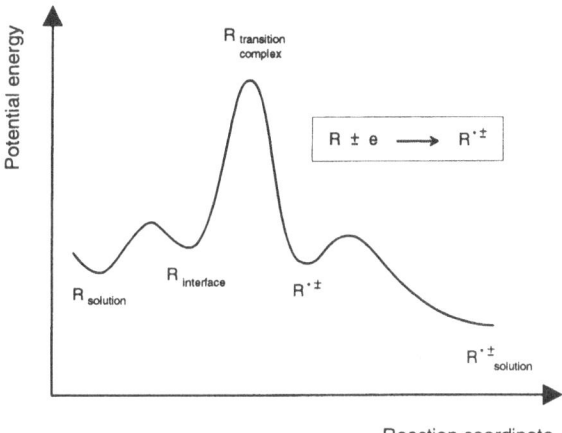

Fig. 1.3. Reaction path of an electrodic reaction, showing possible energy barriers

influenced by the presence of activating groups (electron releasing or withdrawing groups) in the molecular structure. Steric barriers also have effects because such barriers may not allow the electroactive group to approach the electrode site at which the electroreaction occurs. Most organic electrode reactions involve more than one-electron step per molecule. The apparent two-electron step transfers, sometimes observed in practice, are actually two very rapid successive one-electron steps not distinguished by the ordinary experimental techniques in the organic laboratory. Molecular orbital theory [4] considering the repulsion of two electrons when added simultaneously to a certain molecular orbital predicts that addition of a second electron requires an energy of about 5 electron volts. However, solvation of a dianion lowers the separation energy between the first and second ET reaction to only about 0.5 electron volts. Sometimes the second ET occurs more readily than the first, and hence only one voltammetric two-electron wave or peak is experimentally observed in recording voltammetric current-potential curves. Analogously electron abstraction from a molecule is also facilitated by solvation effects [5].

1.2.3 Thermodynamic and Kinetic Aspects

1.2.3.1 Nernstian Equation. Electrode Potential

The fundamental thermodynamic equation in electrodic reactions is the well-known Nernstian equation, given in simple form below:

$$E_e = E_e^{\circ} - \frac{RT}{nF} \ln Q \tag{1.4}$$

$$\text{or } E_e = E_e^\circ - \frac{0.059}{n} \log Q \text{ at } 25°C \tag{1.5}$$

Meaning of electrode potential

In Eq. 1.5, Q represents the ratio of the activities of the products to the activities of the reactants of the electrodic reaction. The value E_c is the equilibrium potential of the electrode, and E_c° is the potential under standard state conditions, that is, when products and reactants are present at unit activities or when, in the general case, the numerical value of the quotient Q is unity, and therefore $E_c = E_c^\circ$. In common practice, concentrations are most often used instead of activities in Eq. 1.5 since activities are not available.

In a physical sense, the electrode potential expresses the energy of the electrons in the electrode in contact with the electrolysis medium [2,3]. When the electrode is negative it has an excess of electrons and when positive an excess of positive charges. Thus the electrode potential affects the energy required to add electrons to the electrode or remove electrons from it. It is important to understand the difference of the sign (− or +) of electrode potential and the sign of the half-cell reaction. The sign in the former is a real physical property whereas the sign of the latter is a defined or conventional quantity and depends on how the reaction is written on paper. We now consider Nernst's equation in connection with a simple *reversible* electrodic organic reaction, as represented in Eq. 1.6

Reversible electrodic reaction

$$O + ne^- \underset{\vec{k}}{\overset{\vec{k}}{\rightleftharpoons}} R \quad \begin{array}{l} \overset{\rightarrow}{-i} \text{ cathodic partial} \\ \qquad \text{current} \\ \overset{\leftarrow}{+i} \text{ anodic partial} \\ \qquad \text{current} \end{array} \tag{1.6}$$

Equation 1.5 may be written in the usual practical form Eq. 1.7

$$E_e = E_e^\circ + \frac{RT}{nF} \ln \frac{C_o}{C_R} \tag{1.7}$$

Equilibrium potential

where C_O and C_R are concentrations of the oxidized and reduced forms of the redox couple O/R in Eq. 1.6. At the equilibrium state the electrode potential attains the steady value E_c. Any equilibrium, E_c, is the relative steady value of the electrode potential that is established in a cell in which the auxiliary electrode is a chosen standard reference electrode (e.g. the NHE) and under conditions that *no current* is allowed to flow in the external circuit. The established equilibrium is *dynamic* since electron transfers take place in both directions at equal rates, as implied by the partial currents −i and +i in Eq. 1.6. By defining the cathodic partial current as negative and the anodic as positive we can write the following expressions:

$$-i + i = 0$$

$$|-i| = |+i| = i_o$$

The symbol i_o represents the *exchange current*. The exchange current is a fundamental characteristic of electrodic processes, in general. Exchange currents have been estimated to be in the range of $10^{-16} - 10^{-3}\,A/cm^2$ at various electrodes at ordinary temperatures. Exchange currents vary with the concentration of the reacting species. For this reason a kinetic parameter, k, independent of concentration is defined [6], as given in Eq. 1.8

Exchange current

$$k = i_o/nFC_o^{\alpha_A} \cdot C_R^{\alpha_C} \tag{1.8}$$

(α_A and α_C are the transfer coefficients for the anodic and cathodic partial reactions)

1.2.3.1a Free Energy of Cell Reaction and Cell Potential

Consider the hypothetical cell reaction

$$A + B \longrightarrow C + D$$

and assume that the free energy, ΔG, of this reaction is positive ($\Delta G > O$). The reaction therefore must be driven by supplying to it electrical energy. This can be done by applying a voltage or potential across the cell. The minimum voltage to be applied for the electrolysis to begin (that is, for the appearance of current in the external circuit) would correspond to the free energy of the reaction as given by the thermodynamic equation, Eq. 1.9

$$\Delta G = -nFE_{cell} \tag{1.9}$$

The value, E_{cell}, represents the theoretical *decomposition* potential, defined as $E_{cell} = E_a - E_c$. Since $\Delta G > O$, E_{cell} is negative by definition. The values E_a and E_c represent the theoretical anodic and cathodic potentials, respectively. In practice the cell voltage will include the anodic and cathodic overpotentials and the (inevitable) potential drop, iR, across the cell. (Note: according to Bockris no electrodic reaction is ever possible without an overpotential) [6a]. The total cell voltage will be distributed across the cell according to Eq. 1.9a. The overpotentials are included in the terms E_a and E_c.

Overpotential

$$E_{cell,total} = E_a - E_c + iR \tag{1.9a}$$

The electrode potentials E_a and E_c are values (in volts) measured with reference to any suitable reference electrode, but the reference electrode must be the same for both the anode and the cathode in order for Eq. 1.9a to be applicable. We may note that absolute electrode potentials, although they exist, can not be measured experimentally [2]. All electrode potentials are relative values referred to a chosen standard electrode [1,2]. In electrochemical synthesis we are primarily interested in the potential of the working electrode. Current-potential curves are usually obtained by plotting current (current densities) versus the potential applied to the working electrode. This potential is measured against a stable reference electrode (such as the SCE, Ag/Ag^+ etc).

1.2.3.2 Kinetic Aspects

Tafel's equation Electrochemical reaction rates are generally expressed as current-potential (i-E) relationships. The most simple and practical current-potential relationship is Tafel's (1905) empirical equation,

$$\eta = a \pm b \log i \tag{1.10}$$

This equation relates the *activation* overpotential, η, to the current i or the rate of the electrochemical reaction, U, since the current is proportional to U, Eq. 1.11

$$i = nFU \tag{1.11}$$

The activation overpotential η is defined as the difference between the applied potential, E, and the theoretical equilibrium electrode potential, E_c, Eq. 1.12

$$\eta = E - E_e \tag{1.12}$$

The constants a and b in Eq. 1.10 are characteristic parameters of the particular electrodic reaction occuring at the anode or the cathode. It must be noted that the Tafel equation is applicable, in a strict sense, only when the slowest or rate determining step (rds) in the overall electrode process is the electron transfer step. Because overpotentials relative to true equilibria potentials, as defined in Eq. 1.12, are not available for most practical electrosynthesis work the Tafel equation is used in the more practical form given in Eq. 1.13

$$E = a \pm b \log i \tag{1.13}$$

In Eq. 1.13, E represents the relative electrode potential measured with reference to any suitable reference electrode. Plots of Eq. 1.13 are useful in selecting electrodes with regard to their electrocatalytic properties and for establishing optimum potentials for a given electrochemical reaction.

Equation 1.10 is a limiting form of the fundamental equation of electrode kinetics, Eq. 1.14, commonly known as the Butler-Volmer equation [2]:

$$i = i_o \left[\exp\left(\frac{\alpha_A nF}{RT}\eta\right) - \exp\left(-\frac{\alpha_C nF}{RT}\eta\right) \right] \tag{1.14}$$

In this equation i is the observed current (or the *net* current), i_o is the exchange current, and α_A and α_C are the anodic and cathodic *transfer*

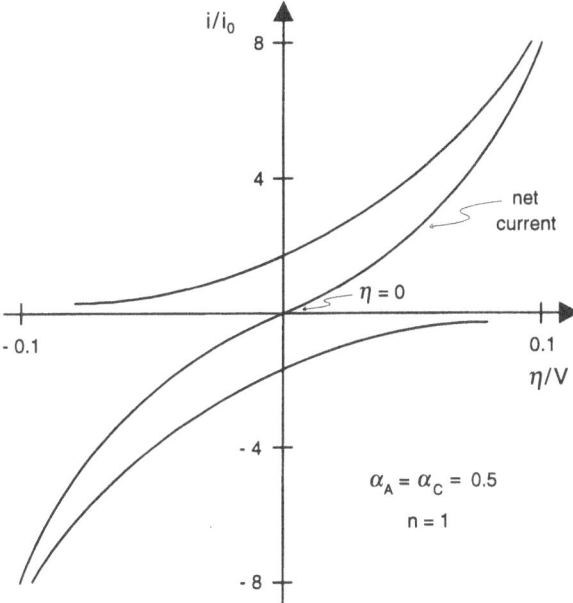

Fig. 1.4. The i-η relationship for a reversible electrodic reaction, $R + e \rightleftharpoons R^{\bar{}}$

coefficients, η is the overpotential and the other symbols have their usual meanings. Equation 1.14 is represented graphically in Fig. 1.4.

In most practical synthetic applications electrolysis is carried out at relatively high positive or·negative electrode potentials, that is at potentials far from the equilibrium potential. Under such potentials, only one of the exponential terms in Eq. 1.14 is practically significant [1,2,7]. Thus at high anodic or cathodic potentials the limiting forms expressed in Eqs. 1.15 are derived from the general equation, Eq. 1.14. These limiting forms represent the detailed Tafel equation as derived from theoretical considerations, and indicate the meanings of the constants a and b in the original empirical Tafel equation.

$$\log i_{anodic} = \log i_0 + \frac{\alpha_A nF}{2.3RT} \eta \tag{1.15}$$

$$\log i_{cathodic} = \log i_0 - \frac{\alpha_C nF}{2.3RT} \eta$$

We further may note that on the basis of experience the sum of the transfer coefficients α_A and α_C is unity and hence only one of them is needed in Eq. 1.14. The α_A in the first exponential term of Eq. 1.14 can be replaced by $(1 - \alpha_C)$. At very low overpotentials another limiting form of Eq. 1.14 is derived by expanding the exponentials, Eq. 1.16,

$$i = i_0 \frac{nF}{RT} \eta \tag{1.16}$$

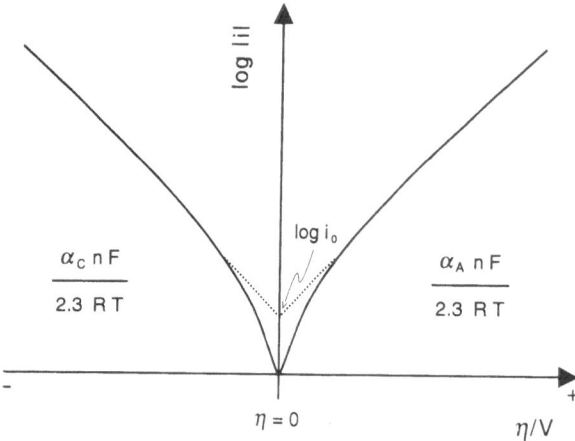

Fig. 1.5. Tafel plots for anodic and cathodic reactions ($\eta = E - E_e$)

It is thus possible to estimate exchange currents from experimental plots of Eqs. 1.15 and 1.16 as illustrated in Fig. 1.5, (provided η is known as $\eta = E - E_e$).

(**Note:** In the kinetic equations given above the symbol, n, refers to the number of electrons involved in the activation step. This number may be ether equal to or less than the number of electrons in the particular half-cell reaction.)

1.2.4 Preparative Electrolysis Under Activation (Kinetic) or Mass-Transfer Controlled Conditions

Consider for example, the simple formal electroreduction, Eq. 1.17

$$R + 2e \xrightarrow{\text{Hg, H}_2\text{O}} RH_2 \tag{1.17}$$

performed electrolytically in aqueous solution using a mercury cathode. In most such electrochemical hydrogenations the reaction is irreversible, even though the first electron-transfer step may be reversible as is illustrated in the following overall reaction scheme:

$$R + e \rightleftharpoons R^{\cdot -} \quad \text{rapid and reversible} \tag{1.18}$$

$$R^{\cdot -} + H_2O \longrightarrow RH^{\cdot} + OH^-$$

$$RH^{\cdot} + e \longrightarrow RH^- \tag{1.19}$$

$$RH^- + H_2O \longrightarrow RH_2 + OH^- \quad \begin{array}{l}\text{may be rapid}\\\text{but irreversible}\end{array} \tag{1.20}$$

$$R + 2e + 2H_2O \longrightarrow RH_2 + 2OH^- \quad \text{overall reaction} \atop \text{irreversible} \qquad (1.21)$$

A current-potential curve of this reaction (in unstirred solution) would resemble curve A in Figure 1.6.

In the rising portion of the curve A the current increases exponentially with the applied voltage E ($i = \text{constant} \cdot \exp(\alpha_C nFE/RT)$). In the potential region corresponding to the plateau (limiting current) region of the curve all the molecules of the substrate R reaching the electrode are reduced. A zone depleted in R is formed near the electrode surface (diffusion zone). The reaction rate (current) at this potential region is mass-transfer or diffusion controlled and independent of the electrode potential. Such limiting currents are common to both reversible (rapid) and irreversible (slow) electrochemical reactions but the shapes of the i-E curves are different as shown in Figure 1.6. A preparative electrolysis may be carried out under activation (kinetic) or mass-transfer control conditions, that is under potentials at the rising portion of the i-E curve or at limiting current or plateau region of the curve. The limiting current, i_l, under mass transfer conditions at a large electrode with an *active* area A is given by Eq. 1.22

$$i_l = nFDAC\delta \qquad (1.22)$$

in which D is the diffusion coefficient (cm^2/s) of the electroactive species. C is the concentration of the species in the *bulk* of the solution, n is the number of electrons per molecule undergoing reaction, and δ the *thickness* of the Nernst diffusion layer, usually 0.1 mm.

For an electrolysis under limiting current conditions (plateau of the i-E curve) the current falls with electrolysis time, t, according to the following

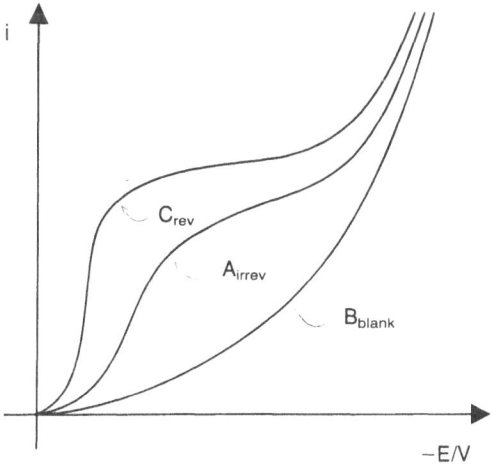

Fig. 1.6. Typical i-E voltammetric curves. Reversible reaction, curve $C_{rev.}$; Inreversible reaction, curve $C_{ir.}$; Curve B, blank, only solvent and electrolyte

i-t relation, Eq. 1.23 where i^o is the starting current (amperes), V is the volume of the

$$\log i = \log i^o - \frac{0.434 \, DA}{V\delta} \, t \tag{1.23}$$

solution (liters), and t is the time of electrolysis (seconds) and the other symbols have the same meanings as in Eq. 1.22. Some other practical equations derived from Faraday's laws are the following (assuming 100% current efficiency):

$$Q = \int_0^t i dt = nFN \tag{1.24}$$

$$m = \frac{M}{nF} \, it \quad \text{constant current} \tag{1.25}$$

$$m = \frac{M}{nF} \int_0^t i dt \quad \text{constant potential} \tag{1.26}$$

In the equations given above the symbols have the following meanings:

Q = total charge consumed (coulombs) for N moles of substrate
t = time (seconds)
i = current (amperes)
n = number of electrons involved in the reaction per molecule reduced or oxidized
m = weight of substrate reduced or oxidized
M = molecular weight of substrate

The following definitions pertain to synthetic electrolytic processes in general:

current efficiency, CE, $= \dfrac{\text{theoretical ampere} - \text{hours}}{\text{actual ampere} - \text{hours}}$

voltage efficiency, VE, $= \dfrac{E_{cell} \text{ theoretical}}{E_{cell} \text{ actual}}$

Energy Efficiency, EE, $= (CE) \cdot (VE)$

(faraday = 96 500 coulombs or 26.806 ampere – hours)

1.2.5 The Electrical Double Layer and its Significance in Electrosynthesis

Adsorption   It has been stated that the electrical double layer or the electrified interface between an electrode and a solution is the "innermost essence" of electrochemistry. Only a pictorial, simple representation of the electrical double layer, Fig. 1.7, can be made here in order to discuss most simply its

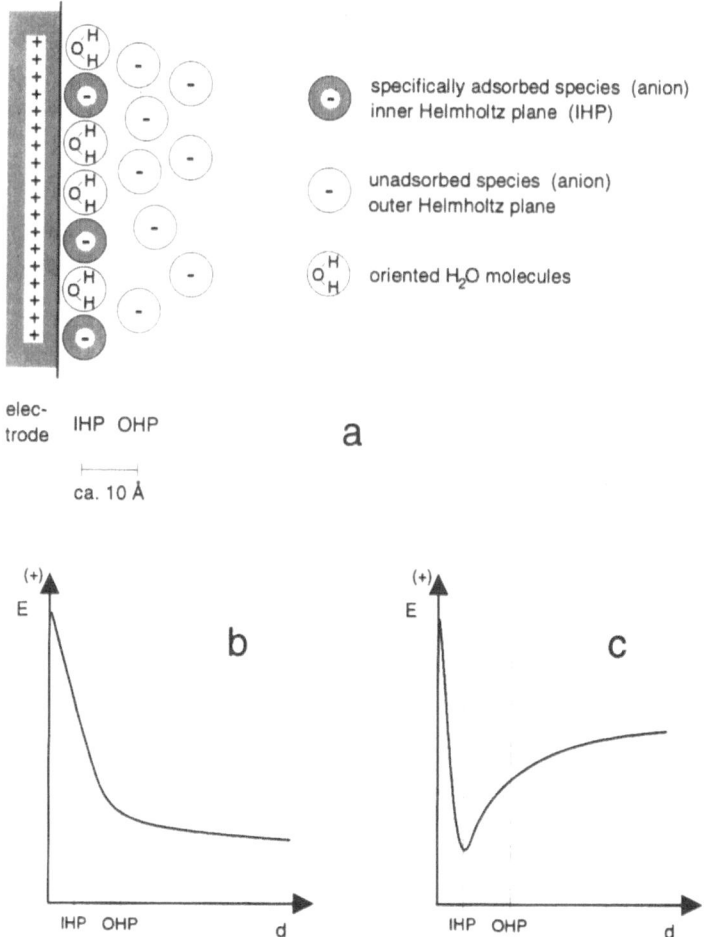

Fig. 1.7. Schematic Representation of the Electrical Double Layer. (b) no specific ad-sorption; (c) specific adsorption; d, distance from electrode surface; E, potential. Thickness of double layer is usually ~10 Angstrums (IHP and OHP refer to inner and outer Helmholtz planes)

practical significance in electrosynthetic work. In Fig. 1.7(a), assuming a positively charged electrode, the electrode surface is shown covered with water molecules (aqueous solution) with the oxygen towards the electrode surface rather than the hydrogens. The opposite would be expected if the electrode was negative charged. This, electrostatically influenced, adsorption or orientation of water at the electrode surface is analogous to solvation phenomena in solutions. It occurs with charged and polar species when the electrode potential is *not* at its characteristic *potential* of *zero*

Helmholtz planes

charge (pzc). However, adsorption of any species may occur by both chemical and physical forces (chemisorption, physisorption). In chemisorption chemical bonds are formed between electrode and adsorbate, whereas electrostatic adsorption occurs when adsorbent and adsorbate are oppositely charged. Chemisorption may occur even when the electrode and the chemical species adsorbed have similar charges. Such adsorption has been called *specific adsorption*. Situations, therefore, arise, as depicted in Fig. 1.7a,b,c, characterised by *inner* and *outer* Helmholtz planes (IHP, OHP).

Location of reacting species

The structure of the double layer is sometimes the most influential factor in the outcome of the reaction, as in the electrohydrodimerization of acrylonitrile, and in stereochemical reactions in which the electrode surface is covered with chiral compounds. The reacting species may be located at the OHP or at the IHP. It is to be noted that it is the potential that an organic molecule experiences that actually affects the electron transfer reaction, rather than the experimentally measured potential of the electrode. (The electric field intensities within the double layer are

Electric field intensities

enormous, as high as 10^7 V/cm). Because the actual loci of the centers of the reacting molecules are usually not known – and it does not matter in most practical synthetic work, reproducible relative potentials are satisfactory. It is, however, important to be *consistent by precisely defining the physicochemical composition of the medium and the potential measurement techniques*. Best electric field uniformity is attainable with parallel plate electrodes of equal dimensions. In cases in which the electroactive organic species is specifically adsorbed it experiences the potential obtaining at the IHP. We should note that organic molecules vary in complexity. As a result they can assume unpredictable orientations when they are adsorbed on electrodes or immersed in the strong electric field near the electrode surface. Electroreactions therefore are often greatly influenced by special orientations of the reacting species. Aromatic molecules can be adsorbed in parallel or vertical positions with respect to the electrode surface. There would be no way to predict how a large biological molecule would fit within the electric double layer and hence how it will react stereospecifically or chemoselectively. A recent review about the electrical double layer has been presented by Parsons [7].

From the aforesaid it would be obvious that the electrode potential, the electrical double layer, and the composition of the electrolysis medium are *very important parameters* in electrochemical reactions. Their relative importance varies, however, and many electrosyntheses, insofar as reproducibility is concerned, require only consistent operational conditions. Constant current electrolysis and indirect electrosyntheses do not require strict potential control in the sense that potentiostatic syntheses do. Furthermore, in practice it is impossible to achieve precise electrode potential control over the entire electrode surface. Also in practice "formal" electrode potentials are used since thermodynamic activities are seldom known and are not necessary for practical synthetic work.

References

1a. Bard AJ (1966) Electroanalytical chemistry. Marcel-Dekker, New York
1b. Bard AJ, Faulkner LL (1980) Electrochemical methods. Wiley-Interscience, New York
1c. Southampton Electrochemistry Group (1985) Instrumental methods in electrochemistry. Ellis Horwood, Chichester
2. Bockris JOM, Ready AKN (1970) Modern electrochemistry, vols 1 and 2, Plenum, New York
3. Eberson L (1987) Electron transfer reactions. Springer, Berlin Heidelberg New York
4. Bard AJ (1971) Pure App Chem 25:379; Saveant JM (1980) Accounts Chem Res 13:323
5. Pysch ES, Yang NC (1983) J Am Chem Soc 85:2124; Fleischmann M, Pletcher D (1973) Adv Phys Org Chem 10:155
6. Ref. 1c, p 26
6a. Bockris JOM (1971) J Chem Educ 48:353
7. Parsons R (1990) Chem Rev 90:813

2 Anodic Reactions

2.1 Fundamental Nature

Anodic or electrooxidative organic reactions generally involve three fundamental intermediates [1,2,7,9,10,11]:

1. Cation radicals, $R^{\dot{+}}$
2. Radicals, $R^{\cdot}$
3. Cations, R^+

These transient species are produced either by direct electron transfer from the organic substrate to the electrode, or indirectly by the action of a mediator [1–6]. It must be noted, with due emphasis, that a large number of anodic reactions proceed by oxidative steps involving continuously generated surface oxides or $OH^{\cdot}$ species adsorbed on the electrode surface. Such oxidative steps are actually a special type of heterogeneous chemical oxidation occurring at the surface of an electronically conducting substrate in contact with an ionically conducting solution. In organic electrochemistry these oxidations are specifically viewed as heterogeneous electrocatalytic reactions [3,5,6]. We might also refer to the formation of cation radicals, by means of one-electron oxidants such as $SbCl_5$, PbO_2 and the organic tris(4-bromophenylaminium) ion. The first of these oxidants oxidizes aromatic hydrocarbons in dichloromethane solutions. The PbO_2 is a most effective one-electron oxidant in CF_3CO_2H/CH_2Cl_2 media. The organic ion is produced anodically (homogeneous electrocatalysis). The generation of cation radicals with PbO_2 and the anode as oxidants is formally shown below in order to provide, perhaps, a first notion of the potential inherent simplicity of an electrodic electron transfer process, as contrasted to a chemical one [7]:

$$2 \; ArH + PbO_2 + 4 \; H^+ \longrightarrow 2 \; ArH^{\cdot +} + Pb^{2+} + 2 \; H_2O$$

$$ArH - e \xrightarrow{\text{anode}} ArH^{\cdot +}$$

The deelectronation of ArH by PbO_2 requires the breaking and formation of bonds and the use of at least stoichiometric amounts of reactants, as shown in the equation. By the direct anodic reaction cation radicals are produced relatively more simply in solutions containing neutral organic molecules, and are especially easily produced if the molecules contain π-electron systems or heteroatoms with unshared π-electron pairs. Extended π-electron systems are most easily deelectronated at the anode. These transient species are best studied by electron spin resonance (ESR) spectroscopy and voltammetric techniques [2,7]. Well-known one-electron voltammetric standards are the compounds, ferrocene, 9,10-diphenylanthracene and dimethylstilbene [1,2].

2.2 Hydrocarbons

Hydrocarbons contain only the two elements carbon and hydrogen in their enormous diversity of molecular structures. Anodic transformations of hydrocarbons, therefore, take place by breaking of carbon-carbon or carbon-hydrogen bonds and making of new such bonds – if, and only if, no other chemical entities are present capable of reacting with the primary electrodic products of the hydrocarbons. Of the three structural types of hydrocarbons, alkanes, alkenes and aromatics, alkanes are distinctly the most resistant to direct anodic oxidation. Positive potentials as high as 3.0 V vs Pd/H$_2$ may be needed for electron release to the anode. At such potentials the solvent or the electrolyte may be, and usually is, more easily oxidized than the alkane molecule. In sharp contrast to alkanes, hydrocarbons with π-electron systems are generally readily oxidized at the anode because the π-electrons are much more available for release to the electrode than the electrons of single C—C or C—H bonds. Synthetically, the most important anodic reactions of hydrocarbons comprise two broad categories:

General aspects

1. Inter- and intra-molecular bond formations (dehydrodimerizations, polymerizations) and
2. Various functionalizations (nucleophilic additions and substitutions).

2.2.1 Saturated Hydrocarbons

It would be quite difficult to conceptualize a plausible mechanism for the direct deelectronation of an alkane molecule at the anode. Yet, under certain, rather severe, conditions even alkanes can be prompted to donate electrons to the anode. In depth studies in this area have been carried out mainly by Pletcher and Fleischmann [1] (Southampton Electrochemistry Group). These studies confirmed that aliphatic hydrocarbons are susceptible to direct oxidation at platinum in FSO$_3$H solution containing electrochemically stable salts as electrolytes [2]. Such salts are those

Oxidation in FSO₃H

whose anions are most difficult to oxidize at the anode (fluorosulfonates, tetrafluoroborates, hexafluorophosphates). In solutions of $FSO_3H/0.1MKSO_3F$ alkanes are oxidized at potentials between 1 and 2.2 V vs Pd/H_2 electrode [1b]. For example, anodic peak potentials of typical alkanes are: isobutane (1.0), cyclohexane (1.1), n-hexane (1.5), n-octane (1.4), n-butane (1.85) and n-propane (2.2). The oxidation of alkanes in FSO_3H apparently proceeds via the protonated form of the alkane molecule. We mention in this connection that theoretical calculations by Kollmer and Smith [3] indicated that the CH_5^+ cation contains a weak $C-H$ bond (9.5 eV) which is much weaker than the average (16 eV) bond in the neutral CH_4 molecule. This possibility, by extention to alkanes, might explain the easier electron release to the anode from the protonated alkane molecule, RH_2^+,

$$RH + FSO_3H \longrightarrow RH_2^+ + FSO_3^-$$

$$RH_2^+ \longrightarrow R^+ + 2e + 2H^+$$

Oxidation in acetonitrile

Alkanes may also be oxidized in acetonitrile and some other solvents such as dichloromethane, nitromethane propylene carbonate and sulfolane. In these solvents the alkane is not protonated, but the solvents are resistant to anodic oxidation. For example, anodic oxidation of octane in acetonitrile affords mixtures of acetamides [1c], following hydrolysis of the intermediate nitrilium ion, Scheme 2.1. The reaction is accepted as proceeding as follows:

$$RCH_2R' \xrightarrow[\substack{Pt \\ MeCN,\ Et_4NBF_4}]{-e} (RCH_2R')^{\cdot+}$$

$$(RCH_2R')^{\cdot+} \longrightarrow RC^+HR' + H^+ + e$$

$$RC^+HR' + MeCN \longrightarrow \underset{RCHR'}{\overset{N=C^+-Me}{|}} \quad \substack{\text{nitrilium ion} \\ \text{intermediate}}$$

$$\underset{RCHR'}{\overset{N=C^+-Me}{|}} + H_2O \longrightarrow RR'CHNHCOMe + H^+$$

$$40 - 50\%$$

Scheme 2.1

Preparative examples

Preparative electrolyses are carried out in acetonitrile saturated with the alkane, using a platinum anode at a potential sufficient to oxidize the alkane (Fig. 2.1).

The anodic transformation of cyclohexane to a ketonic product [1b], as depicted in Scheme 2.2, occurs in $FSO_3H/RCOOH$ medium at a platinum electrode (1.15 V vs Pd/H_2) in a divided cell. Both anolyte and catholyte are the same. The anolyte is saturated with the cyclohexane and is stirred during the electrolysis with dry nitrogen presaturated with cyclohexane.

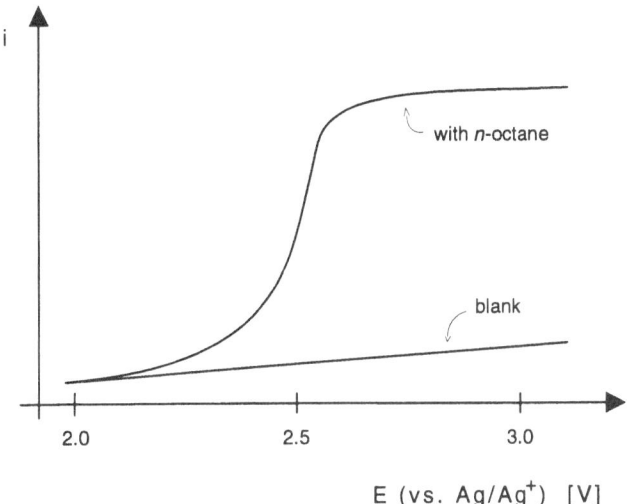

Fig. 2.1. Typical i-E voltammetric curve for the oxidation of an alkane (octane) at a platinum electrode in MeCN/Et_4NBF_4 medium

Apparently the required acyl cation, RCO^+, is generated in the solution by the reaction of FSO_3H with the carboxylic acid which acts as a base in the FSO_3H acid medium.

$$RCOOH + FSO_3H \longrightarrow RCOOH_2^+ + FSO_3^-$$

$$RCO^+ + FSO_3^- + H_2O$$

An excess of FSO_3H shifts the equilibrium as to favor RCO^+ formation, since the excess FSO_3H can bind the produced H_2O.

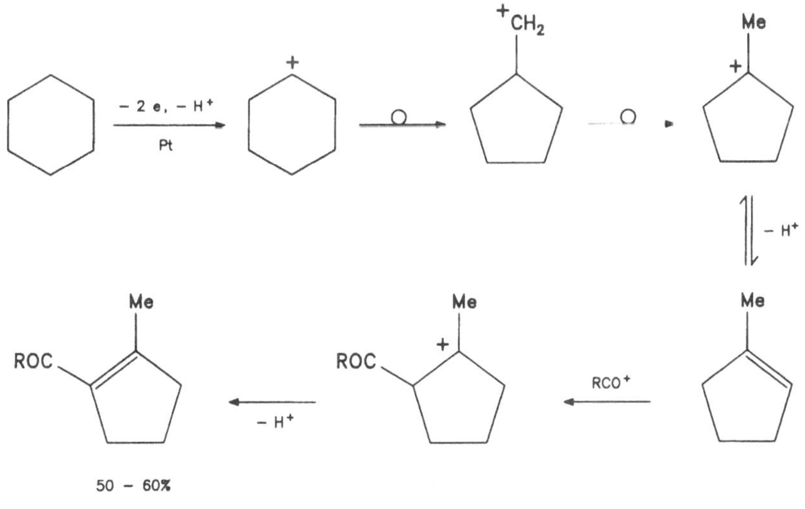

50 – 60%

Scheme 2.2

The products can be recovered from the electrolysis medium by neutral-ization of the acids with Na_2CO_3 (cold solution) and extracting with ethyl ether. We may also consider that alkanes can afford similar ketonic products by chemical reactions in the so-called *superacid* FSO_3H/SbF_5 media containing carboxylic acids [4]. In such media alkanes give cations by loss of hydride ion, H^-.

$$(Me)_3CH + FSO_3H \cdot SbF_5 \xrightarrow{\;-H^-\;} (Me)_3C^+ + FSO_3 \cdot SBF_5 + H_2$$

The anodic method obviates the need of the expensive and difficult to handle SbF_5.

Oxidation of alkanes presence NO_3-ions
Linear alkanes are partially transformed to ketonic products, by anodic electrolysis in *tert*-butanol/H_2O media containing HNO_3 and saturated with oxygen. In this mediated oxidation the nitrate ions are presumably oxidized at the anode to give NO_3 radicals which abstract hydrogen atoms from the alkane. The resulting alkane radicals react with oxygen to form peroxides and are finally transformed to the ketonic products [5].

2.2.2 Unsaturated Hydrocarbons

The presence of π-electrons in unsaturated hydrocarbons allows them to release electrons to the anode readily, and much more easily than the alkanes [6]. The electron release is further assisted by adsorption effects; π-electrons facilitate adsorption, and adsorption usually lowers the over-potential for an ET reaction. Most simple anodic oxidations of unsaturated hydrocarbons, in general, require potentials not higher than two volts vs the SCE, which are easily reached in the organic solvents commonly employed in organic electrosynthesis. The removal of an electron from a π-system in a hydrocarbon results in the formation of a very reactive cation radical which can react directly with a nucleophile or after releasing

Nucleophiles
another electron to the anode. The nucleophile can be an anion species or a neutral species including the solvent (e.g. H_2O, MeOH, CH_3CN). In nonnucleophilic solvents (e.g. CH_2Cl_2, CH_3NO_2, sulfolane) and in the absence of any nucleophiles, dimerizations and polymerizations are very favorable reactions. A solvent ideally suitable for the generation of

Oxidation in SO_2
very active organic cations is liquid SO_2 ($-50\,°C$). This nonnucleophilic solvent can be purified prior to use by shaking it with anhydrous Al_2O_3. Anthracene and 9,9'-bianthryl have been oxidized in SO_2/Bu_4NPF_6 medium and gave their very reactive dications [7]. We further note that alkyl substituted aromatics generally release electrons to the anode more readily as the number of alkyl groups increases: In acetonitrile the deelectro-nations of benzene, toluene, *p*-xylene and pentamethylbenzene occur at the respective potentials 2.0, 1.98, 1.56 and 1.28 vs Ag/Ag^+ electrode. The electron-releasing alkyl groups tend to increase the electron density

on the aromatic ring and thus assist the anodic removal of ring electrons. Fundamental research studies have confirmed that aromatics are generally easily oxidized at the anode in organic solvents to yield inter- and intra-molecular bond products in good to excellent yields. Following are selected typical examples with alkyl and methoxy substituents on the aromatic rings. Durene (**1**) gives the dimer (**2**) as major product upon oxidation at platinum in $CH_2Cl_2/0.1MBu_4NBF_4$ medium by reaction of the generated cation with the durene molecule [9,10]. Two electrons are released from the starting molecule and H^+ is expelled from a methyl group to form the cation. Mesitylene (**3**) dimerizes by two routes [10] to give the bismesityl product:

Oxidation of aromatics

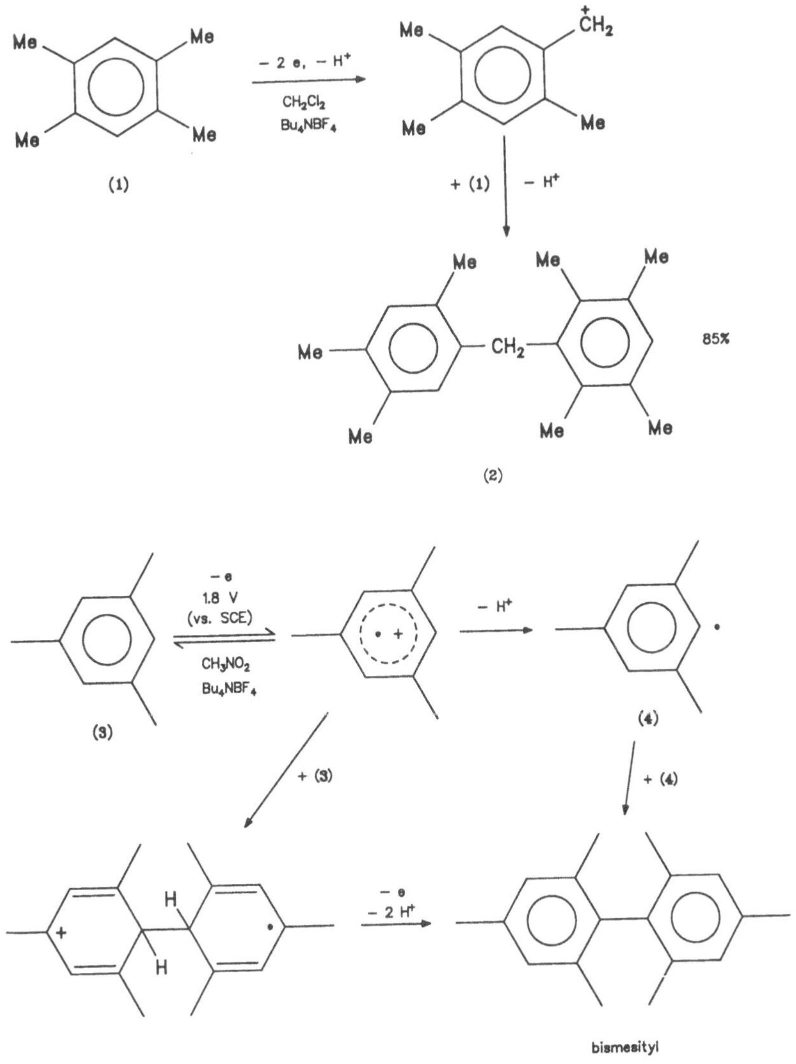

(1)   (2)   85%

(3)   (4)   bismesityl

Scheme 2.3

Although the following oxidation examples are not strictly referable to hydrocarbons they are very illustrative of the π-electron system anodic deelectronation, and will be included here parenthetically. The methoxy (4) and methyl substituted (5) [12] cyclophanes and the isochromanone compound (7) are transformed to the corresponding intra-cyclized products upon oxidation at platinum electrodes in MeCN/Bu₄NClO₄ solution at 1.2 to 1.5 V vs Ag/Ag⁺ electrode [11,12,13]. Anodic intramolecular cyclization of the aromatically substituted *n*-propane compound (9) in acetonitrile-fluoboric acid solution afforded the spirodienone compound (10) in high yield [14].

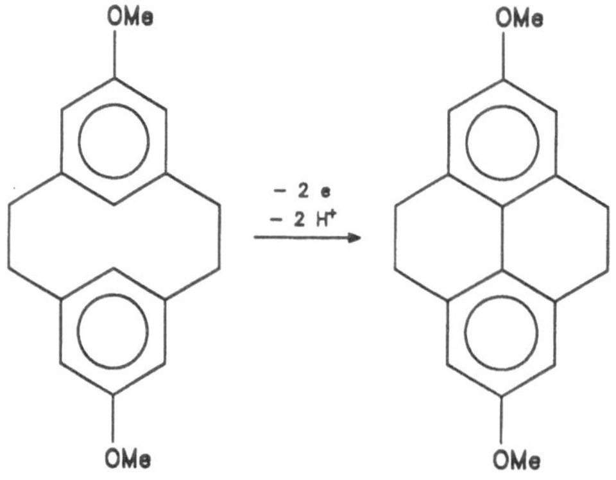

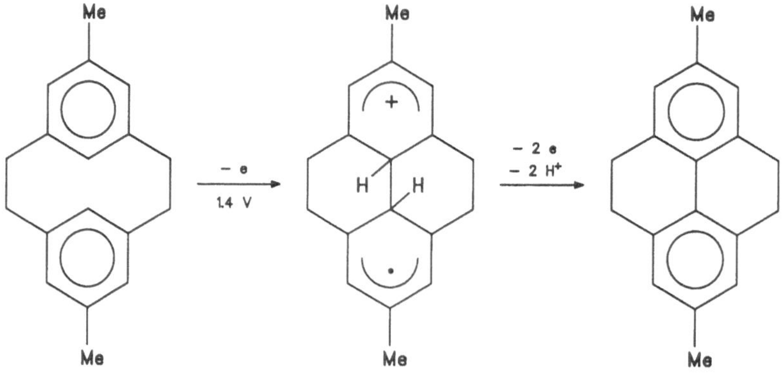

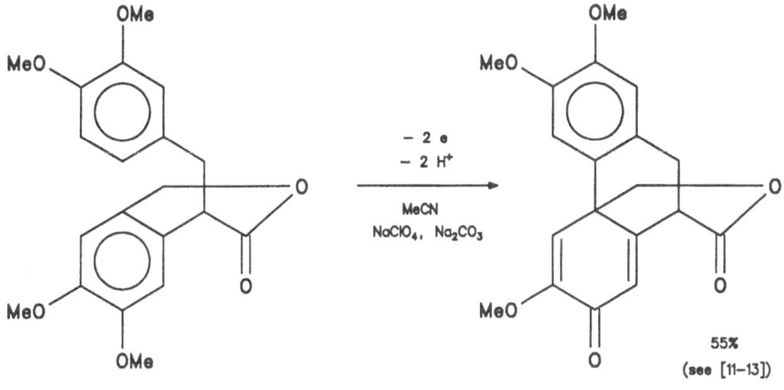

(7)

$- 2\ e$
$- 2\ H^+$

MeCN
NaClO$_4$, Na$_2$CO$_3$

(8)

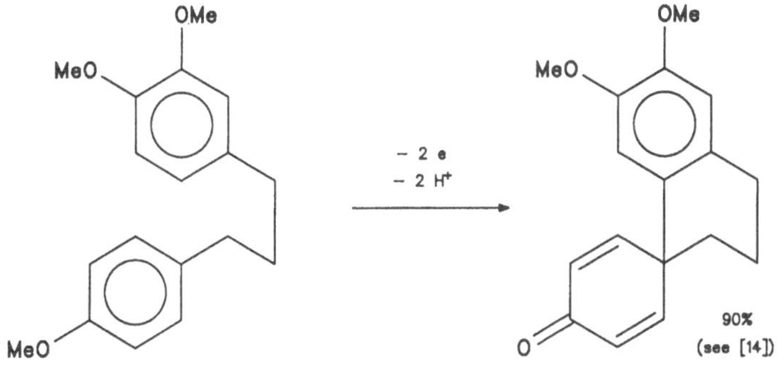

55%
(see [11–13])

(9)

$- 2\ e$
$- 2\ H^+$

(10)

90%
(see [14])

2.2.3 Functionalizations of Hydrocarbons

The initial step in the anodic functionalization of hydrocarbons usually is a one-electron transfer from the hydrocarbon to the electrode whereupon a cation radical is produced. This primary product is attacked by a nucleophile as is symbolically indicated below

$$\text{RE} \xrightarrow{-e} \text{RE}^{\cdot+} \xrightarrow{\text{Nu}} \text{RNu} + \text{E}$$

where Nu and E represent nucleophile and electrophile, respectively. In most cases E is a proton. The nucleophile may be a neutral or negatively charged species (e.g. H$_2$O, MeOH, MeCN, pyridine, RNH$_2$, etc., or

OH^-, RO^-, CN^-, $RCOO^-$, SCN^-, N_3^-, NO_3^-, $I^-\rangle$, $Br^-\rangle$, $Cl^-\rangle$, F^- (nucleophilicity in that order for halide ions).

We would note that a nucleophile bearing a negative charge, e.g. OH^-, would always be a stronger nucleophilic species than its conjugate acid H_2O, a neutral species. Further, we bear in mind that basicity is generally *thermodynamically controlled* but nucleophilicity is *kinetically controlled*, (solvent and other effects). Because most primary electrodic products are transitory, the nucleophilic addition takes place in the electrode's immediate vicinity. If the primary species survives long enough so as to diffuse into the bulk of the solution the chemical reactions can occur homogeneously in the absence of the strong electric field at the electrode vicinity. In cases in which the reactants RE nad Nu are adsorbed on the electrode or if they are oriented in some special position at the electrode surface, stereochemical effects could be expected [17]. Most studies indicate that both aliphatic and aromatic hydrocarbons are adsorbed on the electrode surface [18] prior to the electron transfer to the electrode. Hydrocarbon cation radicals act both as very reactive electrophiles or proton donors, and avidly interact with nucleophiles or bases. Alkyl aromatic cation radicals are extremely strong carbon acids, as for example [19],

Adsorpiom, and stereochemical effects

Carbon acids

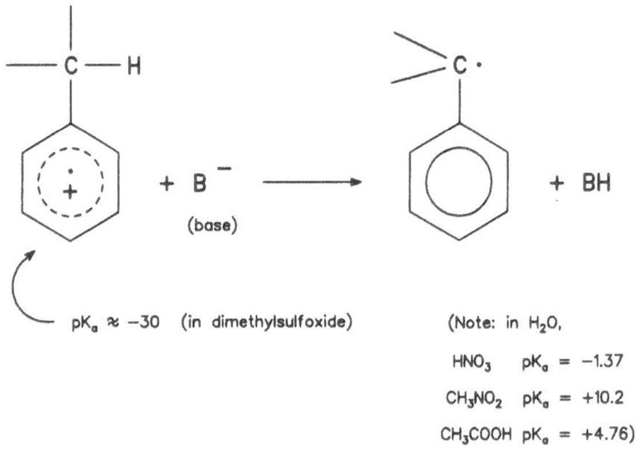

$pK_a \approx -30$ (in dimethylsulfoxide)

(Note: in H_2O,

HNO_3 $pK_a = -1.37$

CH_3NO_2 $pK_a = +10.2$

CH_3COOH $pK_a = +4.76$)

Although cation radicals can dimerize, their tendency usually is to split a C–H bond and to undergo substitution of hydrogen by a nucleophile. A very typical example is the cyanation of methylnaphthalenes in methanol/CN^- solution. According to Yoshida's studies [20] the cyanation occurs as shown in Scheme 2.4.

Cyanation of methylnaph-thalenes

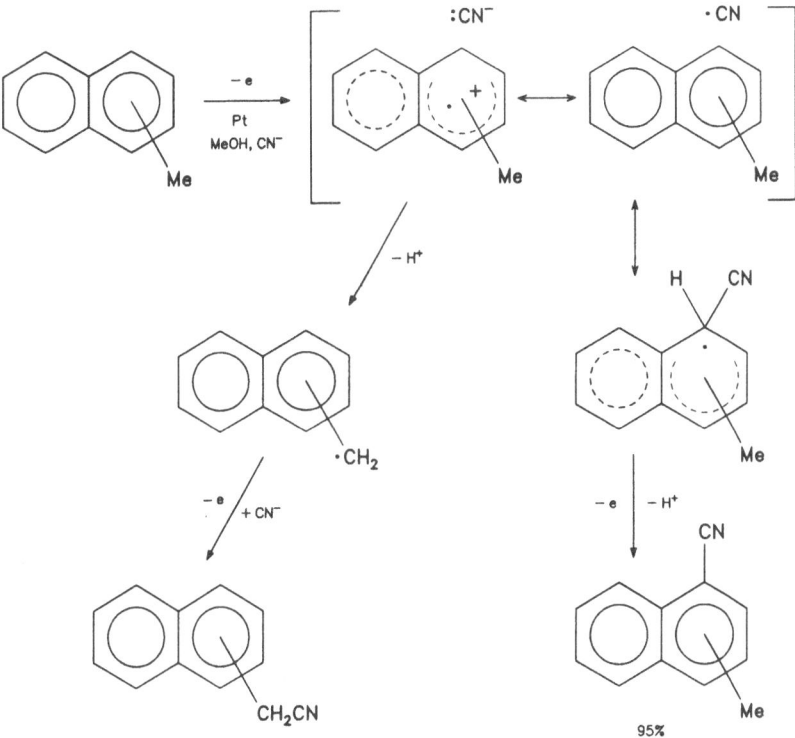

Scheme 2.4

In the above scheme it is seen that two competitive paths lead to two different products. Following the ET step the organic substrate assumes a positive charge which may not be uniformly distributed over the entire cation radical structure. If only electrostatic forces influence the reaction with the nucleophile CN^- the attack by CN^- would be on the carbon with the largest positive charge density. However, because of the influence of other factors substitution of hydrogen by CN^- may occur at carbons other than that with the largest positive charge density in the molecular structure [20]. In anodic substitutions there are usually two mechanistic possibilities to consider, Route 1 and Route 2.

Route 1 : $R \xrightarrow{-e} R^{\cdot+} \xrightarrow{Nu^-} RNu^{\cdot} \longrightarrow$ products

$R^{\cdot+} \xrightarrow{-e, -H^+} R^+ \xrightarrow{Nu^-} RNu$

Route 2 : $Nu^- \xrightarrow{-e} Nu^{\cdot} \xrightarrow{R} RNu^{\cdot} \longrightarrow$ products

R is a hydrocarbon molecule

Most recent opinions, based on experiment, point more to the first route if the organic substrate, R, is oxidized at a less positive potential than the nucleophile Nu^-

2.3 Hydroxy Compounds

General aspects

The electrochemical oxidation of organic substances containing hydroxy groups is a wide and active research field because it attracts interest in both synthetic and energy conversion applications [1-8]. Low molecular weight alcohols and some hydroxy compounds of biomass origin have been studied intensely for energy conversion purposes (obtaining electricity from fuel cells) [9,11,16]. For synthetic applications only partial oxidation of the hydroxy molecule would be desired whereas for fuel cell-type uses complete oxidation to CO_2 would be the goal. For example, glucose when partially oxidized might afford gluconic acid as it does by indirect mediated oxidation (Chapter 4) whereas in a fuel cell the glucose molecule should preferably be oxidized all the way:

$$C_6H_{12}O_6 + 6 H_2O \longrightarrow 6 CO_2 + 24 H^+ + 24 e$$

Oxidation of aliphatic alcohols

Direct anodic oxidations of aliphatic alcohols at pt, Au, and C electrodes are usually not so practical because they require high positive potentials for the alcohols to release electrons to these anodes. Indirect electro-catalytic (heterogeneous or homogeneous via mediators) methods are apparently much more practical [14,16,17]. Electrolysis can be performed in both aqueous and nonaqueous solutions and in mixed organic-aqueous media when the hydroxy compounds are not adequately soluble in water. Acidic, basic and neutral solutions can be used. The nature of the products, however, is influenced by the composition of the medium, the kind of electrode, and cell design. Recently, an interesting electrocatalysis has been reported [18] using a "composite-plated" electrode (Ni/PTFE) for the oxidation of alcohols in H_2O/KOH solutions. The effectiveness of this new type of electrode is attributed to its hydrophobic character (which apparently favors approach of the alcohol rather than water to the anode surface). It is prepared by *composite-plating* a piece of nickel plate in a nickel sulfonate polytetrafluoroethylene dispersion bath hence (Ni/PTFE). With this anode the oxidation of isopropyl alcohol to acetone was carried out with a current efficiency of about 70% and 84% yield. Formally, the anodic partial oxidation of monohydroxy alcohols and their possible oxidation products is conveniently summarized in Scheme 2.5. Oxidation to the aldehyde state requires removal of two electrons and a total of four electrons to the carboxylic acid state. It is usually difficult to stop at the aldehyde state, but this state can be *trapped* in situ by formation of acetals, Eq. 2.4

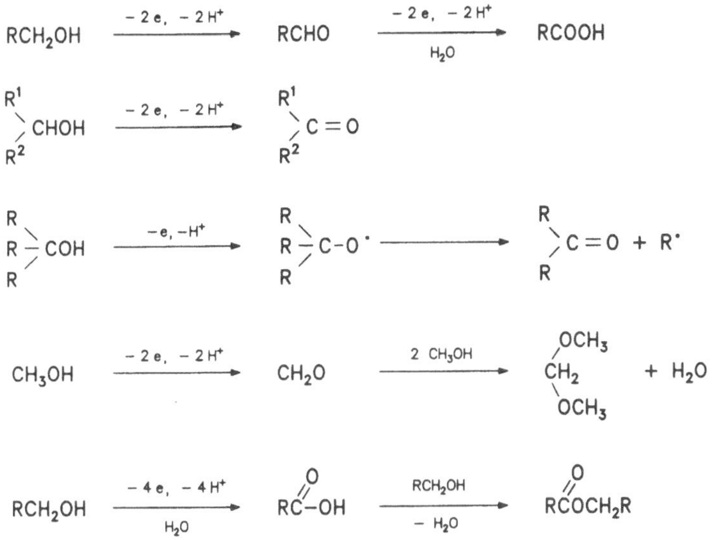

$$RCH_2OH \xrightarrow{-2e, -2H^+} RCHO \xrightarrow[H_2O]{-2e, -2H^+} RCOOH \qquad (2.1)$$

$$\begin{matrix} R^1 \\ \diagdown \\ \diagup \\ R^2 \end{matrix} CHOH \xrightarrow{-2e, -2H^+} \begin{matrix} R^1 \\ \diagdown \\ \diagup \\ R^2 \end{matrix} C=O \qquad (2.2)$$

$$\begin{matrix} R \\ \diagdown \\ R - COH \\ \diagup \\ R \end{matrix} \xrightarrow{-e, -H^+} \begin{matrix} R \\ \diagdown \\ R - C-O^{\boldsymbol{\cdot}} \\ \diagup \\ R \end{matrix} \xrightarrow{\quad} \begin{matrix} R \\ \diagdown \\ \diagup \\ R \end{matrix} C=O + R^{\boldsymbol{\cdot}} \qquad (2.3)$$

$$CH_3OH \xrightarrow{-2e, -2H^+} CH_2O \xrightarrow{2\ CH_3OH} \begin{matrix} OCH_3 \\ \diagup \\ CH_2 \\ \diagdown \\ OCH_3 \end{matrix} + H_2O \qquad (2.4)$$

$$RCH_2OH \xrightarrow[H_2O]{-4e, -4H^+} R\overset{O}{\overset{\|}{C}}-OH \xrightarrow[-H_2O]{RCH_2OH} R\overset{O}{\overset{\|}{C}}OCH_2R \qquad (2.5)$$

Scheme 2.5

Note: High molecular weight alcohols tend to undergo C—C bond cleavage yielding a variety of products [19,20,21].

Because the direct oxidation of aliphatic alcohols at Pt, Au, or C requires high anodic potentials the anions of the supporting electrolyte must be more difficult to oxidize than the alcohols for the process to be synthetically efficient. Such anions are: BF_4^-, ClO_4^-, PF_6^-. At electrocatalytic electrodes the alcohols MeOH, EtOH and $CH\equiv C-CH_2OH$ can be oxidized all the way to CO_2 and H_2O via adsorbed fragments on the anode, the fragments being formed by dissociative adsorption. The oxidative path is probably as represented below:

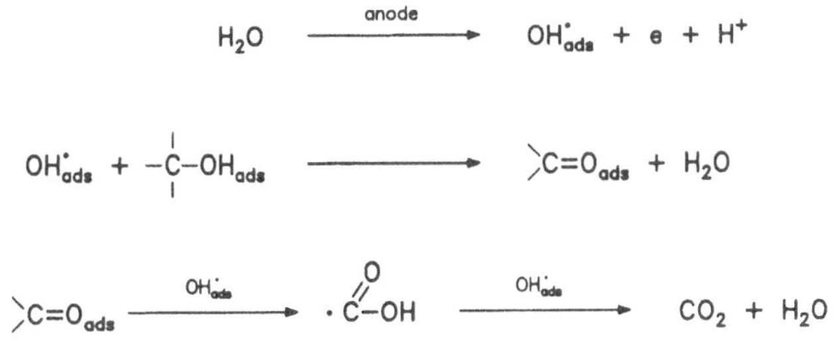

$$H_2O \xrightarrow{anode} OH^{\boldsymbol{\cdot}}_{ads} + e + H^+$$

$$OH^{\boldsymbol{\cdot}}_{ads} + \overset{|}{-\underset{|}{C}}-OH_{ads} \xrightarrow{\quad} {>}C=O_{ads} + H_2O$$

$${>}C=O_{ads} \xrightarrow{OH^{\boldsymbol{\cdot}}_{ads}} {\boldsymbol{\cdot}}\overset{O}{\overset{\|}{C}}-OH \xrightarrow{OH^{\boldsymbol{\cdot}}_{ads}} CO_2 + H_2O$$

In aprotic media with BF_4^- as anion of the electrolyte positive potentials may be reached at which alcohols can be directly oxidized at platinum

Formation of RCH₂O˙ and .CH₂OH radicals

electrodes. In alcoholic/NaOR solutions (R=alkyl) formation of radicals $RCH_2O^{\cdot}$ and $.CH_2OH$ occurs via which alkoxylated and hydroalkoxylated products can be obtained. The electrolyzed solutions must not become too basic since this would lead to aldol-type condensations as the alcohols are oxidized to aldehydes.

2.4 Oxidations at Oxide Covered Electrodes

General aspects

Oxide covered electrodes consist of metals whose surfaces are covered with a thin layer of metallic oxides [1]. These oxides are chemically active oxidizing agents, electrically conductive and insoluble in aqueous or aqueous-organic alkaline solutions. They are continuously electrogenerated during the electrolysis and remain firmly adhered to the electrode surface thus protecting the substrate metal from anodic dissolution. The metals commonly used as oxide covered electrodes are Ni, Ag, Co, Cu and Mn. Possible uses of nickel and silver oxide electrodes have been considered for synthetic [2], electrogenerative and biochemical sensor applications [2,3,4].

Concerning the PbO_2 electrode (used in sulfuric acid media) the generalized oxygen transfer reaction can be expressed as [8]

$$PbO_2 + X{-}R \longrightarrow O{=}X{-}R + PbO$$

$$PbO + H_2O \longrightarrow PbO_2 + 2\ e + 2\ H^+$$

regeneration of PbO_2

The PbO_2 electrode can act both as an electrocatalytic electrode by chemically oxidizing the organic species and as inert electrode for direct ET reactions as in the oxidation of phenolic compounds. Anodic oxidations of alcohols, aldehydes and amines at oxide covered electrodes of nickel or silver take place indirectly (electrocatalytically) by chemical reaction of the surface oxides with the organic substance. These oxidations occur usually at potentials 0.4 to 1.5 V vs SCE, in aqueous or aqueous-organic NaOH or KOH solutions. At these potential ranges the metal surface oxides are formed and are continuously regenerated during the electrolysis. If the organic substrate is not adequately soluble in the alkaline water solution a cosolvent, such as *tert*-butanol, may be used. Two primary oxidation paths are possible:

1. Direct electron transfer from the organic substrate to the electrode.
2. hydrogen atom abstraction from the organic substance by the surface oxide species [5–9].

Fleischmann and Pletcher [7] and Robertson [9] postulated a process of hydrogen atom abstraction by NiOOH formed in situ at a 0.45 V vs

SCE. The oxidation route of an alcohol to carboxylate at a nickel oxide electrode is depicted in Scheme 2.6. Both H˙ abstraction and ET steps are involved in the overall scheme.

Nickel oxide electrode

$$\mathrm{Ni} \xrightarrow[\text{(1 M KOH)}]{\mathrm{H_2O/OH^-}} \mathrm{Ni^{II}(OH)_2 + 2\,e + 2\,H^+} \quad \text{(spontaneous reaction)}$$

$$\mathrm{Ni^{II}(OH)_2 + OH^-} \underset{\text{0.45 V (vs. SCE)}}{\rightleftharpoons} \mathrm{Ni^{III}OOH + H_2O + e} \quad \text{(active oxide formation on electrode surface)}$$

$$\mathrm{Ni^{III}OOH + RCH_2OH} \xrightarrow[\text{rds}]{-\,\mathrm{H^{\cdot}}} \mathrm{Ni^{II}(OH)_2 + R\dot{C}HOH} \quad \text{(hydrogen abstraction step)}$$

$$\mathrm{R\dot{C}HOH} \xrightarrow[\text{anode}]{-\,e} \longrightarrow \longrightarrow \mathrm{RCOO^-}$$

overall anodic reaction:

$$\mathrm{RCH_2OH + 5\,OH^-} \longrightarrow \mathrm{RCOO^- + 4\,e + 4\,H_2O}$$

Scheme 2.6

The abstraction of hydrogen atom is the slowest step in the overall reaction leading to the RCO_2^- product. Electrocatalytic oxidations at these oxide electrodes consume four faradays and five moles of OH^- ion per mole of carboxylate in the product. This is exemplified by the oxidations of glycerol [10] and propargyl alcohol [11] to glyceric and propiolic acids, respectively, and the oxidation of *p*-xylene glycol to terephthalic acid [12], Scheme 2.6a,b,c.

Alcohol oxidation at NiOOH

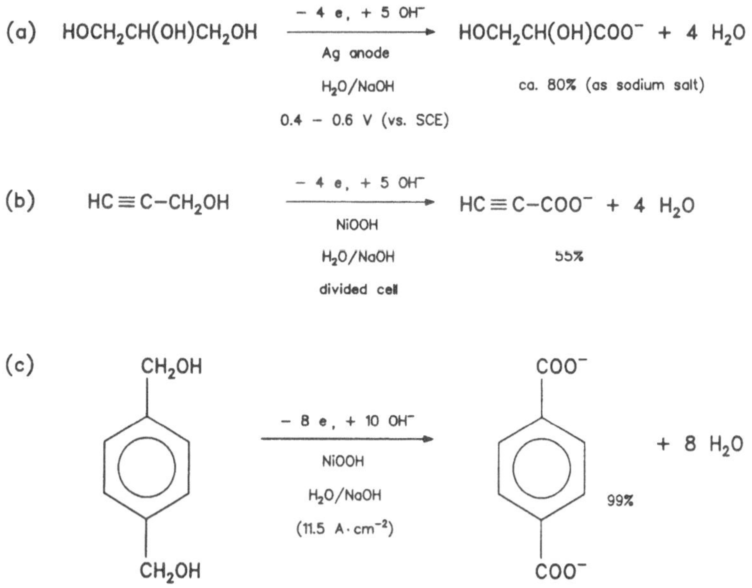
(a) HOCH₂CH(OH)CH₂OH
− 4 e, + 5 OH⁻
Ag anode
H₂O/NaOH
0.4 − 0.6 V (vs. SCE)
HOCH₂CH(OH)COO⁻ + 4 H₂O
ca. 80% (as sodium salt)
(b) HC≡C−CH₂OH
− 4 e, + 5 OH⁻
NiOOH
H₂O/NaOH
divided cell
HC≡C−COO⁻ + 4 H₂O
55%
(c) CH₂OH
CH₂OH
− 8 e, + 10 OH⁻
NiOOH
H₂O/NaOH
(11.5 A·cm⁻²)
COO⁻
COO⁻
99%
+ 8 H₂O

Scheme 2.6a,b,c

Steroid oxidation at NiOOH

Schäfer and Kaulen reported a very efficient oxidation of several alcohols and steroids at a NiOOH electrode in H_2O/NaOH solutions [5]. Typical yields of carboxylic acids obtained from the corresponding alcohols were as follows:

$n-C_7H_{15}CH_2OH$,   acid yield   65%

$n-C_3H_7CH_2OH$,   "   85%

$n-C_{11}H_{23}CH_2OH$,   "   80%

$C_2H_5CH_2OH$,   "   85%

The α,ω-diols afforded the corresponding dicarboxylates. We may note here that chemical oxidations by chromic acid or permanganate are less selective. Also direct anodic oxidation at platinum or carbon electrodes is not selective because of the very positive potentials needed for these oxidations, relative to the much *milder* oxidative conditions at the nickel oxide electrode in alkaline solutions at room temperature. The oxidation of steroids containing 3α- and 3β-hydroxy groups in the presence of 11β- and 20β-hydroxy positions was also found by Schäfer to be "surprisingly selective" at the NiOOH electrode [5]. The Openauer oxidation (a ketone with base as oxidizing agent) gives the 3-lactones, and chromic acid attacks preferentially the 11β-hydroxy group, in contrast to the anodic oxidation at the NiOOH electrode which favors oxidation at the 3β-hydroxy position, – probably because this position is closest to the electrode surface.

36% isolated

These oxidations can be conducted in one-compartment cells with a nickel net to function as anode and a stainless steel net as cathode. Electrolysis media are aqueous 0.1–0.3 M NaOH, or mixtures of water and *tert*-butanol with 0.1–0.3 M KOH, containing the organic substrate to be oxidized. The electrolysis is carried out at constant current at room temperature by applying an anode potential at least 0.45 vs SCE so as to cause the formation of the desired metallic oxide on the nickel electrode. The products are extracted with ether from the acidified (HCl) aqueous solution.

Oxidation of amines

Anodic oxidations of amines occur easily at Ni, Ag, and Cu electrodes covered with their respective oxides in aqueous or aqueous-organic basic solutions. These oxidations transform the amines to the corresponding cyanides or ketones as indicated below. Ethylcyanide and propylcyanide are obtained from propylamine and *n*-butylamine, respectively, in 85% current yields [10].

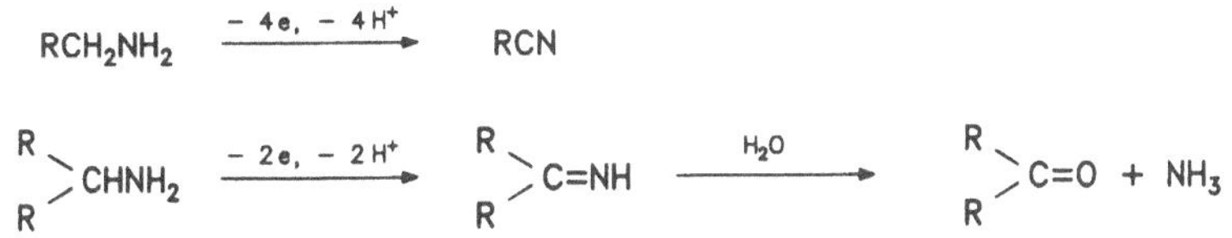

$$RCH_2NH_2 \xrightarrow{-4e, -4H^+} RCN$$

$$\begin{matrix} R \\ R \end{matrix} CHNH_2 \xrightarrow{-2e, -2H^+} \begin{matrix} R \\ R \end{matrix} C=NH \xrightarrow{H_2O} \begin{matrix} R \\ R \end{matrix} C=O + NH_3$$

Hydrophobic aromatic compounds, like the typical methylol compounds shown below, are very efficiently transformed into their carboxylic acid analogs by anodic electrolysis at NiO/NiOOH electrodes in H$_2$O/*tert*-butanol/KOH media [13].

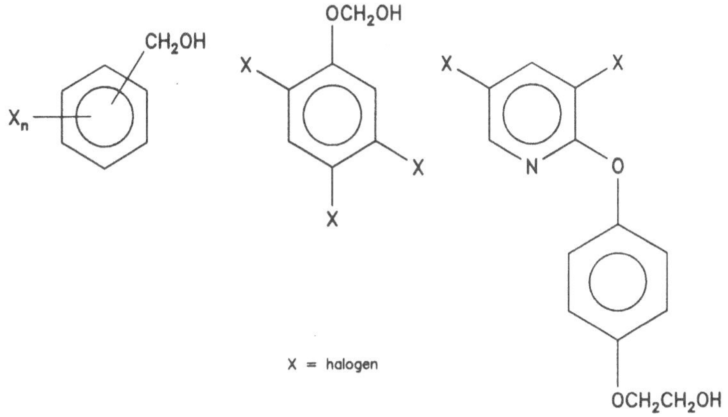

Oxidation of aromatics

X = halogen

The electrolysis is performed at 20–60 °C at anode potentials between 0.5 and 1.0 V vs SCE. The cosolvent, *tert*-butanol enhances the solubility of the hydrophobic compounds. The products are recovered by acidifying with hydrochloric acid, extracting with CH$_2$Cl$_2$ and stripping off the solvent. Identification of products is made by comparing spectroscopic and chromotographic data with those of authentic samples. Yields of 70–90% are obtained.

Note: Usually – but not always necessary, the nickel electrode is activated before the preparative electrolysis by applying short polarity reversals on the electrode immersed in a solution containing some NiSO$_4$, NaOAc and NaOH. This is done until the nickel surface turns black, indicating that the electrode is covered with the active nickel oxides.

The anodic oxidation of propargyl alcohol to propiolic acid at a nickel electrode in aqueous NaOH solution will be described in detail – especially since this oxidative transformation can also be achieved in an electro-generative cell (Chapter 5). Because propargyl alcohol contains an acetylenic bond, a divided cell is necessary to avoid reduction of this bond

Oxidation of propargyl alcohol

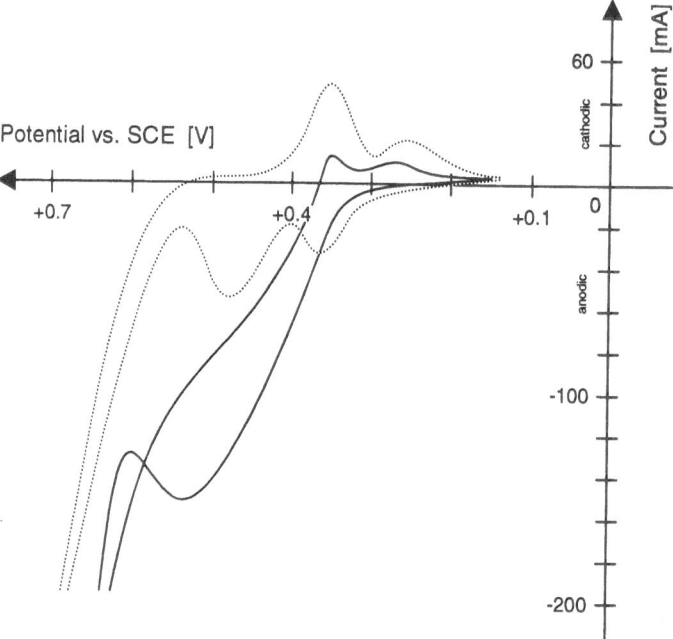

Fig. 2.2. Cyclic voltammogram of propargyl alcohol at a Ni wire electrode (2M NaOH). ....
blank; — with propargyl alcohol, 0.01 M; scan rate 100 mV/s

and formation of acrylic acid instead of propiolic acid. In an undivided cell
most of the propargyl alcohol is converted to acrylic acid. The anodic
potential for this oxidation was chosen on the basis of the cyclic voltammogram, Figure 2.2, using a nickel wire as anode in H₂O/NaOH solution
in the presence and absence of propargyl alcohol. As is apparent from the
blank, formation of the active nickel oxides occurs at 0.4–0.5 V vs SCE.
The oxidation of the alcohol also occurs in that potential range.

Oxidation of Propargyl Alcohol to Propiolic Acid

The electrolysis was performed in an H-type cell with a Nafion
cationic membrane as the divider, Figure 2.3. The potentiostat was
an Electrosynthesis Co. model 415 potentiostatic controller with a
36 V–5 A power supply. The anode was a nickel foil, 50 cm² exposed
area and the cathode a nickel foil or a graphite rod. Typically, a
solution of 2 g (0.03 mol) of propargyl alcohol was dissolved in 50 ml
of 2 M aqueous NaOH solution. The catholyte was 50 ml of 2 M
NaOH. Electrolysis was conducted at 5–8 °C (optimum conditions)
with the cell immersed in an ice bath, and at anodic potentials
between 0.5 to 1.0 V vs SCE (sometimes it was necessary to raise the
potential to 2.0 V for a few seconds in order to restore the activity of

the electrode). Currents of 1 to 0.3 A were passed until the total charge consumed was about 10–20% in excess of the theoretical. At the end of the electrolysis, the cold anolyte solution was slowly acidified to pH 1 with cold concentrated hydrochloric acid. The solution was saturated with sodium chloride and extracted with ether (2 × 50 ml). The combined ether extracts were treated with anhydrous magnesium sulfate or sodium sulfate and the ether was evaporated (vacuum). The liquid residue (1.2 g) was analyzed by chromatography and was identified as propiolic acid by comparing IR and NMR spectra with spectra of an authentic sample. Yield 55%, 99% purity.

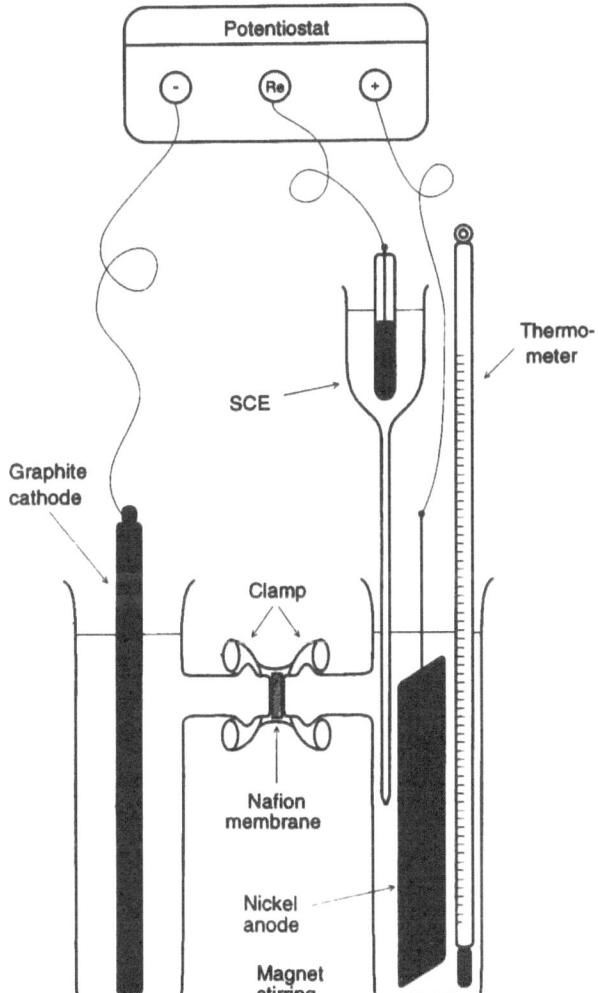

Fig. 2.3. Schematic of H-type cell for the anodic oxidation of propargyl alcohol to propiolic acid

Note: Similar results were obtained by electrooxidation in 1 M aqueous H_2SO_4 solution in a divided cell with a lead anode (PbO_2) and a graphite cathode. This method is used by BASF for the manufacture of propiolic acid.

Oxidation at adatoms

We may at this juncture digress to refer very briefly to the anodic oxidation of lower alcohols at *adatom* electrodes by the underpotential deposition method (UPD) [14]. Adatoms enhance the oxidation rate (of electrocatalysis, heterogeneous type) by preventing adsorption of *poisoning* species on the electrode surface, or perhaps by oxidation of the alcohol by oxygen containing species adsorbed on the foreign adatoms. The UPD adatoms are usually metals of Pb, Bi, Re, Ge, Cd deposited on precious metals, under potentials less negative than thermodynamic potentials. The oxidation of 1,2-propanediol [15], for example, was proposed to take place electrocatalytically at platinum electrodes modified by UPD adatoms of Re or Pb. The first step is a dissociative adsorption process. Oxidation of H_2O generates the active OH˙ species, adsorbed on the electrode surface. These react with the adsorbed fragment of the alcohol, as shown in Scheme 2.7.

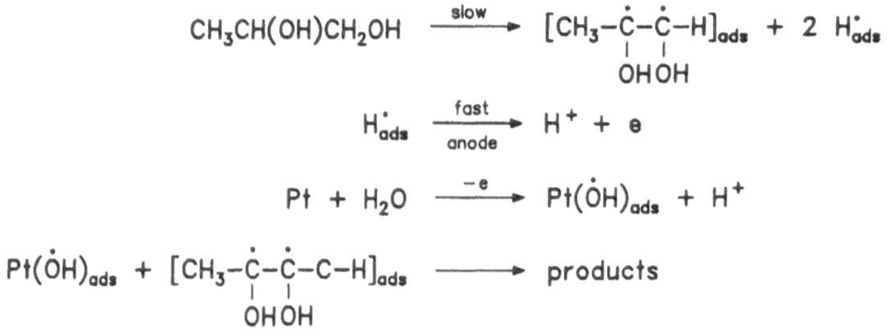

$$CH_3CH(OH)CH_2OH \xrightarrow{\text{slow}} [CH_3-\overset{\cdot}{\underset{|}{C}}-\overset{\cdot}{\underset{|}{C}}-H]_{ads} + 2\ H^{\cdot}_{ads}$$
$$\phantom{CH_3CH(OH)CH_2OH \xrightarrow{slow} [CH_3-}OH\,OH$$

$$H^{\cdot}_{ads} \xrightarrow[\text{anode}]{\text{fast}} H^+ + e$$

$$Pt + H_2O \xrightarrow{-e} Pt(\overset{\cdot}{O}H)_{ads} + H^+$$

$$Pt(\overset{\cdot}{O}H)_{ads} + [CH_3-\overset{\cdot}{\underset{|}{C}}-\overset{\cdot}{\underset{|}{C}}-C-H]_{ads} \longrightarrow \text{products}$$
$$\phantom{Pt(OH)_{ads} + [CH_3-}OH\,OH$$

Scheme 2.7

2.5 Phenolic Compounds

The electrochemistry of phenolic substances is more versatile and also more complex than that of other hydroxy compounds [1]. In phenols the hydroxy group releases electrons to the ring and thus facilitates the release of ring electrons to the anode. The anodic deelectronation of phenol is indicated by Swenton as follows [2]:

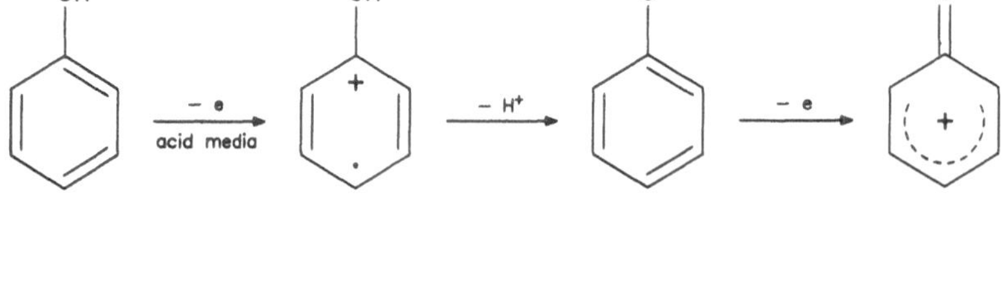

In general, phenols are oxidized by strong oxidizing agents, dichromates, periodates, etc. in very acidic solutions. The anodic method would certainly be a good alternative for oxidation of phenols, especially for large scale production. Most phenols exhibit reversible anodic behavior in the time scale of cyclic voltammetry. As expected, phenols with more than one hydroxy groups are more easily oxidized than the monohydroxy analogs, as is apparent from their voltammetric peak potentials [3]: phenol, 0.53, catechol, 0.14, hydroquinone, 0.02V vs SCE. Alkoxy and Ar—O ring substituents also facilitate oxidation of phenols at the anode. The oxidation of phenols yields cations and neutral species as intermediates, Schemes 2.8 and 2.9, leading to various products in high yields [3]. In the presence of nucleophiles a wide variety of products can be obtained; for example, coumestan derivatives and related heterocyclic compounds [4], Scheme 2.9. Anodic electrolysis of phenols, in general, can be conducted in aqueous and aqueous-acetone solutions, which can be acidic, neutral or basic. Products like those of Scheme 2.9 are physiologically interesting and can be prepared in good yields by the electrochemical method.

Oxidation

Preparation of Coumestan Compounds. General Electrolysis Procedure [4]

Electrolysis is performed in aqueous 0.1–1M sodium acetate solutions containing the starting materials, for example, catechol, A, and 4-hydroxy coumarin, B, Scheme 2.9, in an undivided cell with graphite or platinum electrodes at anodic potentials of 1.0–1.2V vs SCE. Because passivating films may sometimes cover the anode periodic removal of the films is required: Washing the anode with acetone restores the electrode active surface. The products are recovered from the electrolyzed solution after acidification with cold acetic acid whereupon the products precipitate and are then purified by recrystallization from methanol-acetone-ethanol mixed solvent or from acetone alone. Isolated yields are in the range of 90–95%.

Parker and coworkers studied the anodic oxidation of natural phenolic compounds, the α-tocopherol family [4,5].

α-Tocopherol is oxidized reversibly with peak potentials at 0.74V and 0.80V vs SCE in CH_2Cl_2/TFA/Bu_4NF_4 medium. The primary oxidation product is the cation radical which is further transformed, as shown in Scheme 2.10.

Oxidation of natural phenolic compounds

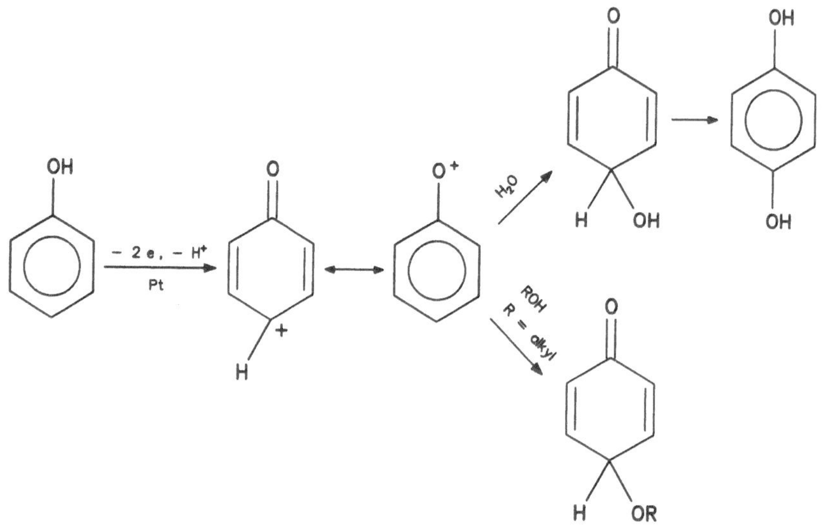

Scheme 2.8

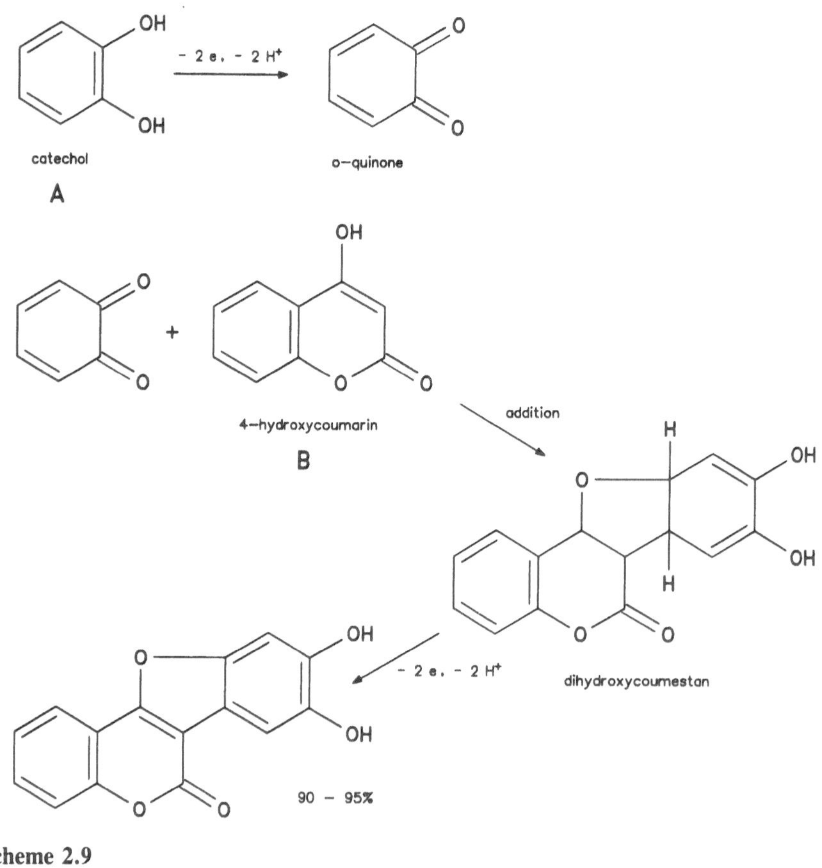

Scheme 2.9

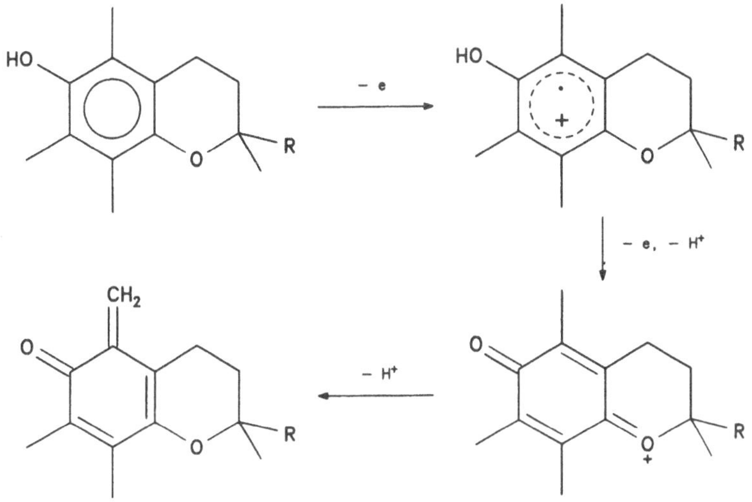

Scheme 2.10

Sterically hindered cresols, when oxidized in basic aqueous solutions on graphite anodes, afford dimers [5]. In acidic solutions phenoxonium ions are formed, Scheme 2.11.

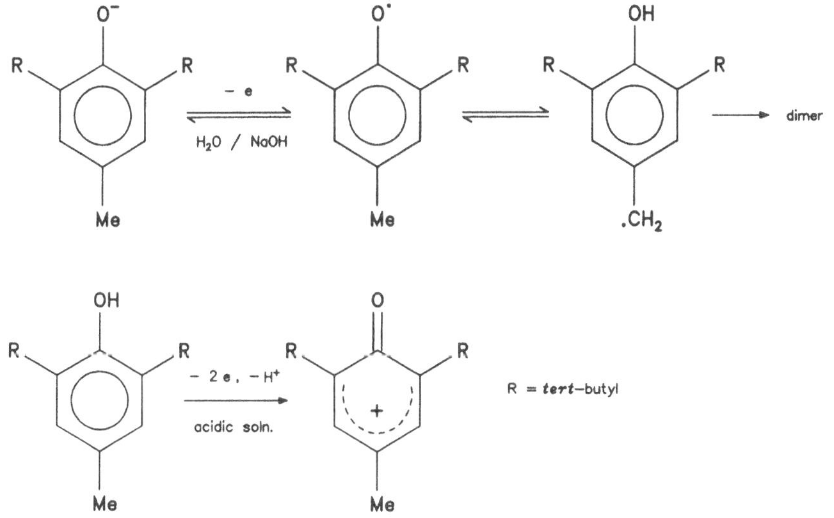

Scheme 2.11

Anodic couplings of methoxy-substituted phenols in methanolic-sodium hydroxide or in acetonitrile-sodium perchlorate solutions in divided cells afford quinones [6]. The oxidation of 2,6-dimethylphenol gave the C—O and the C—C coupled products, depending on the electrolysis medium in

Couplings of phenoes

70–90% yields, and also poly(2,6-dimethyl-1,4-phenyleneoxide (PPO), as indicated in Scheme 2.12.

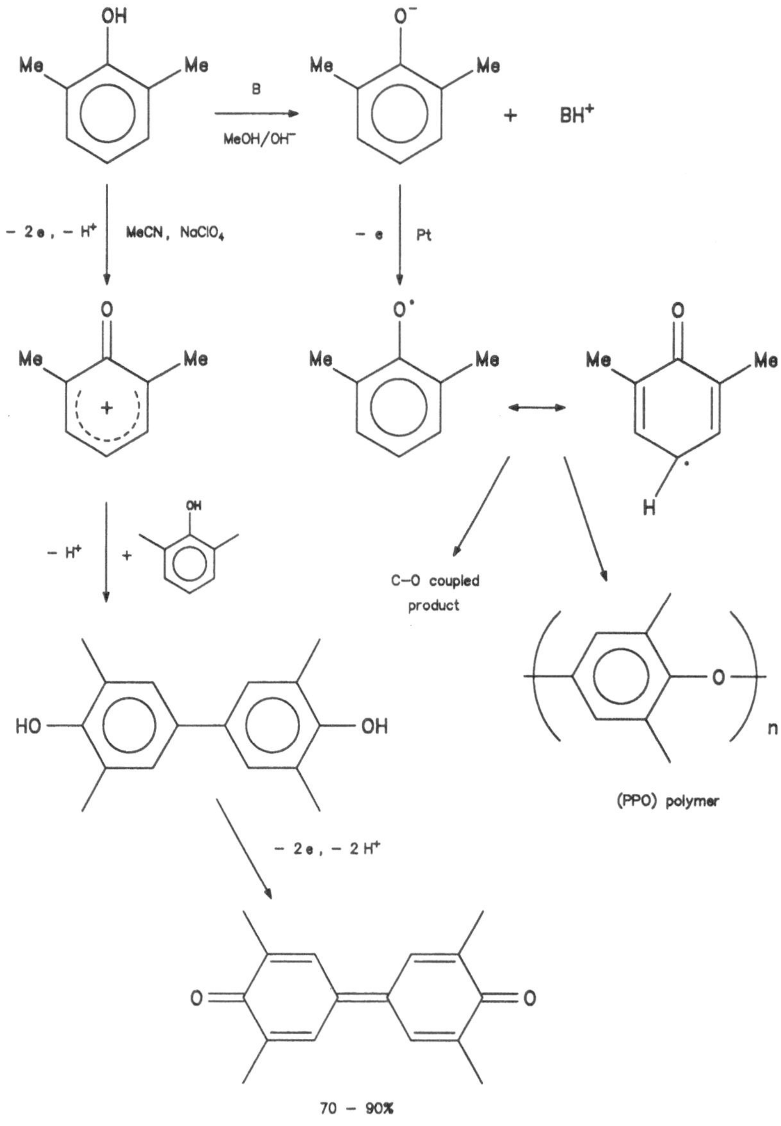

Scheme 2.12

Polymer formation The polymer PPO (via C—C coupling) is most favorably formed in the MeOH/NaOH solution, but the dimer (via C—C coupling) forms in MeCN/NaClO₄ solution by electrophilic attack of the electrogenerated phenoxonium ion on the starting material. The C—O coupling of the phenoxy radicals leading to the PPO polymer proceeds most likely according to the following path:

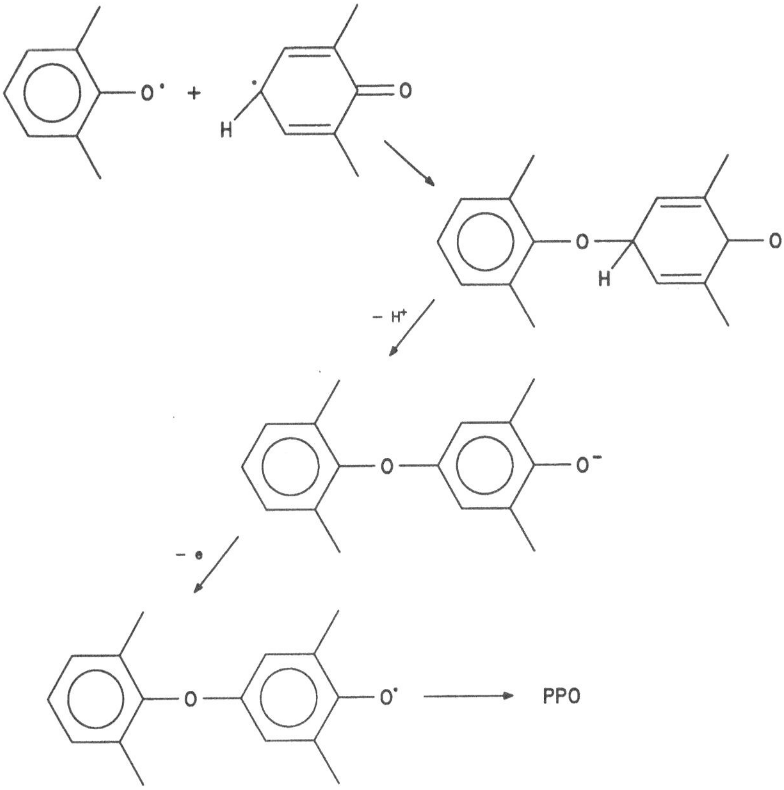

These electrolyses are conducted in divided cells with platinum electrodes under potentiostatic conditions [6]. Recently, Torii and coworkers reported an interesting anodic coupling of phenols [7]. The electrolysis was conducted in a divided cell as if it were a *paired* electrochemical process. The 2,6-*tert*-butylphenol was oxidized at a platinum anode in CH_2Cl_2/MeOH/LiClO$_4$ solution to give the diphenoquinone product which, by changing the current direction, was reduced at the platinum cathode to the biphenol product in high yield, Scheme 2.13.

In this connection we may consider the possibilities of applying electrochemical methods to the degradation of biomass materials such as the lignins, which are ploymeric phenols. According to a recent study by Utley and coworkers [8], the electrochemical oxidation might be a better alternative than the chemical method. The latter requires quite severe reaction conditions, including the use of nitroaromatics as oxidants in concentrated alkali and temperatures of 150–200 °C. The anodic electrolysis achieves

Electro-chemical degradation of biomass materials

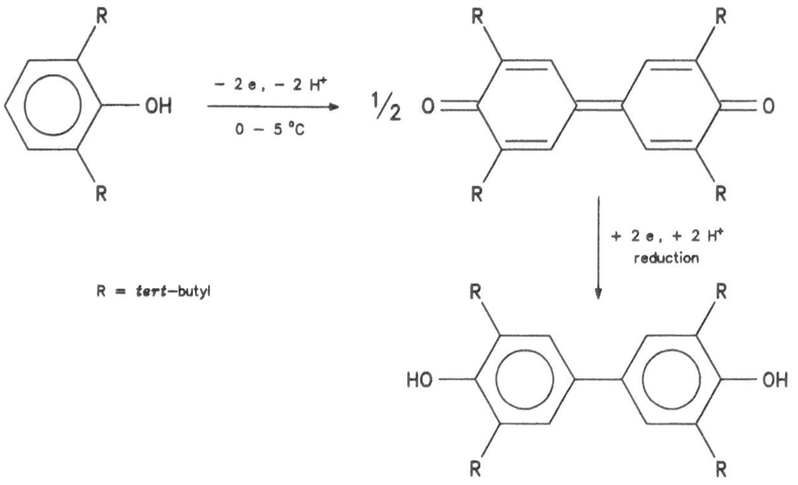

Scheme 2.13

similar results at ambient temperatures with nickel electrodes in aqueous alkaline solution. Thus an organo-sol spruce lignin with some nitrobenzene as additive to facilitate the oxidation afforded vanillin and syringaldehyde in 8–50% isolated yields. Recently, electrochemical oxidation of lignin in 1 N NaOH solution was suggested for the electrolytic production of hydrogen, the lignin acting as anodic depolarizer [9].

Dimerization of Quinones
Quinones and derivatives thereof are easily dimerized by electrooxidation in 50% aqueous acetic acid containing $MgClO_4$ as electrolyte, in divided cells with platinum electrodes [10]. It is noted that chemical methods give dimer yields of less than 20%, as contrasted with 80% by the electrolytic oxidation method:

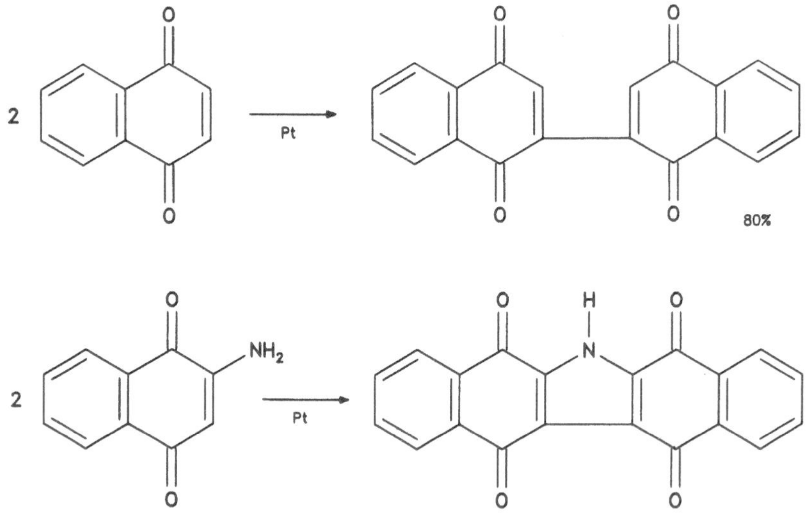

2.6 Carbonyl Compounds

Aldehydic carbonyls are good electroactive groups because they can undergo both reduction and oxidation at the cathode and anode, giving alcohols or carboxylic acids, respectively [1]. Ketonic carbonyls are also electroactive, although to a lesser degree than the aldehydic carbonyls. In principle, aldehydes can be oxidized to the carboxylic acid state by removal of two electrons from the aldehyde, in both acidic and basic aqueous or aqueous-organic solutions, Scheme 2.14. In acidic solutions platinum, graphite and PbO_2 are the usual anodes while in basic solutions oxide-covered electrodes of Ni, Ag, Cu, Co and Mn are the most effective electrodes, especially the first two. Direct anodic oxidation of potentially available aldehydic functions in carbohydrates have not yet been synthetically successful because of concurrent degradative reactions. Indirect (mediated) oxidations have been considerably more successful [2,3,4]. The direct oxidation of monosaccharides in aqueous alkaline solutions has been considered and studied in depth by Bockris and coworkers for fuel cell applications [3].

Direct and indirect oxidation

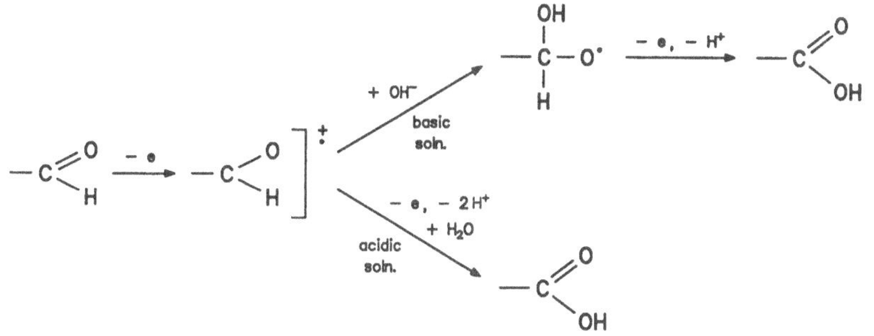

Scheme 2.14

Chiba [4] reported that the aldehydes C_6H_5CHO, $2\text{-}CH_3\text{-}C_6H_4CHO$, $2Cl\text{-}C_6H_4CHO$ and some others are anodically oxidized in MeOH/NaCN solutions at less positive potentials than in MeOH/LiClO$_4$ solutions. Apparently cyanohydrins, formed in situ, are involved in the initial chemical step. This is followed by oxidation of the cyanohydrin to the corresponding acyl cyanides, which then react with methanol to yield methyl esters (73–80%).

Ketones of the general formula RCOR [1] where R is benzyl, or alkyl secondary or tertiary group undergo α-cleavage of the cation radical generated at the anode, in MeCN/LiClO$_4$ solution containing small amounts of water [5,6]. Reactions, as depicted in Scheme 2.15, follow the first step of electron transfer to the anode.

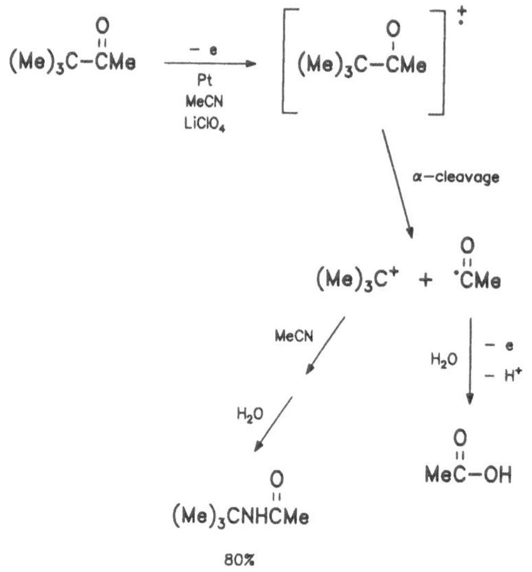

$$(Me)_3C-\overset{\overset{O}{\|}}{C}Me \xrightarrow[\substack{Pt \\ MeCN \\ LiClO_4}]{-e} \left[(Me)_3C-\overset{\overset{O}{\|}}{C}Me \right]^{\cdot+}$$

α−cleavage

$$(Me)_3C^+ \; + \; \cdot\overset{\overset{O}{\|}}{C}Me$$

MeCN

H₂O | −e / −H⁺

H₂O

$$(Me)_3C\overset{\overset{O}{\|}}{N}HCMe$$

$$Me\overset{\overset{O}{\|}}{C}-OH$$

80%

Scheme 2.15

A quite interesting, and rather unexpected, reaction occurs with ketonic substances lacking branching at the α-position. These compounds undergo oxidation by a route involving hydrogen abstraction from a CH_2 carbon remote from the carbonyl group [7], as depicted in Scheme 2.16. The first electrodic step is a one-electron removal step from the ketone. This step is apparently followed by intramolecular attack on the distant C—H bond by the oxygen of the carbonyl group. A second electron is released to the anode resulting in the generation of the carbonium ion and finally the acetamido compound.

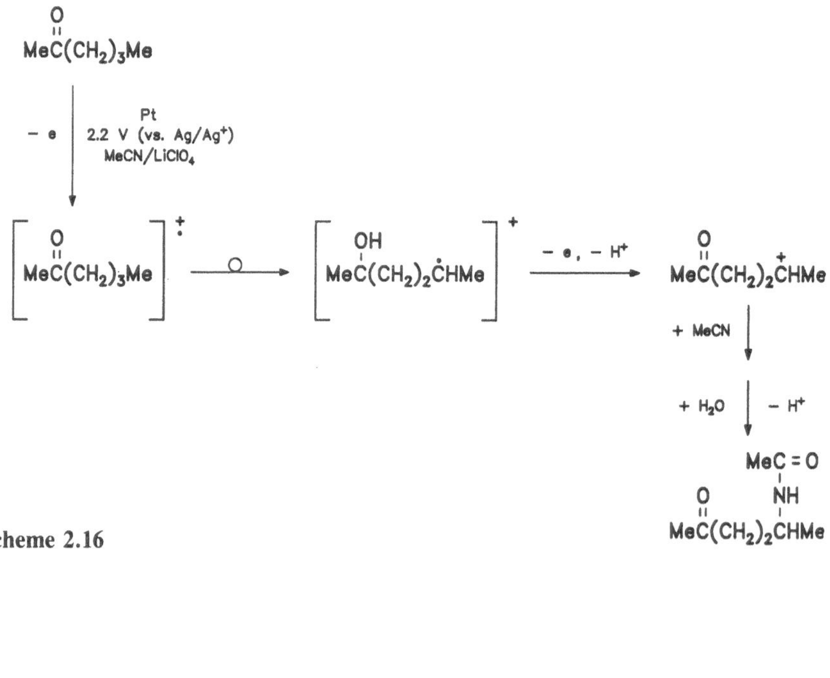

$$Me\overset{\overset{O}{\|}}{C}(CH_2)_3Me$$

−e | Pt 2.2 V (vs. Ag/Ag⁺) MeCN/LiClO₄

$$\left[Me\overset{\overset{O}{\|}}{C}(CH_2)_3Me \right]^{\cdot+} \xrightarrow{\;\;O\;\;} \left[Me\overset{\overset{OH}{|}}{C}(CH_2)_2\dot{C}HMe \right]^{+} \xrightarrow{-e,\,-H^+} Me\overset{\overset{O}{\|}}{C}(CH_2)_2\overset{+}{C}HMe$$

+ MeCN

+ H₂O | − H⁺

$$\begin{array}{c} MeC=O \\ | \\ NH \\ | \\ Me\overset{\overset{O}{\|}}{C}(CH_2)_2CHMe \end{array}$$

Scheme 2.16

Methylcyclohexanones undergo ring opening by electrolysis in EtOH/ H_2O/NaClO$_4$ solutions at platinum electrodes [8]. Rearrangement of the opened ring gives lactones, Scheme 2.17.

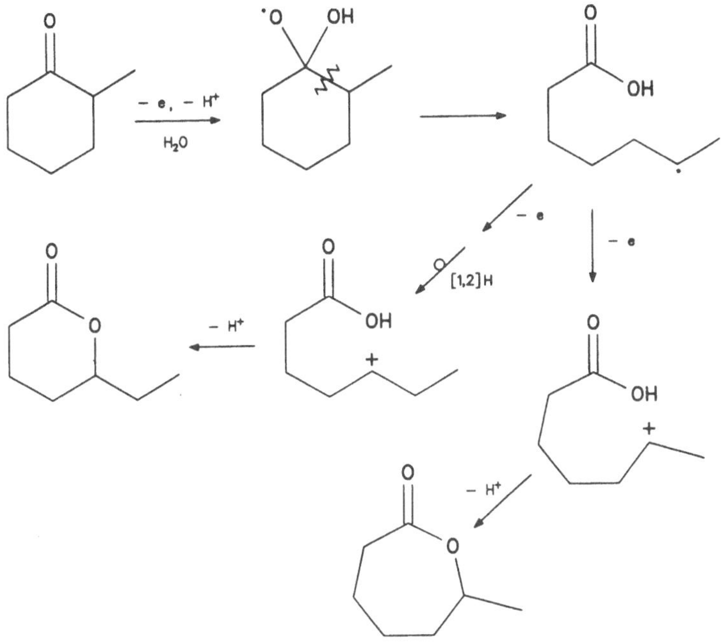

Scheme 2.17

Camphor (**1**) is transformed to lactone (**3**) via the intermediate (**2**) which apparently is produced at the PbO$_2$ anode by OH˙ addition to the carbonyl group, as a first step. [9] Further deelectronation and expulsion of a proton yields the lactone (**3**),

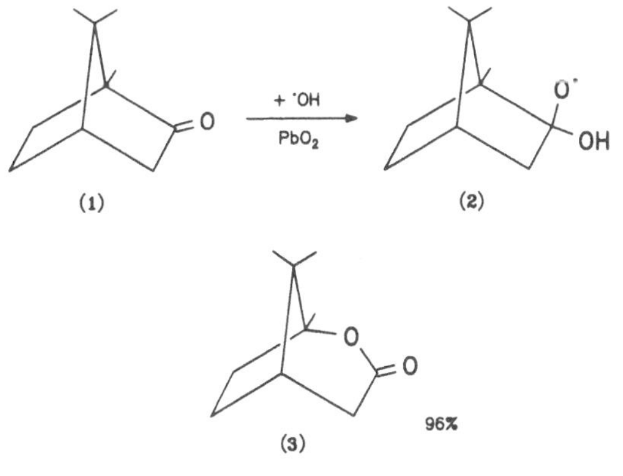

The electrolysis was performed in a mixed solvent consisting of aqueous 1MH$_4$SO$_4$ and acetonitrile in 1:1 volume ratio under a current of 10 mA/cm^2 at 20 °C with PbO$_2$ or Pt anodes. This electrolysis was also

demonstrated with a plate and frame cell divided with a cation exchange membrane. Yields of 62–82% were obtained depending on electrolysis time. Certain carbonyls with structures possessing acidic hydrogens (carbon acids) yield anions by reaction in methanol solutions containing sodium methoxide (formed by adding sodium metal to the alcohol). These anions are readily oxidized at the anode so that substitutions are possible [9,10], as depicted in Scheme 2.18

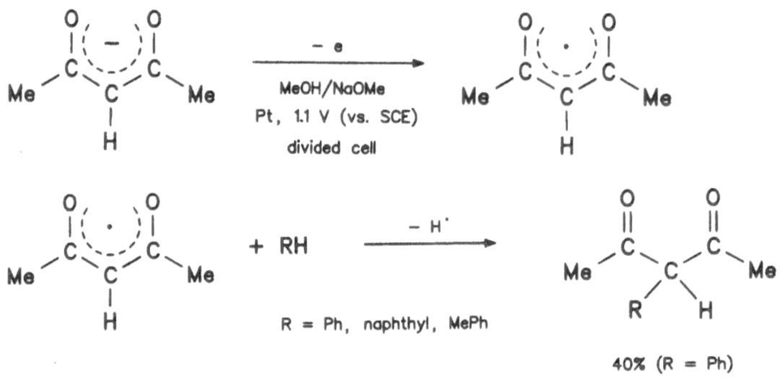

Scheme 2.18

Anodic electrolysis of hydrazine derivatives of aldehydes and ketones in a divided cell, with carbon anode and platinum cathode, in MeOH/NaOAC solution, is an efficient method for the preparation of oxadiazolines. Chiba and Okimoto [11] proposed a reaction mechanism as depicted in Scheme 2.19 for the cyclization of N-acylhydrazones (1) to afford the Δ^3-,1,3,4-oxadiazoline (3). Yields of 30–77% are obtained depending on the type of R^1, R^2, R^3 substituents.

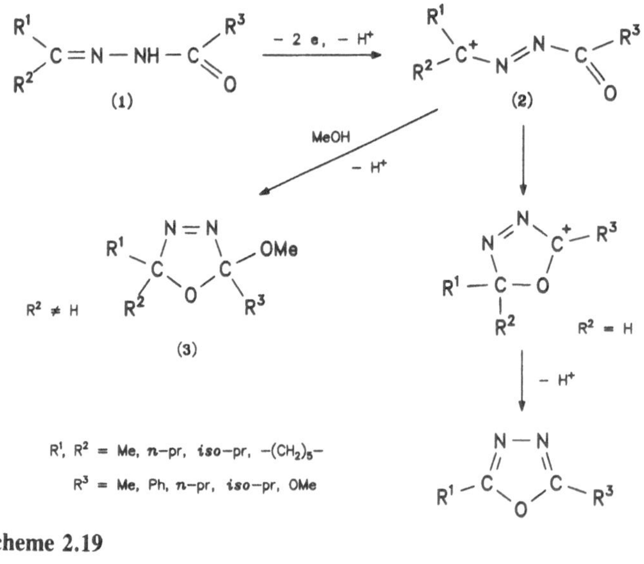

Scheme 2.19

Preparation of Oxadiazolines, (Scheme 2.19)

The electrolysis is conducted at room temperature in a 100-ml cell divided with a glass frit, and equipped with a carbon rod anode and a platinum foil cathode. Typically, 50 mmol of *N*-acylhydrazone is dissolved in 80 ml of methanol containing 15 mmol of sodium acetate. After 3.0 F/mol of charge is passed, at a constant current of 0.05 A, the methanol of the anolyte is evaporated in vacuo at room temperature. Water is added to the residue and the two phases are separated. The organic layer is extracted with ether (20 ml × 3) and the combined extracts are washed with water, dried with anhydrous sodium sulfate and concentrated in vacuo at room temperature. The residue is purified by silia-gel colum chromatography with benzene as eluant (Adapted from (1992) J Org Chem 57:1375. Copyright American Chemical Society, 1992).

2.7 Carboxylic Acids

2.7.1 General Perspective

Electroorganic chemistry, it may be said, began with a typical C—C bond formation – which is the innermost essence of all organic chemistry. This first electrochemical reaction is now widely known as the *Kolbe Synthesis*[1]. In 1834 Faraday first observed the evolution of ethane during electrolysis of aqueous acetate solutions at smooth platinum electrodes. Kolbe later studied the practical possibilities of this electrolytic decarboxylation, although neither he nor Faraday had any idea of the oxidative reaction mechanism as it is accepted today. Kolbe had the plausible idea that it was the nascent oxygen, generated at the anode, that oxidized the acetic acid to give the products, ethane, carbon dioxide and water. The Kolbe reaction is now generally believed to occur directly, by an overall route initiated by an electron transfer from the carboxylate to the anode, as depicted below [2–6]:

Kolbe synthesis

$$RCH_2COO^- \xrightarrow[\substack{Pt \\ 2.2 \text{ V (vs. SCE)}}]{-e} [RCH_2COO]^{\bullet}_{ads} \longrightarrow [RCH_2]^{\bullet}_{ads} + CO_2$$

$$2 [RCH_2]^{\bullet}_{ads} \longrightarrow RCH_2CH_2R$$

Kolbe product

The Kolbe reaction, or more generally the electrochemical decarboxylation of aliphatic carboxylic acids, is a most useful reaction in the organic and bioorganic fields of synthesis. A formal summary of possible reaction routes initiated by the electron transfer step is presented in Scheme 2.20.

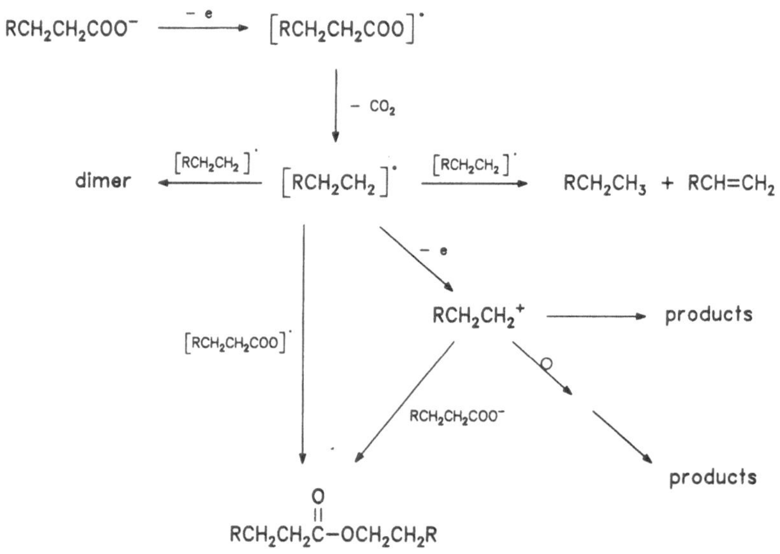

$$RCH_2CH_2COO^- \xrightarrow{-e} [RCH_2CH_2COO]^{\cdot}$$

$$\downarrow -CO_2$$

$$dimer \xleftarrow{[RCH_2CH_2]^{\cdot}} [RCH_2CH_2]^{\cdot} \xrightarrow{[RCH_2CH_2]^{\cdot}} RCH_2CH_3 + RCH=CH_2$$

$$\searrow -e$$

$$RCH_2CH_2^+ \longrightarrow products$$

$$[RCH_2CH_2COO]^{\cdot} \qquad \qquad RCH_2CH_2COO^- \qquad \qquad products$$

$$\underset{\substack{O \\ \parallel}}{RCH_2CH_2C-OCH_2CH_2R}$$

Scheme 2.20

Solvents The most common solvents for Kolbe electrolysis are water, methanol and mixtures thereof. For some special applications absolute methanol, dimethylformamide (DMF) and acetonitrile are employed. The kind of electrode material and the composition of the electrolysis medium generally have profound effects on the reaction paths, and hence on the nature of the final products. The electroactive species is the carboxylate ion. In aqueous solutions 10–15% of the carboxylic acid is neutralized with NaOH or some other suitable base. Usually, the Kolbe electrolysis is performed in one-compartment cells with the anode potential at 2.2 V vs SCE. At this potential oxidation of carboxylate occurs almost at the exclusion of oxygen evolution. Aromatic acids and α,β—unsaturated carboxylates have not yet been oxidized by the Kolbe method to give C—C coupled products in synthetically significant yields. The Kolbe synthesis occurs most efficiently at smooth platinum electrodes. At graphite or PbO$_2$ electrodes the initial radicals are apparently rapidly deelectronated to afford carbonium ions and products typical thereof. Namely, alcohols, esters, ethers, and olefins. Thus some synthetically useful versions of the Kolbe reaction emerged, and are more or less known by their own names:

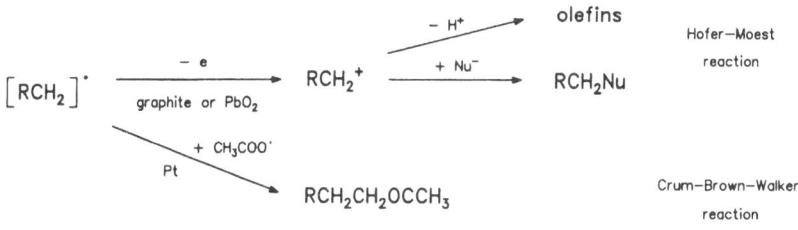

The anodic decarboxylation of aliphatic carboxylic acids is a typical example indicating the overpowering influence of kinetic factors in relation to thermodynamic factors. In aqueous solution the thermodynamically favored reaction

$$H_2O \xrightarrow[\text{anode}]{} {}^1\!/_2\, O_2 \;+\; 2\,H^+ \;+\; 2\,e$$

is almost entirely suppressed by kinetic factors (adsorption effects, etc), as that instead of oxygen evolution, decarboxylation of the organic acid takes place at the anode. Electrode surfaces may be modified in ways that electrocatalytic effects can variously alter the course of decarboxylation reactions and the nature of the final products[6].

The Kolbe synthesis is a versatile and practical method for the synthesis of a great variety of organic compounds. Here only a few typical examples will be given for illustrative purposes. Sebacic acid has been produced on an industrial scale by the Kolbe method. The industrial process consists of three basic steps [7].

1. Esterification of adipic acid with methanol to obtain the monomethyl ester.
2. Electrolytic Koble decarboxylation of the ester to obtain the dimethyl ester of sebacic acid.
3. Hydrolysis of the dimethyl ester to obtain the sebacic acid.

An interesting Kolbe synthesis performed in the laboratory is the conversion of oxolane substituted carboxylic acids to γ-diketones, via the γ-bisdioxolanes produced by oxidation at a platinum anode in methanol solution [8], Scheme 2.21

Unsaturated carbon-carbon bonds are good acceptors of Kolbe radicals, as illustrated in Scheme 2.22, with butadiene as the acceptor hydrocarbon [9]. Low current densities would promote formation of addition products at the expense of radical-radical coupling.

Mixed coupling products and products resulting from the combination of carbocations with certain inorganic anions, NO_3^-, N_3^-, NF_2^-, can be obtained by electrolysis of mixtures of alkanoates, and mixtures of alkanoates with the inorganic ions:

$$RCOO^- \xrightarrow[-\,CO_2]{-\,e} R^\cdot \xrightarrow{-\,e} R^+ \xrightarrow{NO_3^-} RNO_3$$

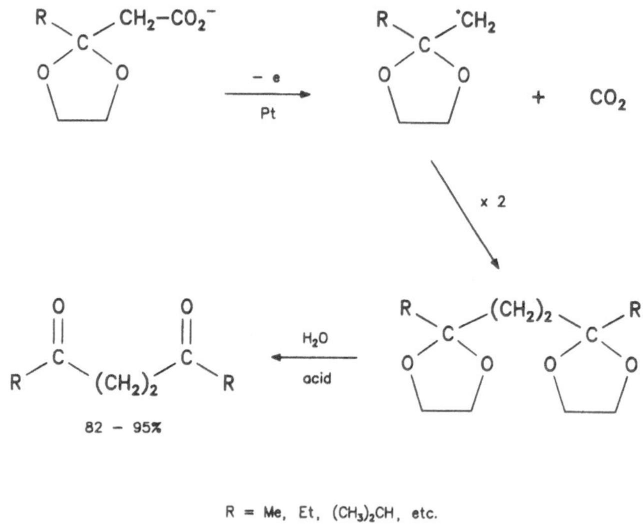

Scheme 2.21

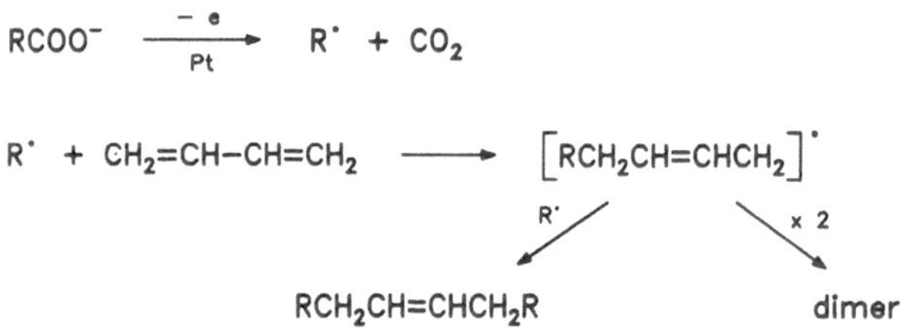

Scheme 2.22

$$R'COO^- \;+\; R''COO^- \xrightarrow[anode]{} R'-R'' \;+\; 2\ CO_2$$

Anodic decarboxylations may lead to aldehydes via one-electron oxidation in the presence of molecular oxygen, and to acetals and acetamides by two-electron oxidation in methanol or acetonitrile solutions [9a]:

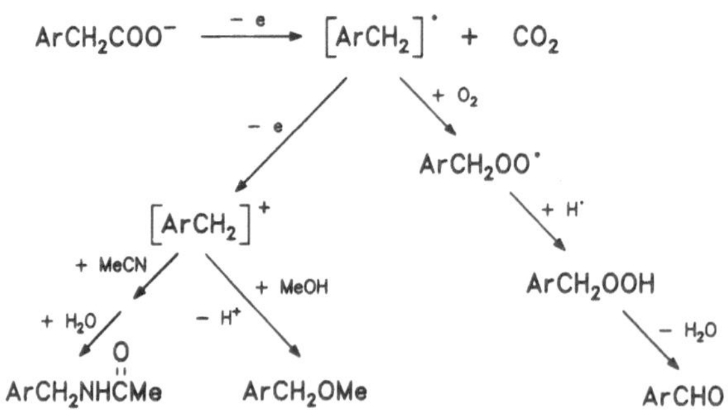

Oxidation of 2-ethoxyphenylacetic Acid. Formation of a Mixed Acetal

The electrolysis is performed in an undivided cell with platinum electrodes [9a]. The solution to be electrolyzed consists of 100 ml methanol, 4 g of carboxylic acid and some sodium methoxide to neutralize about 5% of the acid to form the carboxylate ion, which is oxidized at the anode. A cell voltage is applied that is sufficient to cause a current of $0.15\,A/cm^2$ to pass through the solution while the cell is cooled in an ice bath. Current is passed until the solution turns basic (pH paper) indicating that decarboxylation is complete. The methanol is evaporated under reduced pressure and the product is isolated by distilling the residue (b.p. 205 °C) 70% yield of the benzaldehyde ethyl methyl acetal by two-electron oxidation (Adapted from Ref. 9a, Copyright of American Chemical Society, 1962).

The introduction of the trifluoromethyl group to certain organic compounds is of considerable interest since fluorocarbons are actively studied for possible medicinal and other practical applications. Recently [10], it was found that anodic electrolysis in CF_3COOH (TFA)-MeCN/H_2O media containing compounds with β- and α-carbon-carbon unsaturated bonds results in the incorporation of trifluoromethyl and acetamide groups in the β- and α-carbons, respectively as illustrated in Scheme 2.23.

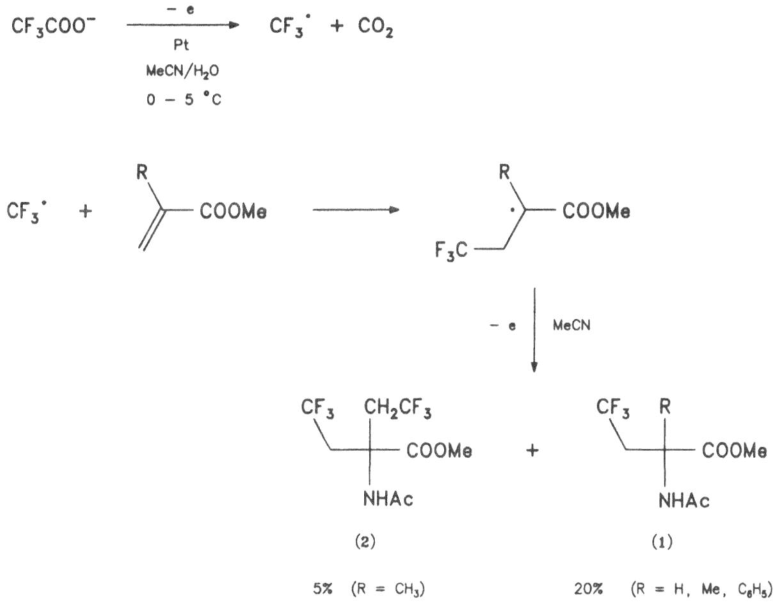

Scheme 2.23

Trifluoromethoxylation of Methylacrylate (Scheme 2.23)

Using methylacrylate (MMA) as a substrate to be trifluoromethylated, the electrolysis was conducted in a one-compartment cell containing 23 ml of mixed solvent MeCN/H_2O, in the volume radio of 20:3, platinum electrodes, 200 mg of MMA, 46 ml of TFA, and 24 mg of NaOH. A current of 24 mA (1 m A/cm^2) was passed for 9 h (or 4 F/mol of MMA) while the cell temperature was kept at 0–5 °C. The electrolyzed solution was neutralized with sodium bicarbonate, the organic layer was separated and the aqueous layer was extracted with ethyl acetate (3 × 4 ml). The combined extracts were washed with saturated sodium chloride solution, dried with sodium sulfate and concentrated in vacuo. The residue was separated by flash column chromatography, and gave 20% of product (1) and 5% of product (2) (Summarized from Ref. 10, copyright to American Chemical Society, 1988).

2.7.2 Oxidations in Fluorosulfuric Acid

Anodic oxidations of carboxylic acids occur at platinum electrodes in fluorosulfuric acid media containing potassium fluorosulfonate [11]. The oxidation of heptanoic, octanoic, 4-methylpentanoic and 5-methylhexanoic acids were thus oxidized and gave lactones and ketones as major products, Scheme 2.24. In this strongly acidic medium the carboxylic acids undergo C-H bond cleavage at carbons distant from the carboxyl group. NMR evidence suggests that the electroactive species is the protonated form of the carboxylic acid, as it happens with the oxidation of alkanes in FSO_3H. Apparently the molecule does not undergo decarboxylation under these electrolysis conditions.

This electrolysis is conducted in a divided cell at anode potentials of 1.8–2.0 V vs Pd/H_2. Short periodic cathodic pulses on the anode may be necessary to minimize possible·electrode passivation. The products are extracted with ether, separated by chromatography and identified by comparing their IR spectra with spectra of authentic samples.

2.7.3 Kolbe Synthesis in SPE Cells

In the fifth chapter, we will consider potential applications of solid polymer electrolyte cells (SPE-Cells) as a new type of cells for electrochemical applications. Ogumi and coworkers [12] demonstrated the feasibility and

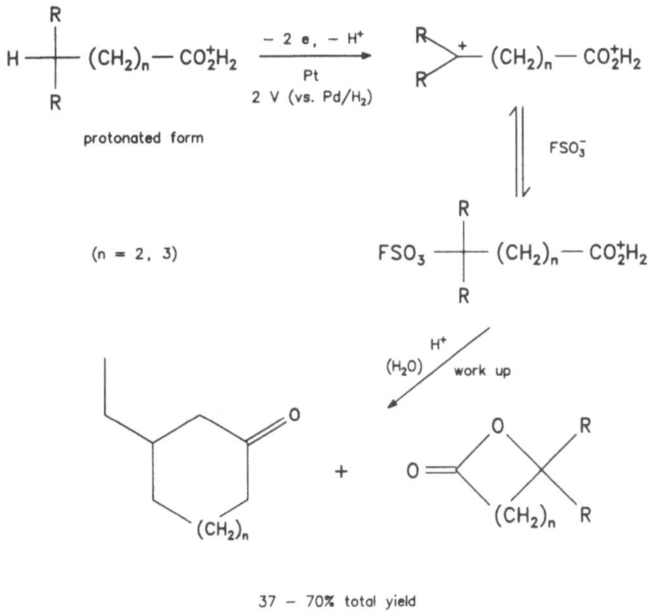

$$H \!-\!\!\overset{R}{\underset{R}{|}}\!\!-\!(CH_2)_n\!-\!CO_2^+H_2 \xrightarrow[\substack{Pt \\ 2\,V\ (vs.\ Pd/H_2)}]{-2\,e,\ -H^+} \overset{R}{\underset{R}{>}}\!\!\overset{+}{}\!(CH_2)_n\!-\!CO_2^+H_2$$

protonated form

$(n = 2,\ 3)$

$$FSO_3\!-\!\!\overset{R}{\underset{R}{|}}\!\!-\!(CH_2)_n\!-\!CO_2^+H_2 \quad\Updownarrow\ FSO_3^-$$

$$\xrightarrow[\substack{(H_2O) \\ work\ up}]{H^+}$$

37 – 70% total yield

Scheme 2.24

practical potential merit of SPE-type electrolysis in performing the Kolbe synthesis of dimethyl sebacate. A cell with a Nafion 415 cation exchange membrane as SPE was used. The SPE functioned also as an effective cell divider. To prepare the *Pt-SPE-Nafion composite* electrode, platinum was deposited on one or both sides of the Nafion membrane by reduction with hydrazine in an aqueous platinum ion-containing solution. In the electrolysis cell the composite electrode (one side plated with Pt) was used as the anode, and a platinum foil was the cathode in the catholyte solution ($0.5\,M\ H_2SO_4$). The anolyte was a solution of 20–60% monomethyl adipate in methanol. (Even net acetic acid undergoes the Kolbe reaction in such a cell but because the IR drop is large, the electrolysis would not be practically interesting.) The current efficiency increases with higher concentrations of the adipate ester from 25 to 60% when the adipate concentration is increased from 20 to 60%, while only little change occurs in the cell voltage. Cell voltages of 5–20 V and current densities of 20–80 mA/cm^2 were used in this study. The anodic reaction is the Brown-Walker type synthesis:

$$2\ CH_3OCO(CH_2)_4COOH \xrightarrow[anode]{Pt-SPE} \begin{cases} CH_3OCO(CH_2)_8COOCH_3 \ + \\[6pt] 2\ CO_2\ +\ 2\ H^+\ +\ 2\ e \end{cases}$$

2.7.4 Photokolbe Synthesis

Some years ago Bard and Kraeutler [1] studied the oxidation of acetate ions in acetonitrile solution under illumination at a rutile TiO_2 crystal electrode (a semiconductor). They found that the decarboxylation could start to occur at potentials as low as -1.6 V vs $Ag/Ag^+NO_3^-$ reference electrode. The n-type TiO_2 electrode was in the form of either a single crystal or in vapor-deposited polycrystalline form. This photo-assisted oxidation on the semi-conductor electrode was explained as follows:

$$CH_3CO_2^- + P_{(TiO_2)}^+ \longrightarrow CH_3^· + CO_2$$

$$2\ CH_3^· \longrightarrow C_2H_6$$

where the symbol P^+ denotes a photogenerated *hole* on the TiO_2 electrode. The electrolysis was conducted in acetonitrile containing tetrabutylammonium acetate and some acetic acid at 0.0 V vs a Ag/Ag^+ reference electrode. The cell was a vacuum type one-compartment pyrex glass vessel fitted with a flat window. The gases C_2H_6 and CO_2 that were produced at the anode were collected and analyzed by mass spectroscopy. (This interesting photoelectrochemical oxidation has the potential of utilizing solar energy for the production of hydrocarbons.)

2.8 Amines and Amides

2.8.1 Amines, in General

Formation of the cation radical

Aliphatic and aromatic amines are readily oxidized at the anode in aqueous and organic solvents. Usually, anode potentials of 0.5–1.5 V vs SCE are adequate for the generation of the primary cation radicals of amino compousdn [1–8]. The primary step is the transfer of an electron from the amino nitrogen to the anode, whereupon a cation radical is produced. Sometimes the first electron transfer step is followed by transfer of a second electron so that an apparent two-electron oxidation step is observed voltammetrically (one wave or peak). Because various difficult-to-control reactions usually follow the primary ET step, the anodic oxidation of aliphatic amines is synthetically less useful than that of aromatic amines which offers many possibilities [2,7,8]. Aliphatic amines give mix-

Aliphatic amines

tures of unstable cation radicals, at platinum or carbon electrodes in various solutions, leading to mixtures of products:

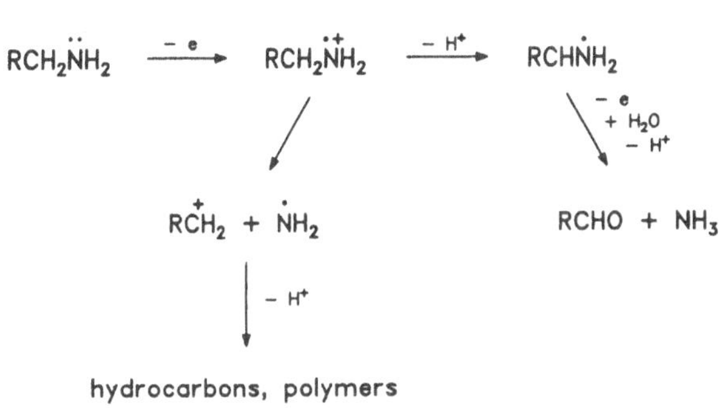

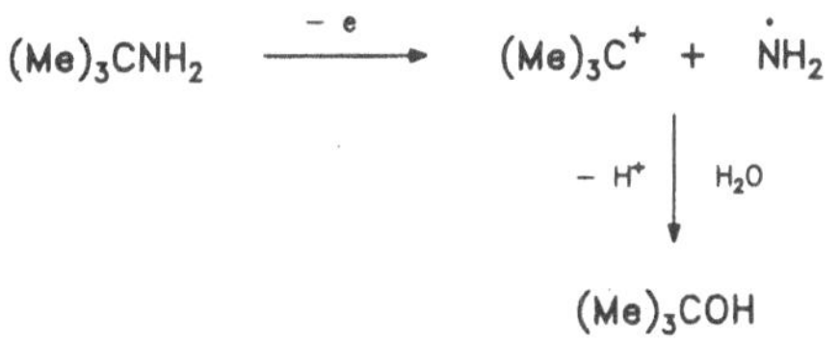

$$RCH_2\ddot{N}H_2 \xrightarrow{-e} RCH_2\overset{\cdot+}{N}H_2 \xrightarrow{-H^+} RCH\dot{N}H_2$$

$$RCH_2\overset{\cdot+}{N}H_2 \nearrow R\overset{+}{C}H_2 + \dot{N}H_2$$

$$RCH\dot{N}H_2 \overset{-e}{\underset{+H_2O}{\searrow}} RCHO + NH_3$$
$$-H^+$$

$$R\overset{+}{C}H_2 + \dot{N}H_2 \downarrow -H^+$$

hydrocarbons, polymers

$$(Me)_3CNH_2 \xrightarrow{-e} (Me)_3C^+ + \dot{N}H_2$$

$$-H^+ \downarrow H_2O$$

$$(Me)_3COH$$

Aromatic amines yield cation radicals that are more stable than those of aliphatic amines. Often dimers are formed via C—C, C—N and N—N bonds, which are more easily oxidized than the starting aromatic monomers [9] (because of enhanced charge delocalization in the dimer intermediates). **Aromatic amines**

Schäfer concluded that the method of anodic oxidation of primary amines at NiOOH electrodes (nickel oxide in alkaline solutions) is generally more effective synthetically than most ordinary chemical oxidation methods [10]. Such anodic oxidations can be performed with nickel electrodes that become covered with a layer of NiOOH continuously regenerated at 0.4–0.5 V vs SCE in MeCN/H$_2$O/KON solutions (see Sect. 2.4):

$$RCH_2NH_2 \xrightarrow[\text{anode}]{NiOOH} RCN$$

Typically, the following amines yielded their corresponding nitriles (% yields shown in parenthesis): **Production of nitriles**

$$n-C_4H_9NH_2 \ (82\%), \quad n-C_6H_{13}NH_2 \ (57\%),$$

$$n-C_{10}H_{21}NH_2 \ (33\%), \quad Ph-CH_2NH_2 \ (85\%)$$

Chemical methods with lead tetraacetate, nickel peroxide or iodine pentafluoride achieve about equal conversion but require at least equivalent amounts of these oxidizing agents.

In this connection we mention the conversion of diphenylamine(1) to the nitrile compound(2) by anodic cyanation of the primary cation radical of (1) in MeCN/Et$_4$N$^+$CN$^-$ solution, as proposed by Serve [11], Scheme 2.25.

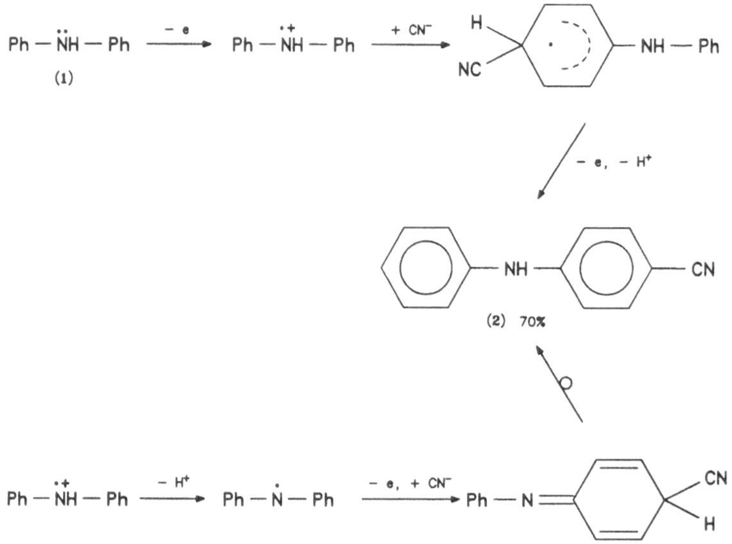

Scheme 2.25

Production of azobenzenes   Anodic oxidations of anilines in MeCN/pyridine solutions afford azobenzenes [12]. The aniline monomer is polymerized in 0.5 M H$_2$SO$_4$, forming electrically conducting films [13].

Azobenzenes, it may be noted, can be produced via the chemical diazonium salt method if a strong electron releasing group (e.g. OH) is on the ring to facilitate electrophilic attack by ArN$_2^+$.

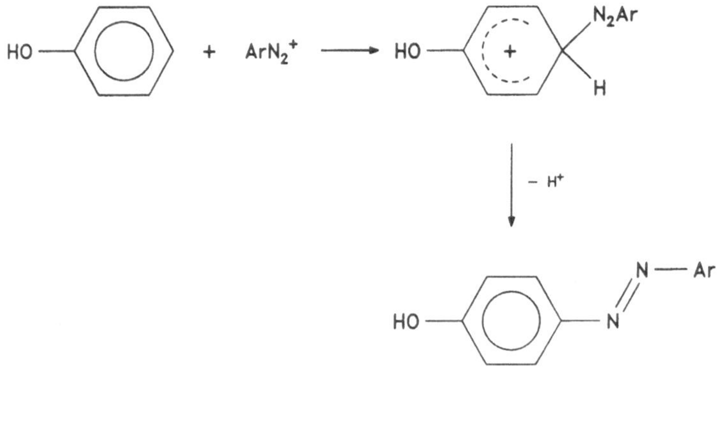

The anodic oxidation of aniline to the azobenzene can readily take place in water dispersions or in H_2O/DMF and also in $H_2O/MeCN/pyridine$ solutions [14]. The type of dimers produced by anodic oxidation of anilines is influenced greatly by the pH of the solutions [12].

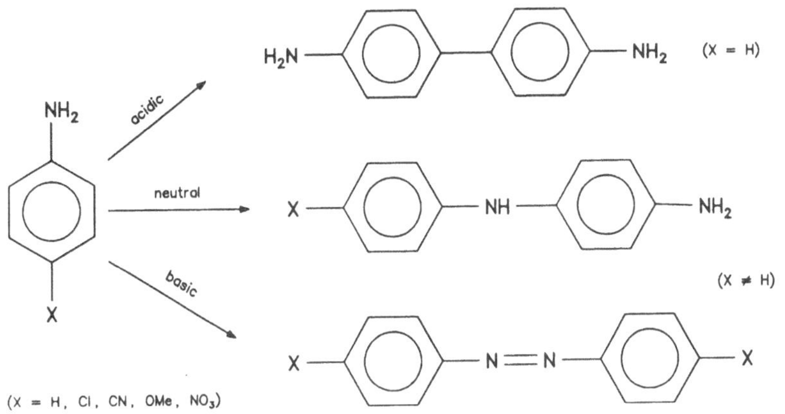

pH effect

Some aromatic amines when oxidized yield stable salts of their cation radicals with perchlorate or bromide as negative counterions (Wuster salts), both as solids and in solution [15]. Triarylaminium salts are best known as very effective one-electron oxidants. One such cation radical used in indirect electrolysynthesis is the tri-*p*-bromophenylaminium ion $Ar_3N^{\cdot+}$. Anodically produced stable organic salts are rare. They are usually sparingly soluble in the medium and tend to adhere to the electrode surface. They may require mechanical removal for the electrolysis to continue.

Production of organic salts

2.8.2 Amides

Amides are a special type of substituted amines. In general, they are oxidized at the anode, but at considerably more positive potentials than the amines. The carbonyl group delocalizes the lone electron pair of the nitrogen atom and thus makes these electrons less available for release to the anode:

An amide, upon surrendering an electron to the anode, is transformed into a cation radical which may abstract a hydrogen atom from the environment:

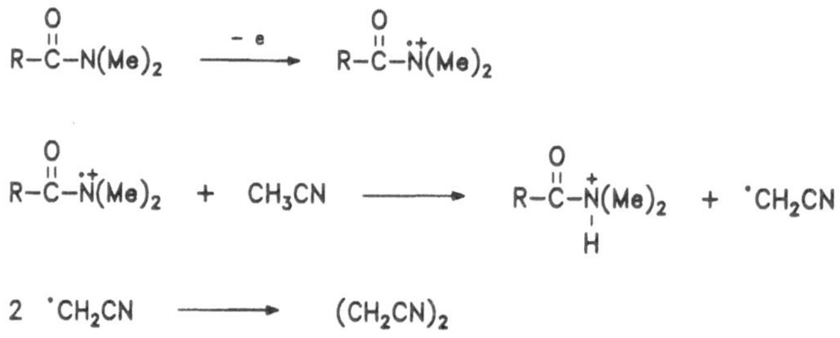

$$R-\overset{\overset{O}{\|}}{C}-N(Me)_2 \quad \xrightarrow{\;-e\;} \quad R-\overset{\overset{O}{\|}}{C}-\overset{\cdot+}{N}(Me)_2$$

$$R-\overset{\overset{O}{\|}}{C}-\overset{\cdot+}{N}(Me)_2 \;+\; CH_3CN \quad \longrightarrow \quad R-\overset{\overset{O}{\|}}{\underset{H}{C}}-\overset{+}{N}(Me)_2 \;+\; \cdot CH_2CN$$

$$2 \;\cdot CH_2CN \quad \longrightarrow \quad (CH_2CN)_2$$

Dimethylformamide, if oxidized at a platinum anode in MeOH solution, yields polyalkoxylated products in 60–70% total yields, but much lower yields at graphite electrodes [16,17]. Tertiary amides are oxidized in MeOH in one-compartment cells under constant current electrolysis and give products from which α-acyliminium ions can be obtained [17]:

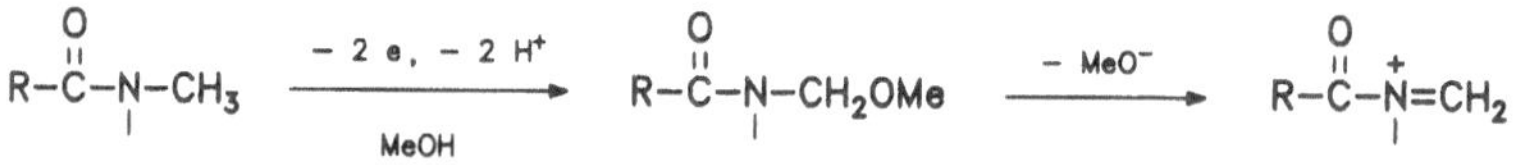

$$R-\overset{\overset{O}{\|}}{\underset{|}{C}}-N-CH_3 \quad \xrightarrow[\text{MeOH}]{-2e,\,-2H^+} \quad R-\overset{\overset{O}{\|}}{\underset{|}{C}}-N-CH_2OMe \quad \xrightarrow{-MeO^-} \quad R-\overset{\overset{O}{\|}}{\underset{|}{C}}-\overset{+}{N}=CH_2$$

Tertiary amides have been studied as electrochemical models [18] for the oxidative N-dealkylations of amines by the cytochrome P-450 enzyme. The oxidation yields N-methylamides in high yields via an ECE process involving amminium intermediates.

N-N,-dialkylamides and urethanes undergo anodic methoxylations affording synthetically useful intermediates [19]. For example, the methoxylation of the cyclic amide takes place very efficiently in MeOH/BF_4^- solution:

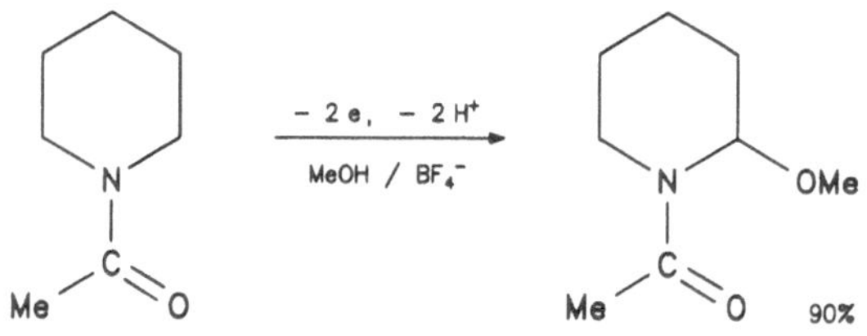

Urethanes thus methoxylated afforded intermediates that were chemically converted to the alkaloids, hydrogline, sedamine and others [17]. Urea is oxidized to the simplest oxidation products at platinum anodes [20]. Urethanes are produced in high yields by electrolyzing formamides in undivided cells with graphite anodes and halogenides as electrolytes [21]. Swenton and coworkers [22] recently studied the anodic oxidation of p-methoxyanilides (1) for the purpose of developing a method for the preparation of acylated quinone imine ketals (5). These compounds are intermediates for the construction of a number of alkaloids. A very efficient and "reasonably general" method was developed and a plausible reaction mechanism (EEC_rC_p) was proposed for the anodic reaction, Scheme 2.26.

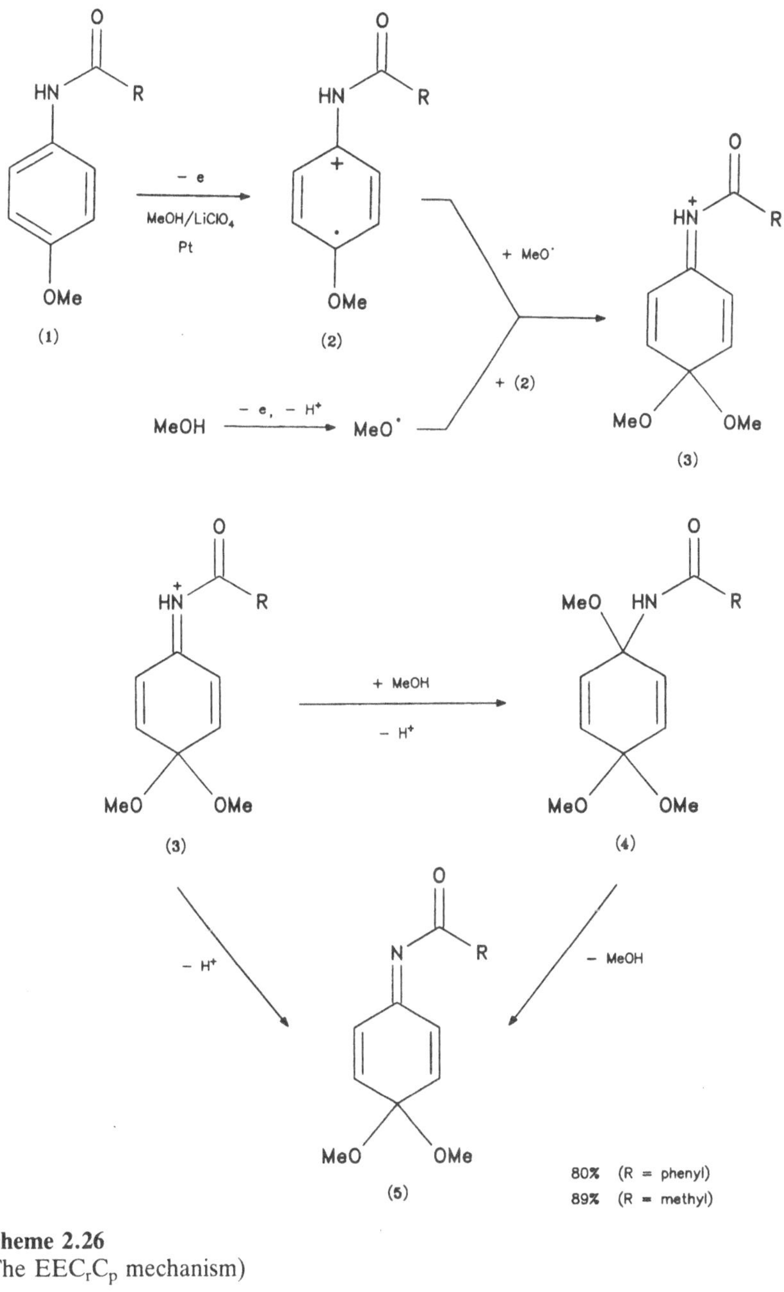

Scheme 2.26
(The EEC_rC_p mechanism)

EEC$_r$C$_p$-type Electrolysis (Scheme 2.26)

This interesting anodic electrolysis is carried out at 0 °C under constant current in MeOH/LiClO$_4$ solution containing the anilide in the presence of either sodium bicarbonate or 2,6-lutidine in a single-cell apparatus with platinum electrodes. The anodic potential is such that both the anilide and the methanol are oxidized by electron transfers to the anode while lithium ion is reduced at the cathode, and the metallic lithium reacts with methanol to liberate hydrogen. The electrolyzed solution is concentrated in vacuo at about 10 °C, and the residue is extracted with methylene chloride (3 × 40 ml) after adding 40 ml of water. After work-up of the organic phase the acylated quinone ketals are recovered in yields of 80–89% (Adapted from Ref. 22, Copyright of the American Chemical Society, 1989).

The notation EEC$_r$C$_p$, Scheme 2.26, implies two electron steps (EE), to generate the radical cation and the radical species; C$_r$ refers to methoxy radical CH$_3$O· combining with cation radical (**2**), and C$_p$ refers to the polar step, the proton removal from (**3**) to give (**5**) or addition of methanol to (**3**) to give (**4**) which loses methanol to give (**5**).

2.9 Ethers and Esters

Aromatic ethers Aromatic ethers are generally directly oxidized at the anode. In contrast, aliphatic ethers are extremely resistant to anodic oxidation, and this property allows them to be employed as electrolysis media, alone or as *cosolvents*, for certain electrosynthetic applications. Aromatic-alkoxy ethers are deelectronated quite easily because the alkoxy substituents tend to release electrons to the aromatic ring, from which the electrons are removed, and they also stabilize the produced cation radicals [1]. Aryl ethers, such as anisoles, couple readily upon oxidation in solvents containing trifluoroacetic acid (TFA). This type of anodic coupling can be followed by cathodic reduction of the *stable in the TFA medium cation radicals* of the coupled intermediates [2,3]. Miller reported an unexpected ring-ring coupling in the anodic oxidation of 2-methoxy-1,4-bis(trimethylsiloxy) benzene [4]:

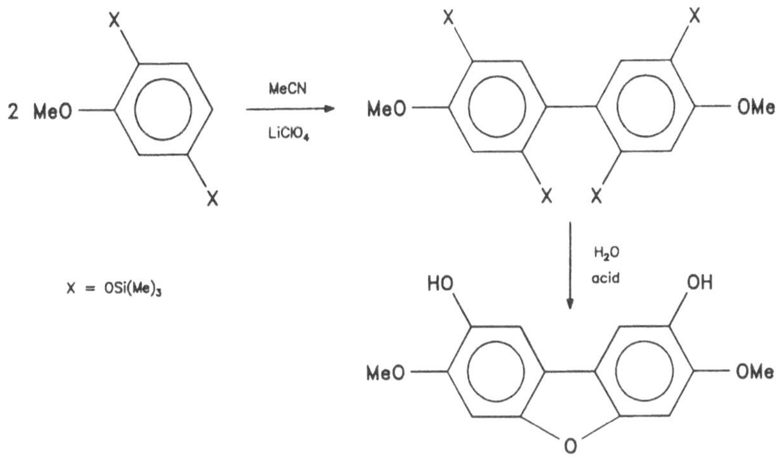

2 MeO —

MeCN
LiClO₄ → MeO —

X

X

— OMe

X = OSi(Me)₃

H₂O
acid

HO

MeO —

OH

— OMe

O

Apparently the presence of the methoxy group promotes anodic coupling rather than the expected cleavage at the OSiMe₃, a reaction that would have led to the quinone product. Anodic oxidations of aryl ethers in aqueous methanol-potassium chloride solutions, afford quinone monoketals in 65–78% yields, and dimerized products by electrolysis in acetonitrile containing Et₄NClO₄ as electrolyte [5]:

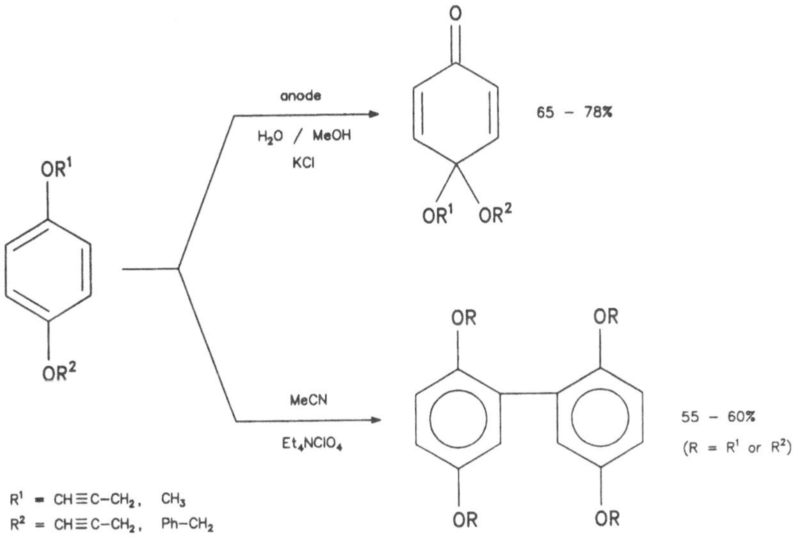

OR¹

OR²

anode
H₂O / MeOH
KCl

O

OR¹ OR²

65 – 78%

MeCN
Et₄NClO₄

OR

OR

OR

OR

55 – 60%
(R = R¹ or R²)

R¹ = CH≡C–CH₂, CH₃
R² = CH≡C–CH₂, Ph–CH₂

Saturated aliphatic ethers are indirectly methoxylated in MeOH/KOH solutions by reacting with $CH_3O^{\cdot}$ radicals generated from the oxidation of CH_3O^- ions [6]. Phenolic ethers are methoxylated directly, that is, via the generated cation radicals of the ethers, since these ethers are more easily deelectronated at the anode [2] than the CH_3OH or CH_3O^- ion. Enol ethers are easily oxidized at the anode in methanolic solutions to give methoxylated dimer products [8]:

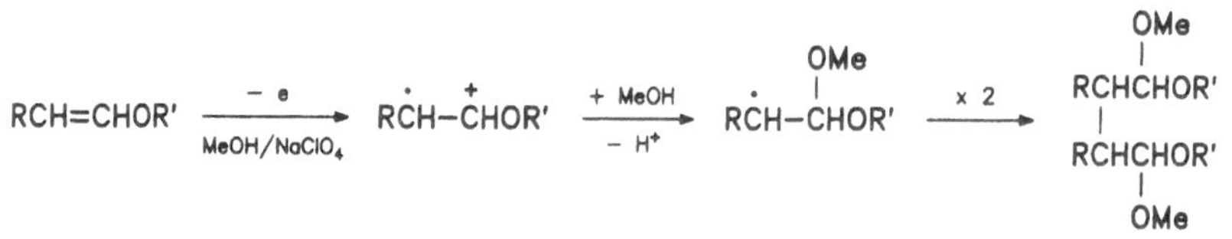

$$RCH=CHOR' \xrightarrow[MeOH/NaClO_4]{-e} R\overset{\cdot}{C}H-\overset{+}{C}HOR' \xrightarrow[-H^+]{+MeOH} R\overset{\cdot}{C}H-\overset{OMe}{\underset{|}{C}HOR'} \xrightarrow{x2} \begin{array}{c} OMe \\ | \\ RCHCHOR' \\ | \\ RCHCHOR' \\ | \\ OMe \end{array}$$

This reaction has been found useful for the synthesis of octaethylpor-phyrins.[9] The coupled products from the oxidation of aldehyde enol ethers in methanol, gave by reaction with benzyl carbamate N-carboxybenzyloxy-pyrroles, which after deprotection and reaction with formaldehyde gave the porphyrins. Electrolysis of zinc-octaethylporphyrins in DMF/CN⁻ media gave the cyanoprophyrins [9]. Bis-enol ethers are synthetically interesting electron-rich olefinic compounds which can undergo efficient intramolecular coupling reactions by anodic oxidation [10]. Electrolysis can be carried out under constant current in an undivided cell with a platinum anode and 10% MeOH in MeCN in the presence of 2,6-lutidine as proton scavenger. Two typical examples are shown in the following:

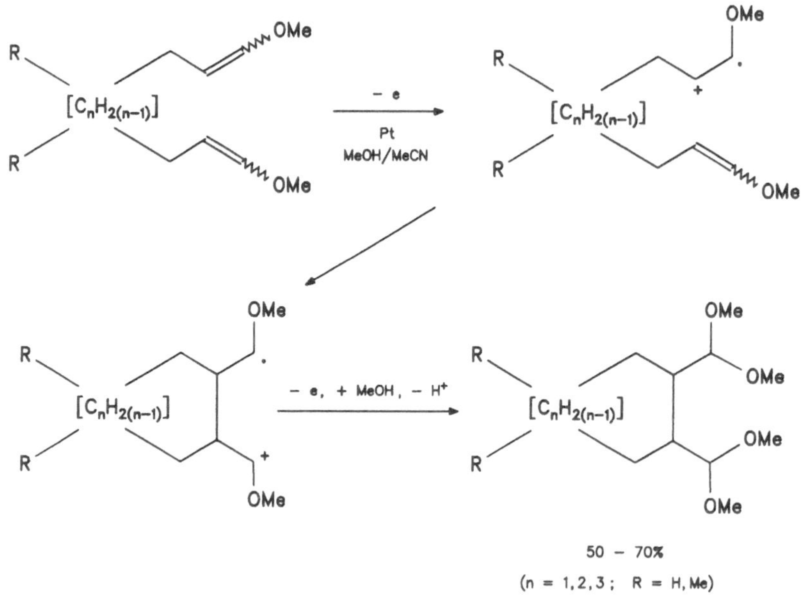

50 – 70%

(n = 1, 2, 3 ;   R = H, Me)

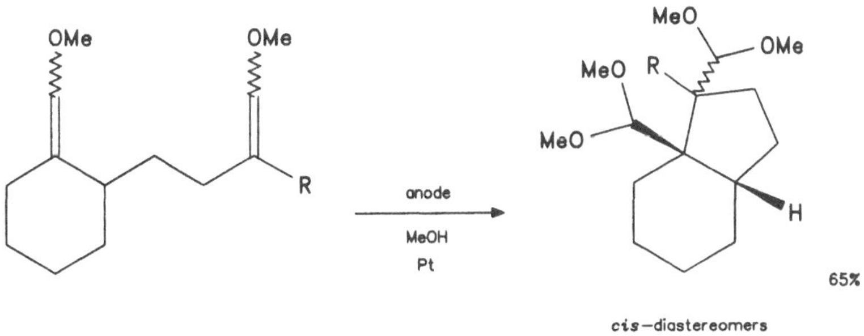

cis—diastereomers

Construction of rings as represented above is of interest in the synthesis of natural products. The oxidation of dibenzyl ether [11] in acetonitrile yields N-benzylacetamide as one of the products via splitting of the intermediate carbonium ion Ph $\overset{+}{C}$HOCH$_2$Ph. Esters undergo some anodic reactions depending on the kind of ester and the molecular structure. Straight chain esters undergo ω-1 substitutions [12]. Half esters, when oxidized in methanol at carbon electrodes and at low current densities, yield methoxylated products, and acetamidated products when oxidized in acetonitrile solvent [13].

2.10 Organic Halides

2.10.1 Alkyl Halides

Anodic synthetic applications of organohalides are limited by comparison with the very wide range of their versatile cathodic applications. Oxidation of alkyl iodides can take place at platinum electrodes in MeCN/LiClO$_4$ solutions. An alkyl iodide yields, as primary anodic product, the organoiodo cation radical which by expelling an iodine atom is transformed to an organocation [1,2]. The cation radical may react directly with a nucleophile before expulsion of I· as depicted in Scheme 2.27. Alkyl iodides can also be oxidized indirectly by the iodonium ion, I$^+$, which can be prepared externally by oxidizing iodine in acetonitrile solution:

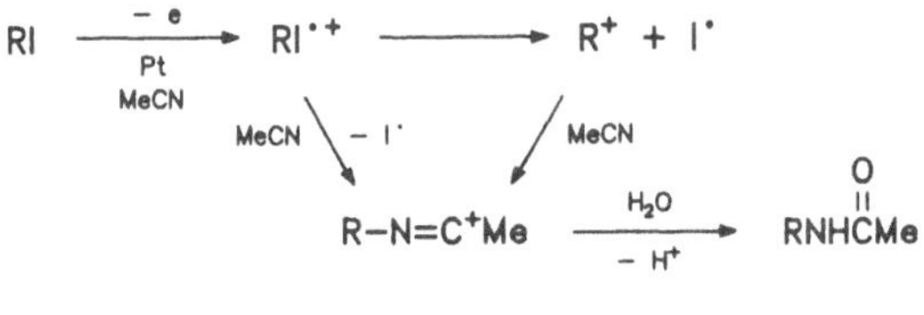

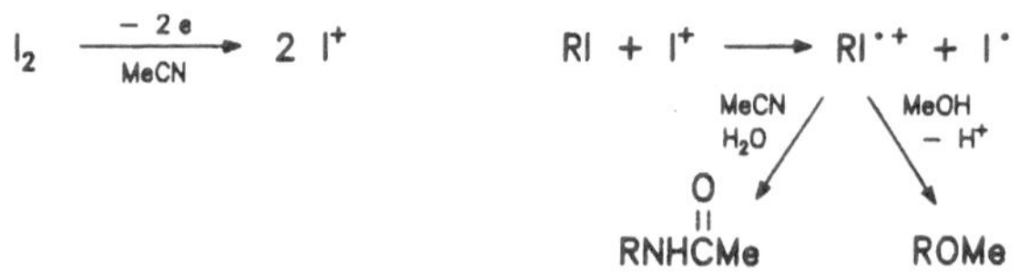

$$I_2 \xrightarrow[\text{MeCN}]{-2e} 2\ I^+ \qquad RI + I^+ \longrightarrow RI^{\cdot+} + I^{\cdot}$$

$$\overset{\text{MeCN}}{\underset{H_2O}{\diagdown}} \qquad \overset{\text{MeOH}}{\underset{-H^+}{\diagdown}}$$

$$\underset{RNHCMe}{\overset{O}{\overset{\|}{}}} \qquad ROMe$$

Scheme 2.27

Primary and tertiary bromoalkanes are directly oxidized at platinum electrodes in acetonitrile solutions containing $LiClO_4$ or Et_4NBF_4 electrolytes. Becker [5] proposed a reaction mechanism similar to that for the oxidation of iodoalkanes. An electron is released to the anode from the nonbonding orbital of bromine and a very unstable cation is generated, as would be expected from the quite high potential required for the electron release to occur (2.3–2.4 V vs Ag/Ag$^+$). We also note that anodic oxidation of bromofluoroalkanes in fluorosulfuric acid transforms them to the fluorosulfate analogs [6]. Bromine is released at the anode and is substituted in the starting molecule by the OSO_2F group. The process is monoelectronic and the released bromine acts as a catalyst (mediator) for the oxidation of the bromofluoroalkane.

$$CF_3(CF_2)_5Br \xrightarrow[\text{HFSO}_3]{-e} [CF_3(CF_2)_5]^+ + Br^{\cdot}$$

$$[CF_3(CF_2)_5]^+ + HFSO_3 \xrightarrow{-H^+} CF_3(CF_2)_5OSO_2F$$

Polyfluoroalkyl iodides are oxidized to hypervalent iodanyl radicals which by releasing iodine result in the nucleophilically substituted product [7]:

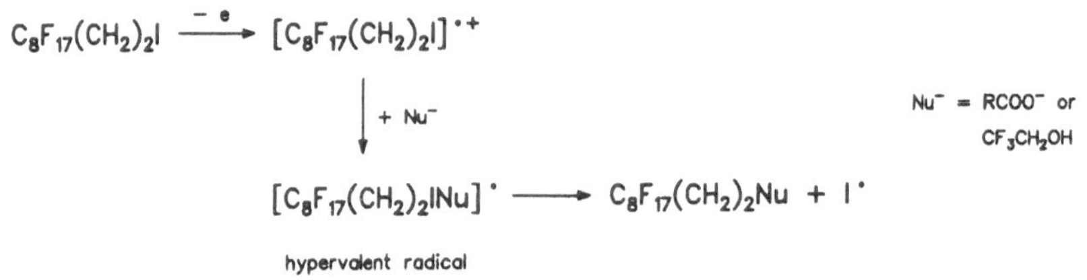

$$C_8F_{17}(CH_2)_2I \xrightarrow{-e} [C_8F_{17}(CH_2)_2I]^{\cdot+}$$

$$\downarrow + Nu^- \qquad\qquad Nu^- = RCOO^- \text{ or } CF_3CH_2OH$$

$$[C_8F_{17}(CH_2)_2INu]^{\cdot} \longrightarrow C_8F_{17}(CH_2)_2Nu + I^{\cdot}$$

hypervalent radical

2.10.2 Aromatic Halides

It has not yet been possible to achieve carbon-halogen bond cleavage of aromatic halides by direct anodic oxidation. Instead of carbon-halogen bond cleavage, the deelectronation is followed by hydrogen expulsion, which often leads to new C—C bond formations. A rather unique anodic oxidation is that of iodobenzene in the presence of benzene in acetonitrile [8].

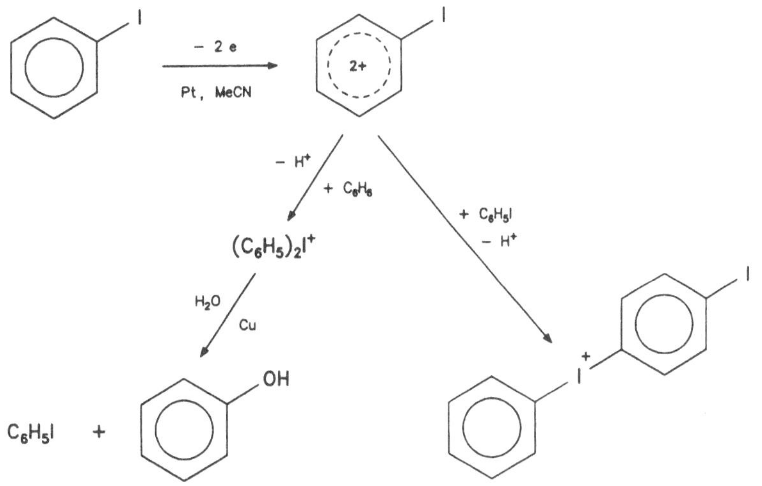

The iodobenzene may in effect be a recyclable mediator for the indirect oxidation of benzene to phenol. This is possible because benzene reacts easily with the $(C_6H_5I)^{2+}$ species to give the dimer $(C_6H_5)_2I^+$ from which the phenol and the starting iodobenzene are obtained.

2.11 *N*-Heterocyclics

The electrochemistry of *N*-heterocyclic compounds is of very special interest because this class of organic substances has profound biological importance [1]. Here the discussion will be limited only to some highlights that are representative of the anodic electrochemistry of these materials, since an adequate presentation of this subject would require a separate monograph. Simple *N*-heterocyclics readily give their cation radicals or dications which are rapidly attacked by a nucleophile, e.g. a CN⁻ ion, or enter into various coupling reactions, as in the following examples [1,2]:

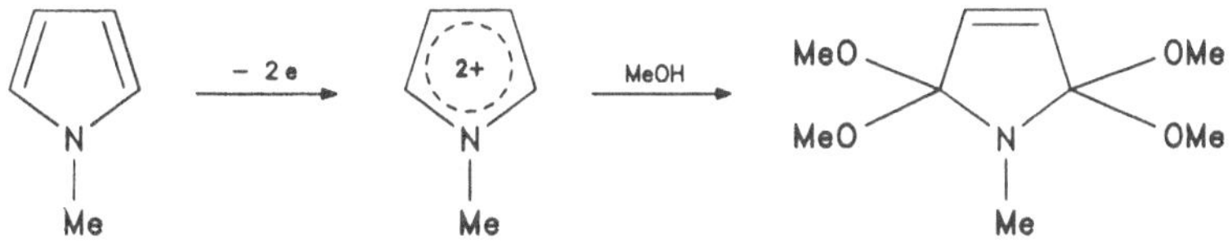

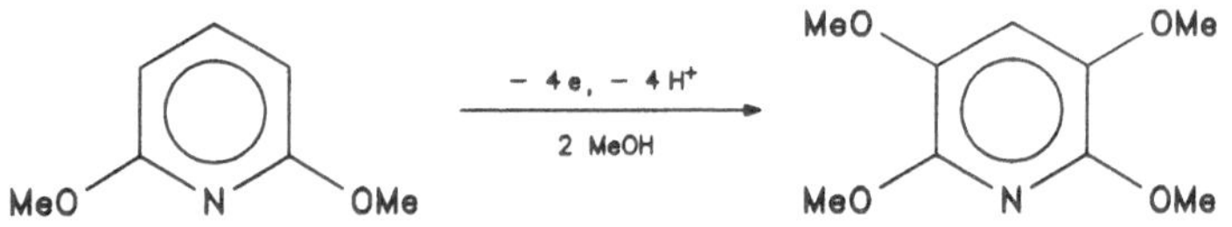
MeO
N
OMe
− 4e, − 4H+
2 MeOH
MeO
OMe
MeO
N
OMe

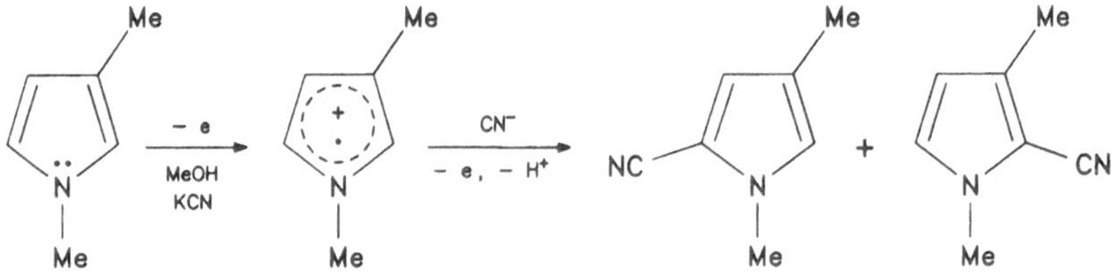
Me
N
Me
− e
MeOH
KCN
Me
+
N
Me
CN−
− e, − H+
Me
NC
N
Me
+
Me
CN
N
Me

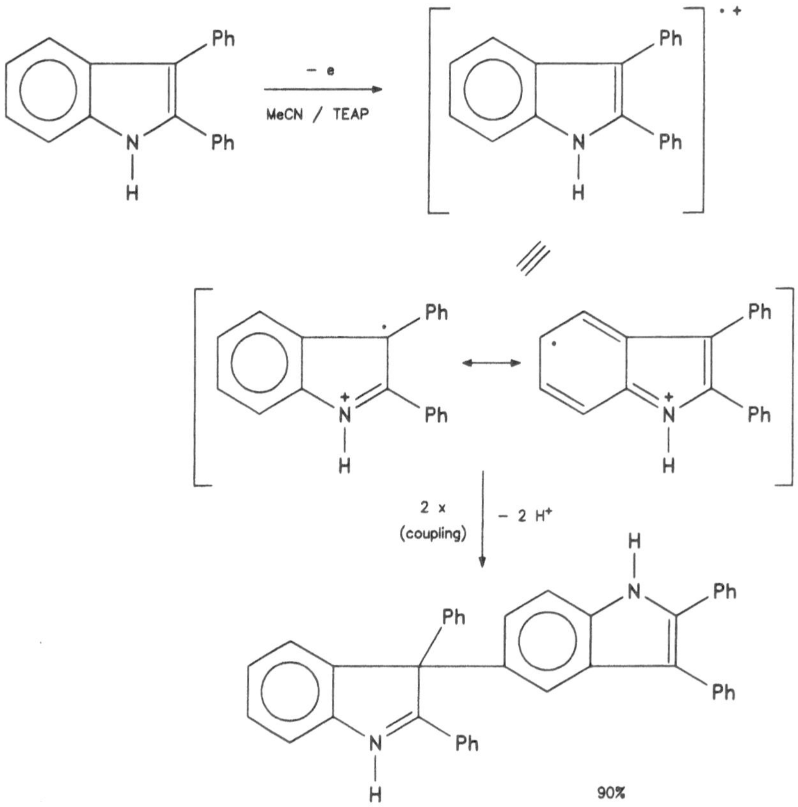
Ph
N
H
Ph
− e
MeCN / TEAP
Ph
N
H
Ph
• +
Ph
N
H
Ph
Ph
N
H
Ph
2 x
(coupling)
− 2 H+
H
N
Ph
Ph
Ph
N
H
Ph
90%

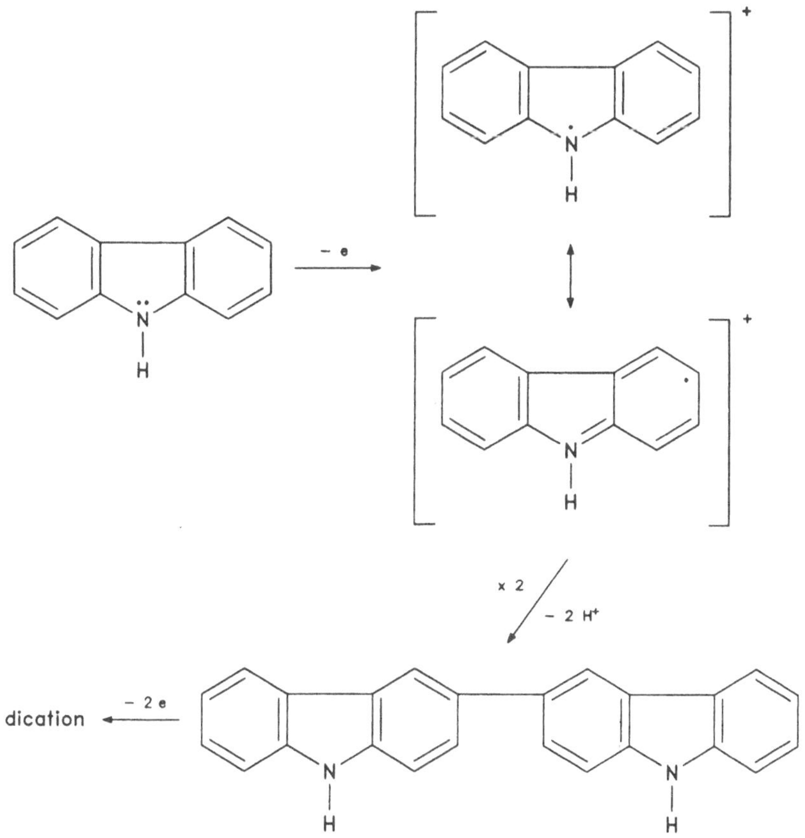

Callot and coworkers [3] studied the electrochemical oxidation of porphyrins and metalloporphyrins with relation to substituent effects. Cyanation at the meso (methine bridge) was achieved efficiently at the anode while chemical procedures failed practically. Zinc octaethylporphyrins are easily prepared, and are cyanated in a divided cell in DMF/Et$_4$NCN solution using platinum electrodes at the easily reached potentials of 0.4–0.9 V vs SCE. Meso-, mono-, di-, tri-, and tetra- cyanoporphyrins are thus produced in 35–95% yields. Cyanations of aromatic *N*-heterocyclics take place by mechanisms similar to those of aromatics, in general. Cyanations occur either by replacement of aromatic hydrogen or of hydrogen on the side chain [4].

Anodic oxidation of *N*-acylprolines and *N*-acylpipecolic acids is a very efficient method for the synthesis of *N*-acyl-*N*-*O*-acetals [5], Scheme 2.28. Methoxylation is carried out in MeOH/NaOMe solution, and hydroxylation in H$_2$O/THF/KOH solution. Undivided cells with graphite electrodes are used and the electrolysis is performed under constant current of about 200 mA/cm^2. The dimethoxylated product is obtained by electrolysis in MeOH/Et$_4$NOTs solution.

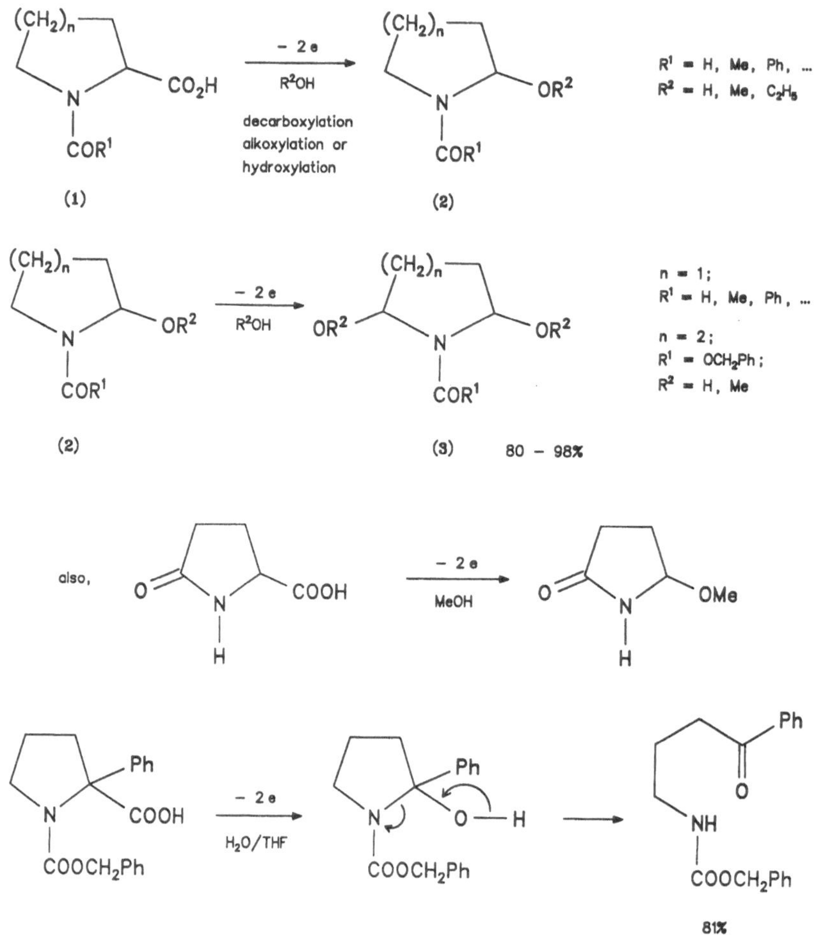

Scheme 2.28

Heteroaromatics in general, usually exhibit more than one voltammetric oxidation step forming first cation radicals. These, in many cases, are distinctly colored and stable enough to display reversible voltammetric behavior. For example, phenothiazine forms an orange-colored cation radical. This is further oxidized to a dark red dication which is stable for several days in the absence of air and moisture [6].

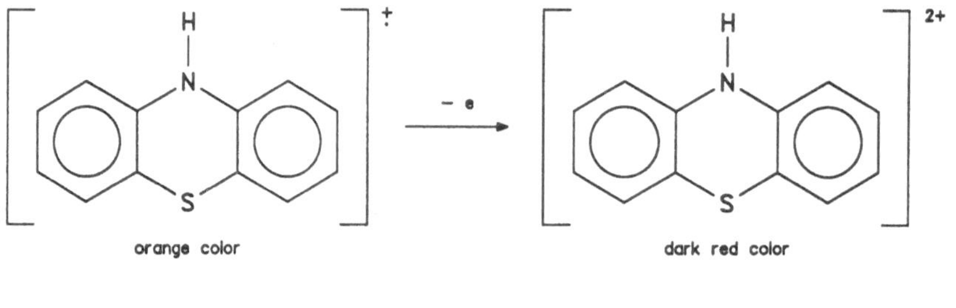

Cauquis studied the anodic oxidation of heterocyclic amidines which are precursors of biologically active materials [7]. Apparently the NH_2 group of these molecules is more easily deelectronated than the ring nitrogen. The anodic —HN—NH— and —N=N— couplings of aminothiazole compounds are explained by mechanisms as given in Scheme 2.29.

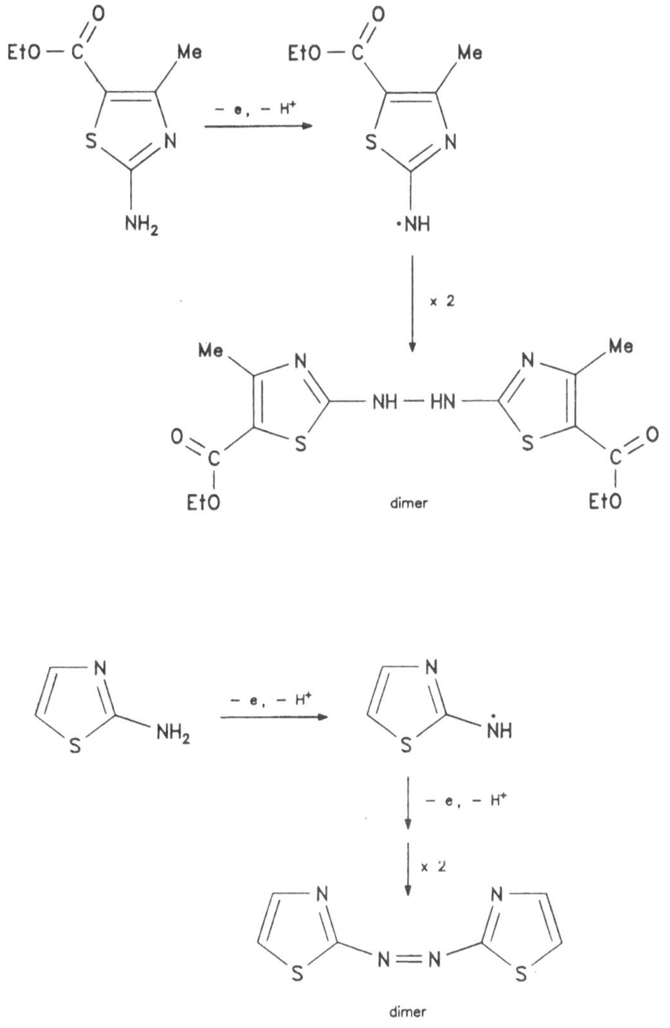

Scheme 2.29

Anodic oxidation of aminopyrimidines preceded by electroreduction in aqueous media affords open ring products [8] at the anode, as depicted in Scheme 2.30.

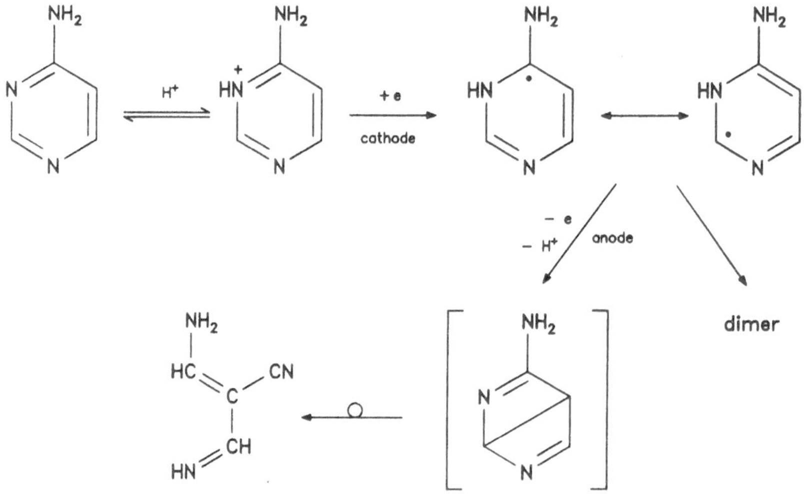

Scheme 2.30

In the anodic oxidation of 1-phenyl-3-aminopyrazoline in acetonitrile the amino group remains intact while dimer formation takes place by joining the phenyl rings [9]:

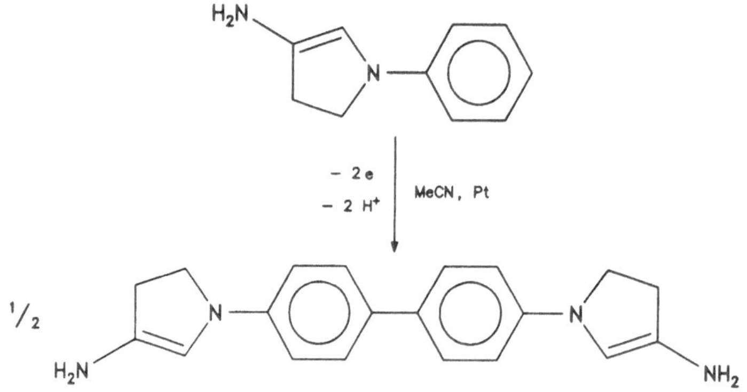

The dimer is further oxidized at more positive potentials by releasing a total of six electrons and four protons to give the dication,

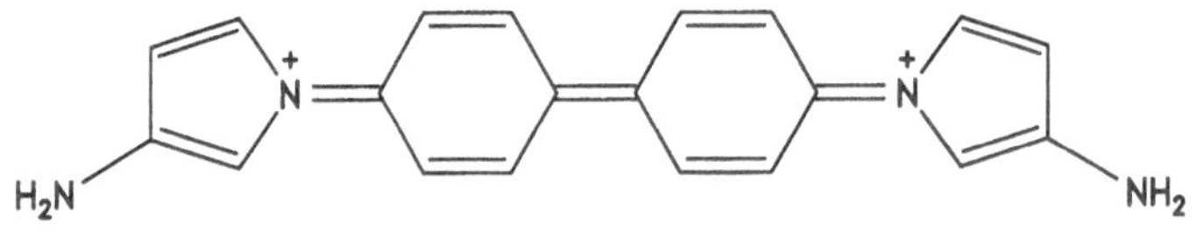

Couplings of *N*-heterocyclics with nitrogens external to the ring may take place via N—N bonds by more than one path. Pyrazoles involve both ring and external to ring nitrogen, as indicated in the following example [10]:

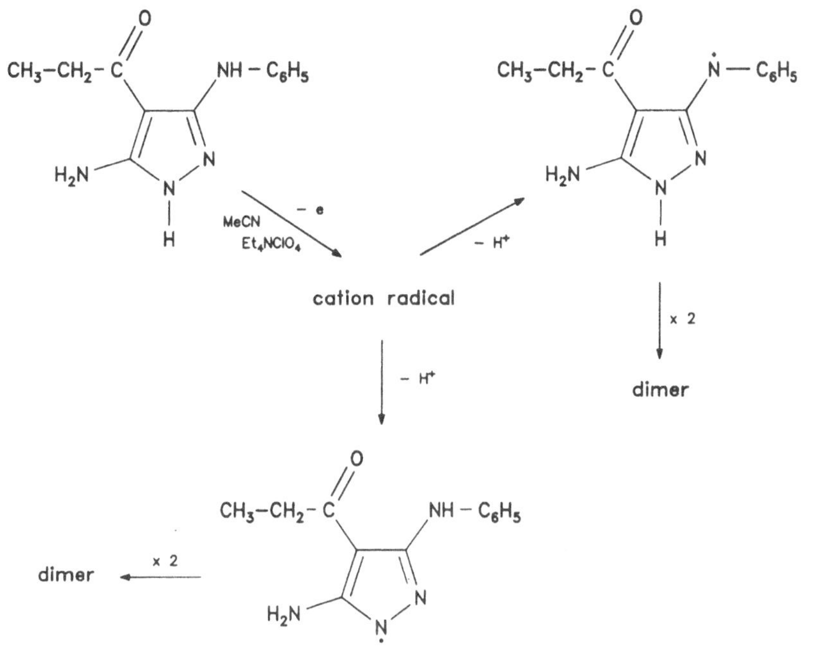

In the anodic oxidation of 1-phenylazolidin-3-ones under controlled potential, two products were formed the yield ratio of which was influenced by the kind of the anion of the electrolyte [10]. While perchlorate favored formation of the dimer (**2**) chloride ion suppressed formation of the dimer in favor of product (**1**), Scheme 2.31.

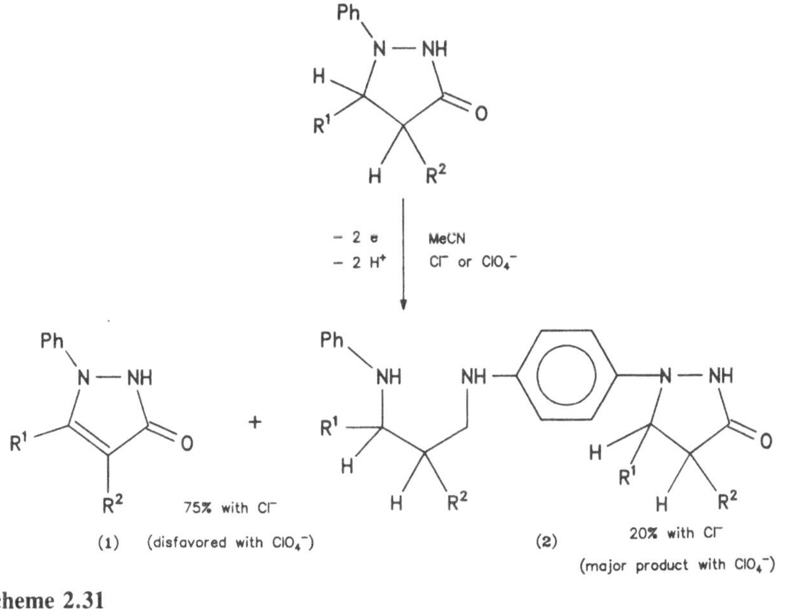

Scheme 2.31

The electrolyte effect was attributed to the fact that Cl⁻ ion behaves like a base in acetonitrile (HCl is 99% associated in this solvent), and this apparently affects the reaction course. Methoxylation of pyrrolidinols leads to the syntheses of *cis*-3-hydroxy-L-proline and *trans*-3-hydroxy-D-proline [11]. The methoxylation on the *N*-methyl group of antipyrine has been studied in connection with antipyrine metabolism [12]. The electrolysis was performed in MeOH/KOH solution. The methoxylation occurs on the methyl group attached to nitrogen, as shown in Scheme 2.32.

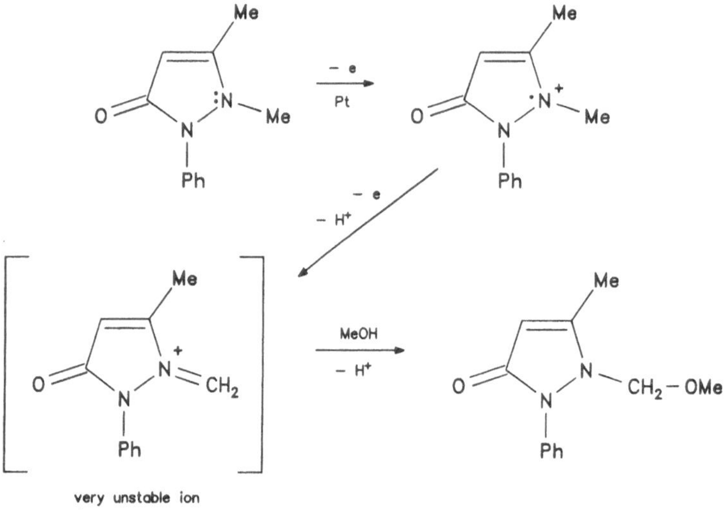

Scheme 2.32

Dryhurst [13] and Moiroux [14] independently studied the electrochemical oxidation of the biologically important *N*-heterocyclic substance 5-hydroxytryptamine (5H-T). The electrooxidation of 5H-T in acetonitrile containing perchloric acid at platinum electrode yields the 3,4'-linked indole-indolenine dimer in 80% yield [13]:

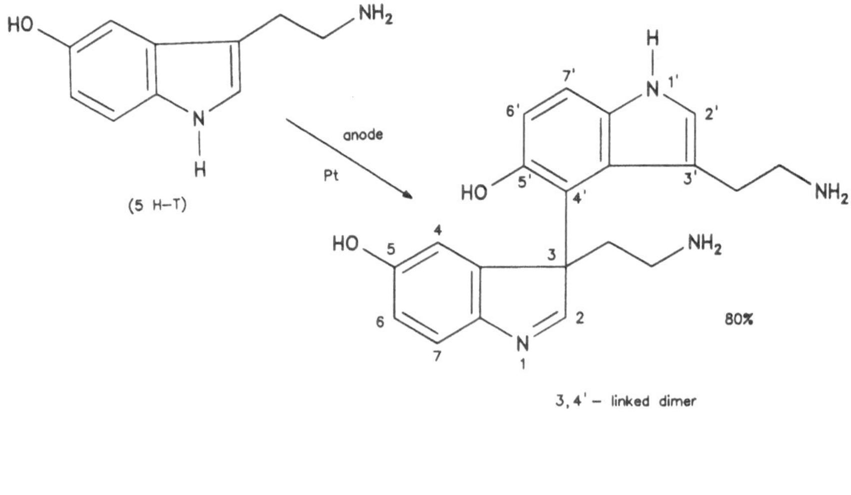

Moiroux also obtained 80% yield of the same dimer, and proposed the following reaction scheme:

$$AH \quad \underset{}{\overset{-\ e}{\rightleftharpoons}} \quad AH^{\cdot+} \quad \xrightarrow{+\ AH} \quad DH_2^{\cdot+} \quad \xrightarrow{AH^{\cdot+}} \quad DH_2^{2+} \ + \ AH$$

$$DH_2^{2+} \quad \underset{rds}{\overset{-\ H^+}{\xrightarrow{\hspace{1cm}}}} \quad DH^+$$

AH = 5H–T

$DH_2^{\cdot+}$ = dimer of AH with $AH^{\cdot+}$

Scheme 2.33

Oxidation of 5-Hydroxytryptamine [13,14] (5H-T) (Scheme 2.33)

Electrolysis of 5H-T in aqueous solution (pH 2) gave predominantly the 4′,4-linked dimer along with the 4′,3, 4′,2 and 4,6′ dimeric products. It could not be concluded which of the two types of electrolytic oxidations best mimic the biological oxidation of 5H-T.

These microscale electrooxidations were performed in divided cells at 22–25 °C under anodic potentials near 0.5 V vs SCE, with 50 ml of anolyte solution into which a total of 30 mg of 5H-T was added incrementally as electrolysis progressed and the solution became deep purple. Electrolysis was ended when more than 95% of the starting compound was oxidized, as determined by periodic analysis of the solution by high performance liquid chromatography (HPLC). The eluted products were collected and identified by spectroscopic techniques.

The alkaloid-*N*-heterocyclics are readily oxidized at the anode, but because their molecular structures are very complex the oxidation mechanisms are not easily, if at all, elucidated. Picolines are clearly transformed to their carboxylic acids and may act also as nucleophiles [15,16].

The anodic oxidation of *N*-aminophthalimides affords tetrazane and tetrazene systems [17] apparently by a 2-electron deelectronation process and deprotonation of the NNH₂ group to give the intermediate amino-nitrines which dimerize. Yields are less that 40%, and they seem to vary with the kind of electrolyte. Tetrazanes are converted to tetrazenes very readily.

Dihydropyrimidines are oxidized at a carbon anode in MeCN and are "self-protonated" in the process by the nascent protons from the starting molecule as it is deelectronated and deprotonated at the anode [18].

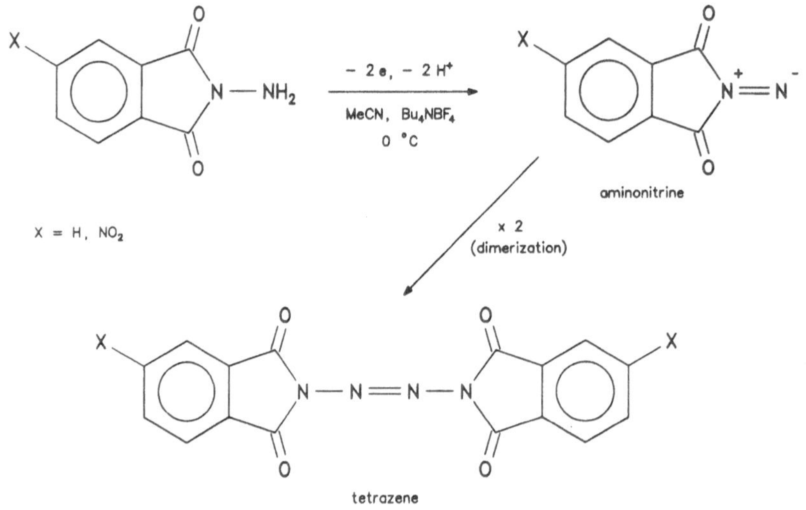

Scheme 2.34

2.12 Organosulfur Compounds

Sulfur containing organic substances are studied electrochemically with special interest because, apart from synthetic applications, the sulfur heteroatom is present in many biologically important substances.

2.12.1 Sulfides

The anodic oxidation of organic sulfides in the presence of water proceeds clearly by two-electron successive steps to afford sulfoxides and sulfones [1] in good yields.

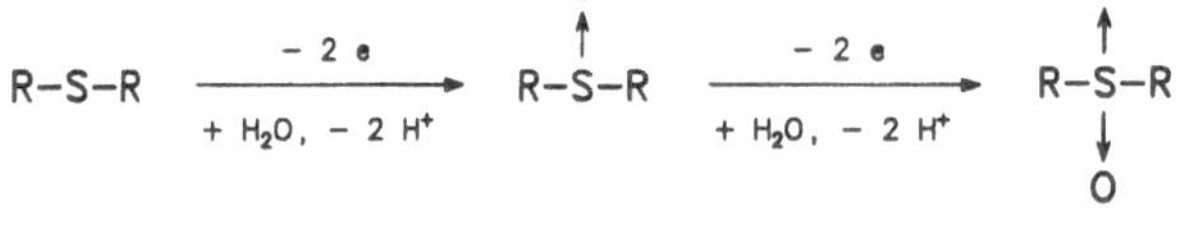

Anodic oxidation of sulfide salts, RSNa, afford disulfides by one-electron oxidation of the RS^- anion.

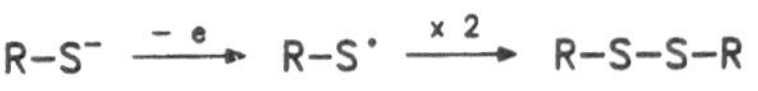

Thianthrene and thioanisole are readily oxidized and act as oxidative mediators, as in the following example in which the thioanisole is regenerated in the presence of base [2].

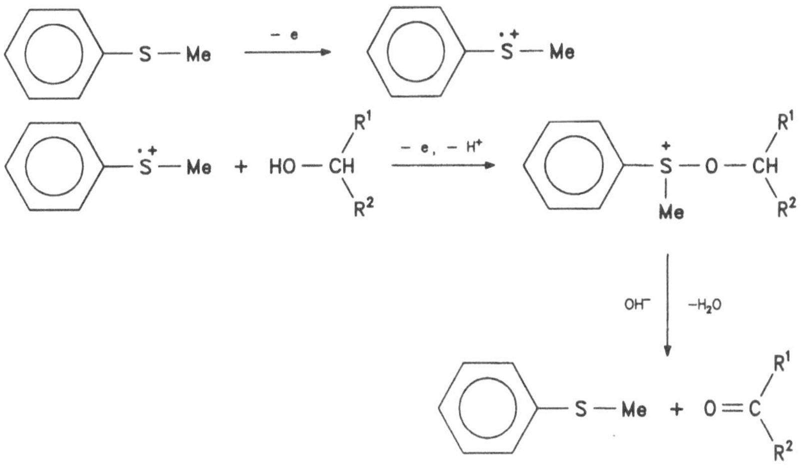

Sulfur compounds possessing the structure R—N—C—S— are commercially interesting. Anodic oxidation leads to dimers, as in the industrial production of Thiourem M. Cuttler summarized the overall cell reaction for thiourem M as follows [4]:

$$2\ CS_2\ +\ 2\ (CH_3)_2NH\ +\ 2\ NaOH\ \xrightarrow[-\ 2\ H_2O]{}\ 2\ (CH_3)_2NCSSNa$$

$$\downarrow{\scriptstyle -\ 2\ e\quad -\ 2\ Na^+}$$

$$[(CH_3)_2NCS]_2S_2$$

thiourem M

cathode $2\ H_2O\ +\ 2\ Na^+\ +\ 2\ e\ \longrightarrow\ 2\ NaOH\ +\ H_2$

overall
cell reaction $2\ CS_2\ +\ 2\ (CH_3)_2NH\ \longrightarrow\ thiourem\ M\ +\ H_2$

Scheme 2.35

This is an industrial example of *paired electrosynthesis*. Both half-cell reactions generate the required products for the desired overall cell reaction.

2.12.2 Thiolates and Dithianes

Thiolates and dithianes are electrochemically easily oxidized. This is of interest in some protective group-synthetic applications. Thiolates are prepared for carboxyl group protection. Deprotection of the carboxylic acid may be done by anodic indirect oxidation in aqueous acetonitrile in the presence of bromide ion as mediator [5]. Aldehydes and Ketones protected as 1,3-dithiane derivatives are deprotected by anodic electrolysis in acetonitrile in divided cells. The partial regeneration of the protected benzaldehyde is depicted below [6]:

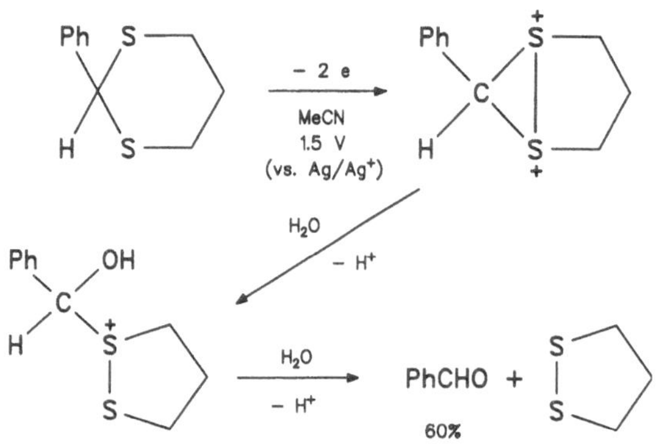

In the presence of pyridine, the 1,3-dithiane (**1**) was oxidized at a platinum anode and gave the tetrathiofulvaline (**2**) in 40% yield after 3 F/mole were passed [7]. The electrolysis was performed in MeCN with Bu₄NF as electrolyte. (Compound (**2**) is of interest as a one-dimensional electronic conductor.)

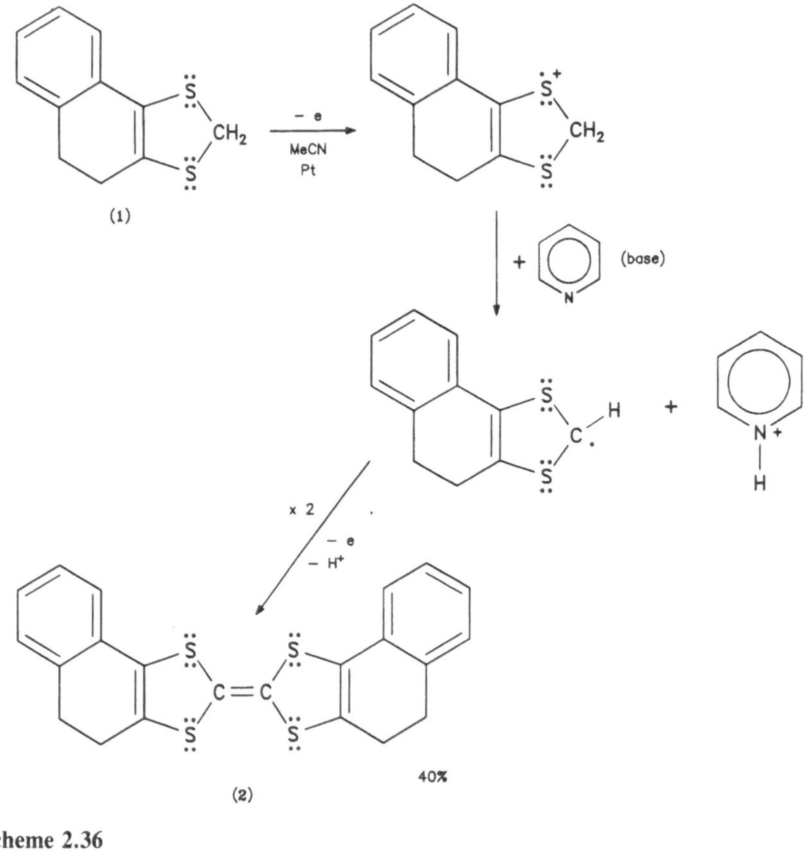

Scheme 2.36

Dithianes are transformed to dithiolanes by an overall four-electron process in 20–70% yields [8]. Such oxidative transformations could probably be effected by indirect electrooxidation with ceric ion salts as mediators.

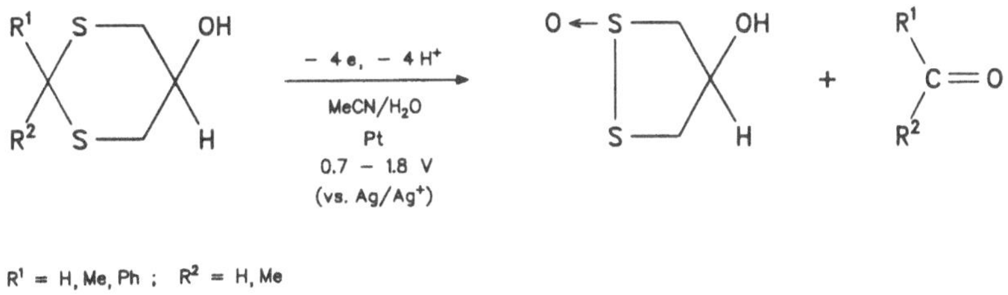

R^1 = H, Me, Ph ; R^2 = H, Me

2.12.3 Sulfonium Salts

Oxidation of some sulfides in acetonitrile with small amounts of water and lithium perchlorate electrolyte affords both sulfoxides and sulfonium salts. Thus diphenyl sulfide is oxidized to diphenyl sulfoxide, 1,4-bisphenylthiobenzene and the sulfonium salt. The salt is formed by reaction of the anodically generated cation radical with the starting sulfide [9]:

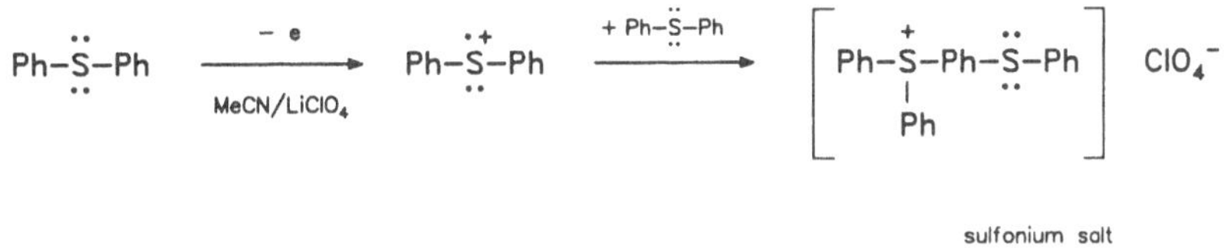

sulfonium salt

Thianthrene is an interesting molecule. It acts as an oxidizing mediator and also forms sulfonium salts, as shown becow, by anodic oxidation in acetonitrile in the presence of anisole or *tert*-butylamine [10]:

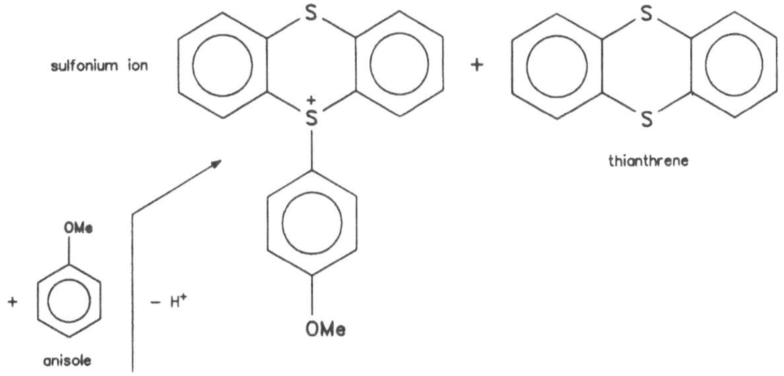

Thianthrene

sulfonium ion

thianthrene

OMe

anisole

– H$^+$

OMe

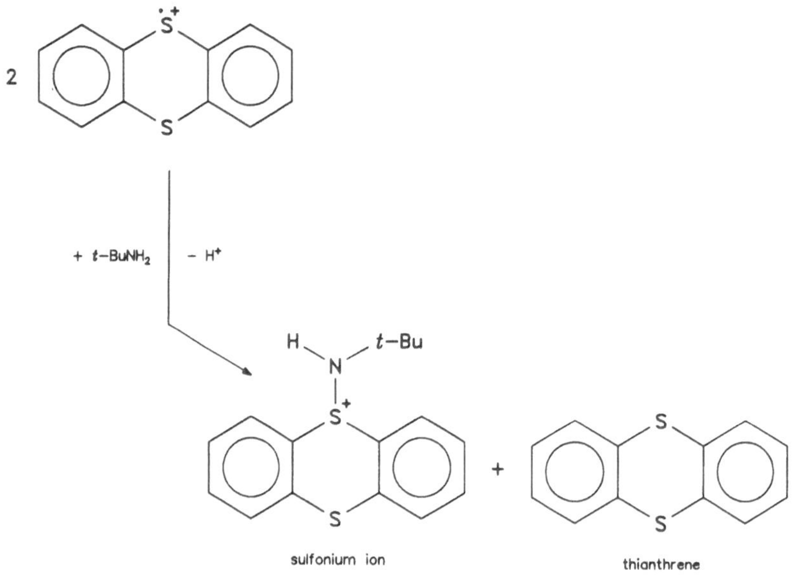

Scheme 2.37

The thianthrene molecule releases an electron to the anode reversibly to give a cation radical which in the presence of anisole is further oxidized to a dication by the following mechanism, as proposed by Parker and Eberson [12].

$$Th \quad \underset{}{\overset{-e}{\rightleftharpoons}} \quad Th^{\cdot +}$$

$$Th^{\cdot +} \quad + \quad AnH \quad \underset{k_{-1}}{\overset{k_1}{\rightleftharpoons}} \quad (Th/AnH)^{\cdot +}$$

$$(Th/AnH)^{\cdot +} \quad + \quad Th^{\cdot +} \quad \underset{k_{-2}}{\overset{k_2}{\rightleftharpoons}} \quad (Th/AnH)^{2+} \quad + \quad Th$$

$$(Th/AnH)^{2+} \quad \overset{k_3}{\longrightarrow} \quad (Th/An)^+ \quad + \quad H^+$$

Th = thianthrene ;   AnH = anisole ;   (Th/An)$^+$ = sulfonium ion

The (ThAnH)$^+$ is more easily deelectronated than the Th$^+$ thus by reaction with Th$^+$ gives the dication (Th/AnH)$^{2+}$ and finally the sulfonium ion (Th-An$^+$). The thianthrenium ion forms salts with aromatics containing electron donating groups, and also with amines and olefinic and acetylenic compounds [13]. In trifluoroacetic acid the thianthrenium cation radical survives for weeks, as is evident from cyclic voltammetry (ratio of anodic and cathodic peaks = 1.0). The trifluoroacetate salt can be obtained as a crystalline purple solid after evaporation of solvent at low temperatures [14].

2.13 Anodic Cyanations

Anodic cyanations are well established reactions by means of which
carbon-carbon bonds are formed and the synthetically valuable nitrile
intermediates are obtained. A wide literature has been developed since
the detailed report of Andreades [4] (1969) on anodic cyanation of
aromatic compounds [1–17]. It might be noted that aromatic rings are not
easily cyanated by chemical procedures. Chemical methods are usually
very laborious and require at least two major steps:

Comparison with chemical methods

1. Introduction to the aromatic ring of a suitable leaving group, and
2. displacement of this group by the cyano group, chemically or
   photochemically.

Cyanation agents that have been used for the cyanation of aromatic
hydrocarbons and phenols include organometallics in the presence of
KCN, BrCN, and Cl_3CCN. The electrochemical method is much simpler.
The electrocyanation can be carried out in acetonitrile or methanol solutions containing KCN or NaCN or Q_4NCN salts which provide the CN^-
ion and also function as electrolytes. Sometimes emulsions of CH_2Cl/H_2O
are employed to overcome solubility difficulties. Usually, divided cells
with platinum electrodes are used with the SCE or $Ag/Ag^+NO_3^-$ as
reference electrodes.

Because the CN^- ion is easily oxidized at the anode, it is sometimes
difficult to elucidate the cyanation mechanism. This may be further complicated by the ability of the CN^- ion to reduce the generated cation
radical,

Cyanation mechanism

$$ArH^{\cdot+} \ + \ CN^- \ \longrightarrow \ ArH \ + \ CN^{\cdot}$$

We note that in water the CN^- ion is oxidized at 0.6 V vs SCE but in
methanol the anodic limit extends beyond 1.6 V. Most recent studies
suggest that cyanations take place after the organic substrate is deelectronated to yield the cation radical RH^+ or the cation R^+. These species
are then attacked by the very powerful nucleophile CN^- as shown in
Scheme 2.38.

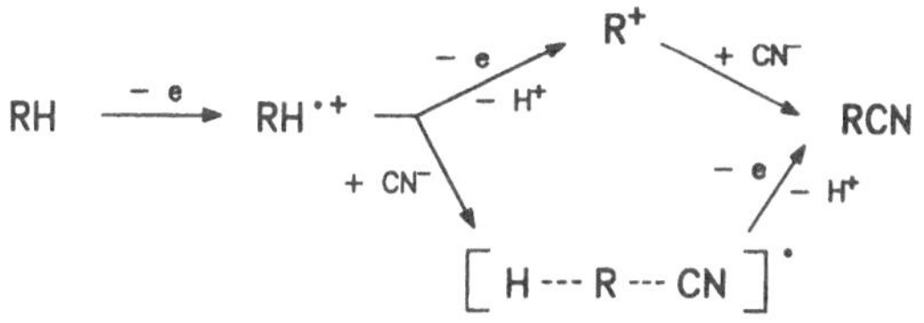

Scheme 2.38

Initial electrochemical cyanations   In the initial electrochemical cyanations by Andreades and Zahnow [4] the electrolyses were carried out in anhydrous acetonitrile solutions with large concentrations of tetraethylammonium cyanide and selected aromatic substrates. Both divided and nondivided cells with smooth platinum electrodes were employed. In the undivided cells cylindrical electrodes concentrically arranged were employed, and a SCE reference electrode was placed with the tip near the anode. Potentials, for preparative synthesis, in the range of 0.9–2.0 V were applied. The anode was rotated during electrolysis so as to minimize concentration polarization effects at the anode. Typical reactants and product yields at 50% conversion of the aromatic reactants were: anisole, anisonitrile (5%), p-dimethoxybenzene, anisonitrile (95%), veratole, o-methoxybenzonitrile (94%), 1,2,3-trimethoxybenzene, 2,6-dimethoxybenzonitrile (86%). (Caution: Cyanations should always be carried out in well ventilated hoods.)

The cyanations of p-dimethoxybenzene and tertiary amines take place by displacement of a methoxy group in the former and cyanation of an alkyl group in the latter [4]:

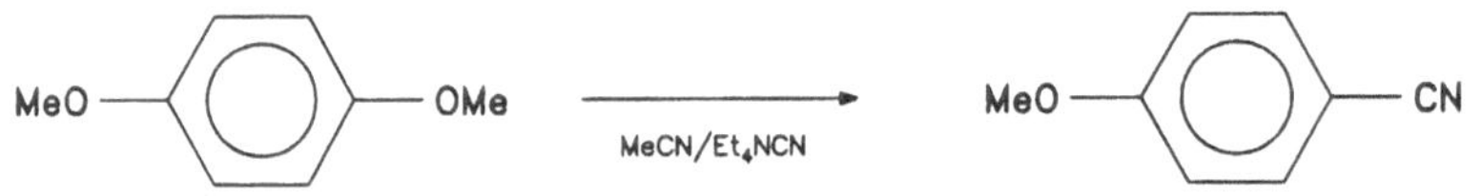

Cyanation of naphthalines   Yoshida [14] studied in depth the cyanation of various naphthalines. Naphthalenes with methyl or ethyl substituents form preferentially naphthonitriles instead of side-chain cyanation products. Chemically naphthonitriles are obtained by reaction of their corresponding halides with cuprous cyanide under quite severe conditions. According to Yoshida's observatins (voltammetric i-E curves) the anodic cyanation occurs by an ET from the naphthalene molecule to the anode. This first step is followed by a rapid attack by CN$^-$ ion on the cation radical (see Sect. 2.2, Scheme 2.4). Yoshida [18] also found that the anodic cyanations of pyrroles and **Pyrroles and Indoles** indoles is a very efficient synthetic reaction and it seems to be a general reaction for cyanations of N-heterocyclic compounds. The chemical methods for the cyanation of pyrroles via Manish bases (RCO(CH$_2$)NH$_2$) or

aldoximes RCN are complicated procedures, and give lower yields than the electrochemical procedures.

Cyanation of 1-Methylpyrrole

The anodic cyanation of 1-methylpyrrole yielding thc 2-carbonitrile analog is carried out in MeOH/NaCN in a divided cell with a platinum anode set at 1.0 V vs SCE (the reaction mechanism is depicted in Scheme 2.39). The electrolyzed solution is subjected to distillation to remove the methanol and the residue is extracted with ether from a saturated NaCl aqueous solution. The ether extract is treated with $MgSO_4$, and after filtration the ether is evaporated. The oily residue is distilled under vacuum to obtain the product (64%) (Adapted from Ref. 18, Copyright of the American Chemical Society, 1979).

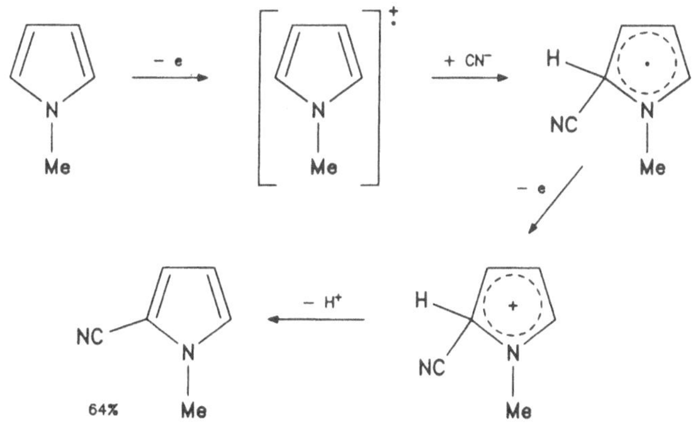

Scheme 2.39

Cyanation of *N-n-Propylpyrrolidine* [19]

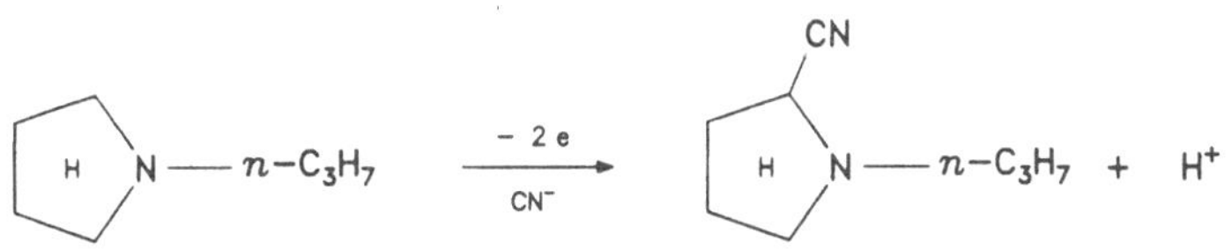

The electrolysis is performed in a 100-ml divided cell, using a porous cup for the catholyte. The anode is a cylindrical platinum net (4.5 cm in height, 11.0 cm in circumference, 55 mesh) and the cathode is a platinum *coil*. The reference electrode is the SCE. The electrolysis is carried out at 3 °C with constant stirring by a magnetic bar. The anolyte contains 0.1 mol of the amine and 0.15 mol of sodium cyanide in 75 ml of 1:1 methanol-water solution. The catholyte is an aqueous methanol solution of sodium cyanide. Current is passed for 8 hours at an anode potential in the range of 1.1 to 1.5 volts so as to maintain a constant current of 0.5 A during the entire electrolysis (0.15 F). At the conclusion of the electrolysis the anolyte is concentrated under reduced pressure at 35 °C on a rotary evaporator. The remaining liquid is saturated with anhydrous potassium carbonate. The organic layer is separated and the aqueous layer is extracted twice with 30 ml portions of ether. The combined extracts are dried with potassium carbonate and then analyzed by GLC. The distillation of the crude product at 114–119 °C (5.3 kPa) gives 8.2 g of colorless liquid, N-n-propyl-2-cyanopyrrolidine, as confirmed by IR, NMR, and elemental analysis of the picrate derivative. The yield and current efficiency are 59 and 79 %, respectively (Adapted with permission from Ref. 19, Copyright American Chemical Society, 1977).

Methoxylation versus cyanation

In methanol/cyanide solutions some substrates methoxylate rather than cyanate while others do the opposite. Yoshida [20] reported that benzene, toluene, p-xylene and tetralin prefer to methoxylate, and naphthalenes, anthracenes, biphenylacetylenes and bibenzyl undergo predominently cyanation. The cyanation of trans – stilbene (**1**) and its dimethoxy derivative was carried out in H_2O/CH_2Cl emulsion [21]. Several products were obtained besides the desired product (**2**) which was favored at lower current densities.

Comparison of cyanation methods

The cyanation of dibenzofuran was recently carried out by electrochemical, chemical and photochemical methods [22]. Isomer distributions were different in these methods. The anodic method (MeOH/NaCN) gave mostly the 3-isomer of dibenzofurancarbonitrile in 71 % of the total yield (40 %). The chemical method (BrCN/AlCl_3/CS_2) and the photochemical (NaCN/MeOH/hv) method gave the 3-isomer in only 15 % of the total yield (90 %) and the photochemical method in 64 % (total yield 85 %). The electrochemical cyanation occurs via the ArH_·^+ + CN^- paths, the chemical via the CN^+ electrophilic path, and the photochemical via the ArH_·^+ +

CN⁻ path. The anodic electrolysis was performed at constant current (0.5 A) in a nondivided cell with platinum electrodes.

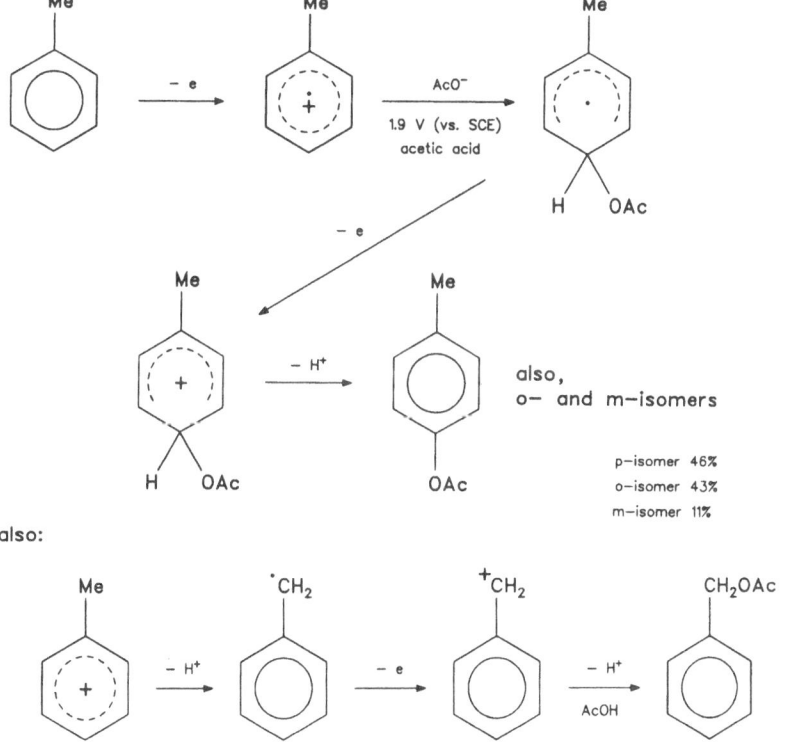

2.14 Acetoxylations and Acetamidations

2.14.1 Acetoxylations

Acetoxylations, by the anodic method, represent a general type of electro-organic reactions for synthesis involving formation of C—O bonds. They have been extensively studied by Eberson and coworkers [1–3] and several other electroorganic researchers [4–8].

The generally accepted acetoxylation mechanism is illustrated with toluene as a substrate, Scheme 2.40.

Scheme 2.40

The anodically generated cation radical is rapidly attacked by the AcO⁻ nucleophile or by the weaker neutral nucleophile CH_3COOH as indicated above for the acetoxylation via the benzyl radical of the toluene molecule. Electrochemical acetoxylations usually yield products similar to those obtained by chemical oxidations with metal oxidants, Pb(IV), Ce(IV), Co(III), Mn(III), Cu(II), Cu(III) and Ag(II) in acetic acid solutions. The electrochemical method avoids the need to these oxidants and associated process problems, (product isolation, environmental problems). However, these oxidants may be considered as mediators for indirect electrosynthesis. It must be noted that the composition of the electrolysis medium influences the nature of the acetoxylated products [1,3,4], as is seen in Scheme 2.41.

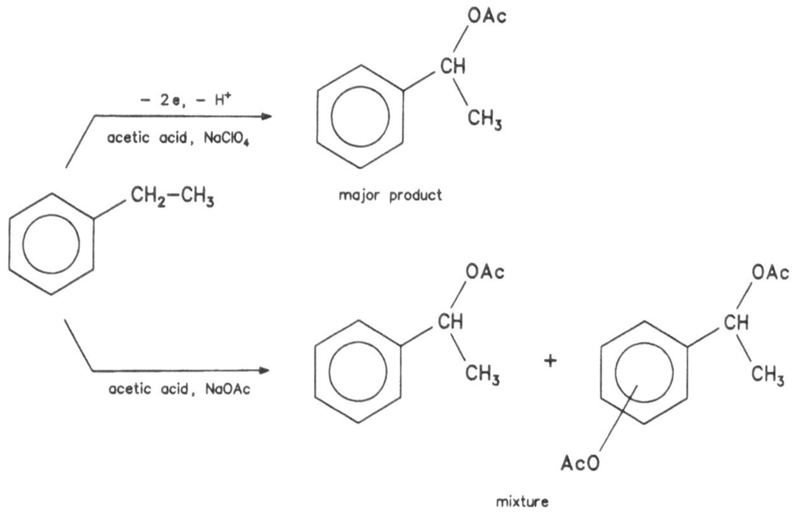

Scheme 2.41

In performing an acetoxylation the most suitable anode potential is best selected from an i-E curve. It should be set at a value at which only the substrate is deelectronated, while the acetate or acetic acid remain free to react nucleophilically with the cation radical or cation produced at the anode. We consider here some interesting acetoxylations and trifluoroacetoxylations. The α,β-diacetoxylation of N-substituted piperidines was reported by Shono [9] as a functionalization method of a less reactive methylene group. Yields of acetoxylated products of 55–80% are obtained, Scheme 2.42

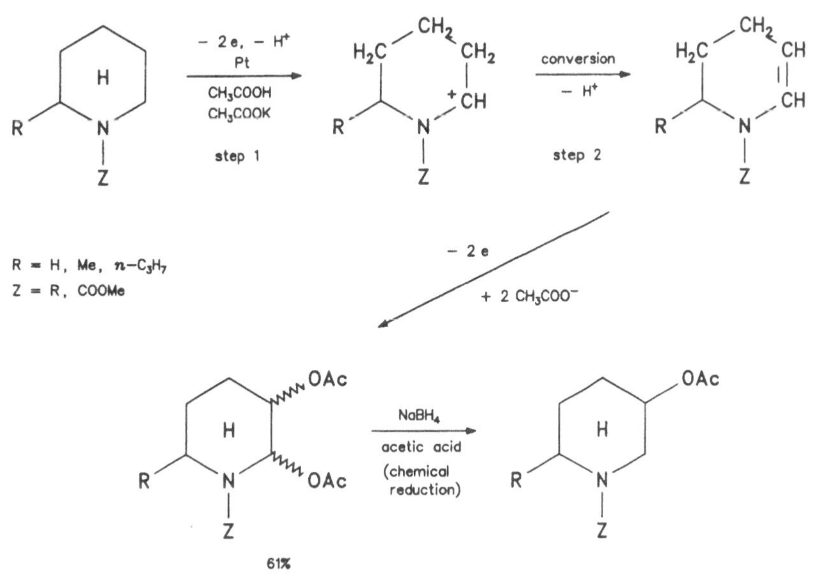

Scheme 2.42

Acetoxylation of Piperidines (Scheme 2.42)

The electrolysis was performed with 2 g substrate in 50 ml acetic acid/potassium acetate in a glass beaker with a platinum plate anode and a carbon rod cathode. A terminal cell voltage of 35 V was needed to pass a current of 0.4 A through the solution. The products were extracted with CH_2Cl_2 from a cold aqueous neutralized ($NaHCO_3$) solution. The extracts were dried with $MgSO_4$, and the residue, after solvent removal (vacuo), was chromatographed on silica gel (ethyl acetate:hexane 1:1) to obtain the α,β-diacetoxy-N-(methoxycarbonyl) piperidine (61%). The chemical reduction step with $NaBH_4$ was done in acetic acid. From this mixture the β-acetoxy product was obtained by CH_2Cl_2 extraction and silica gel chromatography (Adapted from Ref. 9, Copyright of American Chemical Society, 1987).

Selective nuclear acetoxylation of alkylaromatics was achieved in an undivided cell utilizing a Pd-carbon catalyst [2]. This interesting system allows the hydrogen produced at the cathode to react with the benzylic acetate product and to cleave it. And thus, by continuously regenerating the starting substrate, it augments the yield of the nuclear-acetoxylated product. This product is apparently resistant to hydrogenolysis. Isodurene, p-xylene, mesitylene and durene were thus acetoxylated in acetic acid-potassium acetate media, giving yields of 75–98%. The anodes were

Selective nuclear acetoxylation

graphite cylinders and the cathodes stainless steel foils. Cell voltages of about 17 volts were needed for a current of 1.4 A to pass through the solution at 17–18°C.

Trifluoro-acetoxylations

Analogous to anodic acetoxylations are the trifluoroacetoxylations by means of which the trifluoroacetoxy group is introduced to organic molecules. Trifluoroacetoxylations are performed in organic media containing trifluoroacetic acid (TFA) or in net TAF. Such media are: TFA/CH_2Cl_2/Bu_4NPF_6, TFA/acetone/Bu_4NBF, TFA/Bu_4BF_4. *trans* Dichlorohexane and fatty acids were trifluoroacetoxylated [10], the first affording the 3,4-dichlorohexyl trifluoroacetate and the latter the ω-2 and ω-4 trifluoroacetoxylated products (~50%). In TFA solutions alkanes are oxidized to give cations which react with TFA to give trifluoroacetoxy products in 50–70% current yields [11].

$$RCH_2R \xrightarrow[\text{TFA/Bu}_4\text{NBF}_4]{-\ e} [RCH_2R]^{\cdot +} \xrightarrow{-\ e,\ -\ H^+} R\overset{+}{C}HR \xrightarrow[-\ H^+]{+\ TFA} RCH(OCOCF_3)R$$

Trifluoroacetates are very easily hydrolyzed to afford the hydroxy products. Trifluoroacetoxylations, therefore, provide a very convenient indirect route for the hydroxylation of aromatic hydrocarbons [8]:

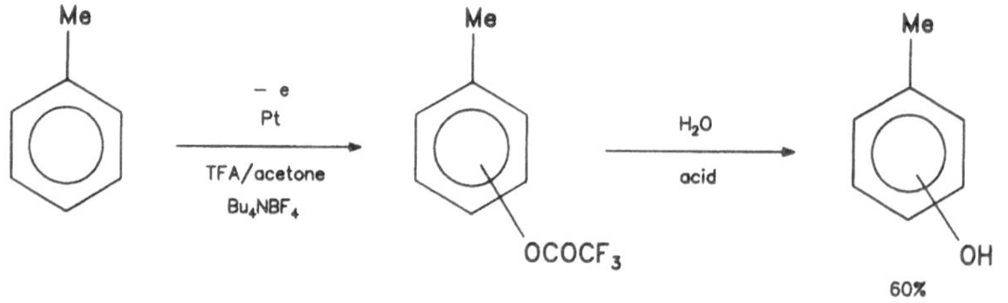

2.14.2 Acetamidations – Electrochemical Ritter-Type Reactions

Anodic acetamidations are reactions of a general type taking place in acetonitrile solution, as depicted symbolically in Scheme 2.43

$$RH \xrightarrow{-\ 2\,e,\ -\ H^+} R^+ \xrightarrow{+\ MeCN} R-N\overset{+}{=}C-Me \xrightarrow[-\ H^+]{+\ H_2O} RNHCOMe$$

nitrilium ion

Scheme 2.43

For the acetamidation to be efficient the carbocation R^+ should first be produced, and the solution should not contain nucleophiles that compete with acetonitrile. Eberson and Nyberg, and Baizer and Wagenknecht, independently established this useful version of the Ritter reaction [12,13]. The electrochemical Ritter reaction takes place with alkyl benzenes, alkyl halides, aliphatic ketones, olefins and even saturated hydrocarbons. In the presence of H_2O competition with MeCN should be expected, that is, hydroxylation versus nitrilium ion formation. This would result in a mixture of hydroxylated and acetamidated products, in ratios that would be affected by the MeCN/H_2O ratio and also the reactivity of the carbocation R^+. A striking example indicating the effect of the kind of electrolyte ions on the acetamidation in competition with hydroxylation is shown below [13a]:

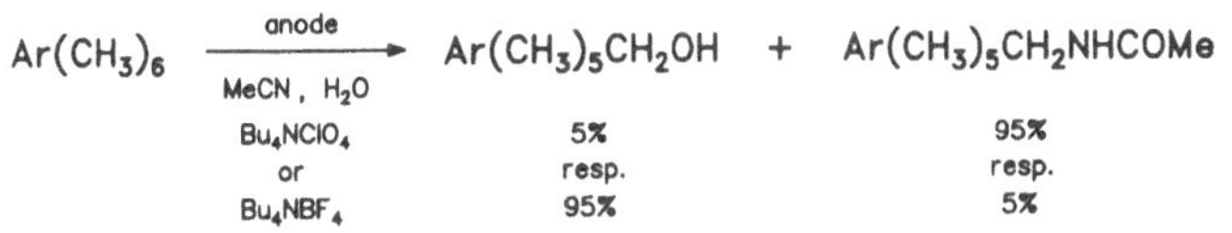

$$Ar(CH_3)_6 \xrightarrow[\substack{MeCN,\ H_2O \\ Bu_4NClO_4 \\ or \\ Bu_4NBF_4}]{anode} Ar(CH_3)_5CH_2OH \ + \ Ar(CH_3)_5CH_2NHCOMe$$

	5%	95%
	resp.	resp.
	95%	5%

It is seen that the product distribution is reversed by changing the electrolyte (these are probably double layer effects). Chemically, the Ritter reaction as represented below requires a strong acid to protonate the aldehydic carbonyl for the generation of the needed carbocation:

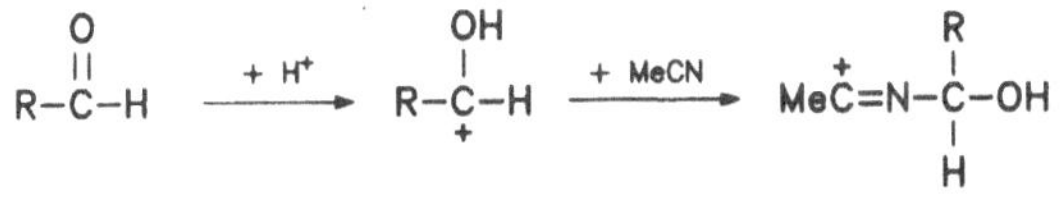

$$\underset{R-C-H}{\overset{O}{\|}} \ \xrightarrow{+\ H^+} \ \underset{R-\overset{+}{C}-H}{\overset{OH}{|}} \ \xrightarrow{+\ MeCN} \ Me\overset{+}{C}=N-\underset{\underset{H}{|}}{\overset{\overset{R}{|}}{C}}-OH$$

Acetamidations occur by substitution on the aromatic ring and also on side chains. Sometimes very dry acetonitrile may be required. In such cases trifluoroacetic anhydride (TFAA) is a most effective drying agent. Thus in acetonitrile/TFAA solution anthracene gave the 9-acetamido product in 80% yield [14]. Saturated and unsaturated hydrocarbons afford mixtures of acetamidated products by electrolysis in MeCN/0.1 M Bu_4NBF_4 at smooth platinum electrodes [15], in 55–80% yields. *tert*-Butylcyclohexane undergoes cleavage to give the C—N coupled products [16]:

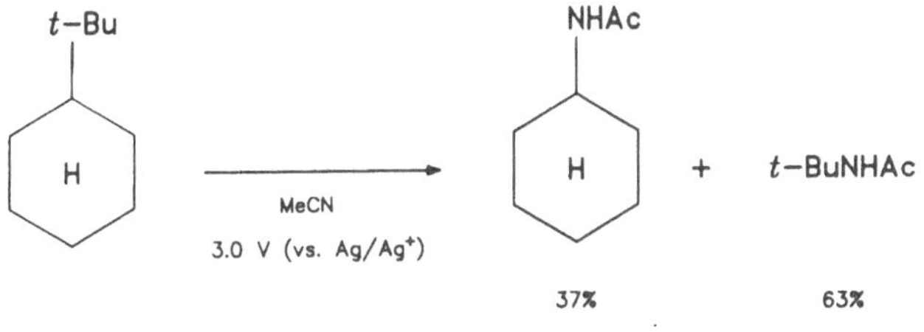

Anodic electrolysis of propylene in wet acetonitrile gives the *N*-allylacetamide [17,18]. Side chain acetamidations may sometimes lead to amines [19] through further hydrolysis:

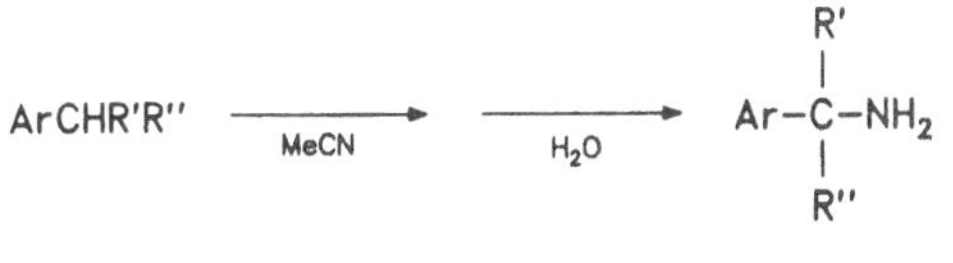

2.15 Anodic Halogenations

Halofunctionalizations take place mostly by indirect electrolysis in solutions containing a halide salt and the organic substrate. That is, the latter is halogenated by the anodically produced halogen species [1–4]. Fluorinations and some chlorinations are possible by direct oxidation of the organic substrate as a first step. This anodic step is followed by an attack on the oxidized substrate by the fluoride or chloride ion, as in the formal fluorination example:

$$RH \xrightarrow[\substack{MeCN \\ Et_4NF \cdot 3HF}]{-e} [RH]^{\cdot +} \xrightarrow{+F^-} [RHF]^{\cdot} \xrightarrow[\substack{-e \\ -H^+}]{} RF$$

$$RH \xrightarrow{-e, -H^+} R^+ \xrightarrow{+F^-} RF$$

Large scale anodic fluorination   Anodic fluorination of hydrocarbons is performed on a large scale in liquid anhydrous hydrogen fluoride containing the organic material to be fluorinated (steel tanks). Cell voltages below that required for the oxidation of F^- ions are applied. Hydrogen gas is evolved at the cathode. The Simon's process [1] employs a divided cell (porous diaphragm) with an iron cathode and a nickel anode immersed in a liquid HF/KF mixture which dissolves a moderate amount of the starting hydrocarbon. At the nickel electrode a coating of K_2NiF_6 is produced which apparently donates fluorine to the organic substrate and is continuously regenerated to repeat the fluorination cycle (heterogeneous electrocatalysis). Following are a few examples indicating that all hydrogens are replaced by fluorine in the products, which are obtained in high yields [1].

$$CH_3(CH_2)_7SO_2Cl \xrightarrow[2-3 \text{ V (cell)}]{HF/KF} CF_3(CF_2)_7SO_2F \xrightarrow{H_2O} CF_3(CF_2)_7SO_3H$$

$$CH_3(CH_2)_6COOH \xrightarrow[2-3 \text{ V (cell)}]{HF/KF} CF_3(CF_2)_6COF \xrightarrow{H_2O} CF_3(CF_2)_6COOH$$

$$(C_4H_9)_3N \xrightarrow[2-3 \text{ V (cell)}]{HF/KF} (C_4F_9)_3N$$

Chlorination of aniline   Chlorination of aniline occurs directly in DMF or dimethylacetamide solvents. In these solvents the generated aniline cation radical is stable enough to be attacked by the chloride ion. Platinum, carbon and the DSA-type electrodes are suitable anodes for this reaction [5].

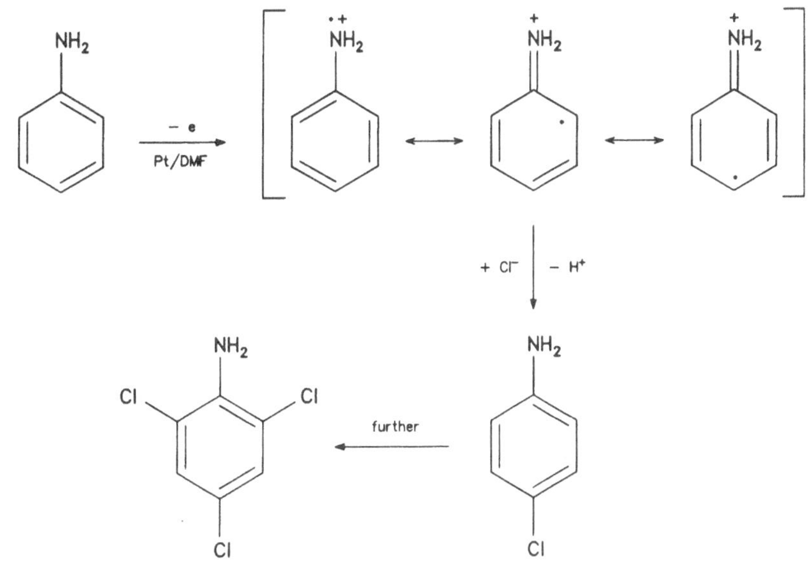

Anthracene can be brominated in MeCN/Et$_4$NB$_r$ at 1.2 V vs Ag/Ag. It must be noted that both the anthracene molecule and the bromide ion are oxidized at this anode potential. The product 9-bromoanthracene is obtained in 48% current yield. Similarly naphthalene gives the 1-bromo analog with 70% current efficiency [6].

Bromination of anthracene

Shono [7] reported a convenient method for chlorination of certain unreactive methylene groups in cyclic amines by electrolysis in aqueous acetonitrile containing ammonium chloride. N-methoxycarbonyl cyclic amines afforded the α-hydroxy-β-chloro analogs in good yields. We note that there are no general effective methods for such chlorination:

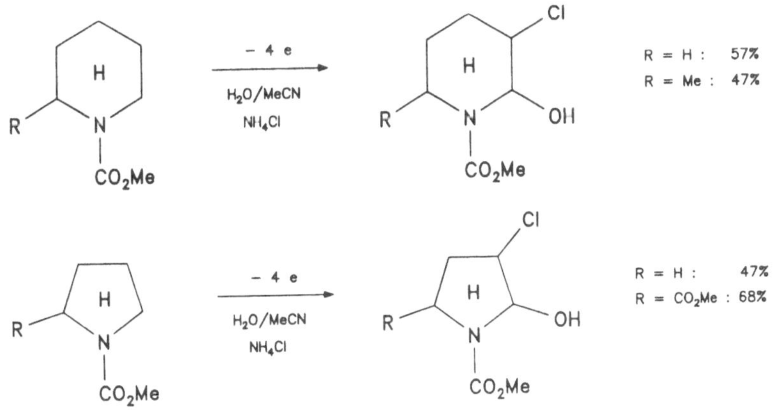

Chlorenation of methylene groups in cyclic amines

The α-OH group is easily removed by reduction with NaBH$_4$ and thus the β-chlorosubstituted products are obtained in 80% yields.

Chlorination of naphthalene butadiene and methoxy- and ethoxy benzenes

Chlorination of naphthalene in H_2O/CH_2Cl_2 emulsion is facilitated by the presence of $ZnCl_2$ in the emulsion. It was proposed that a complex $ZnCl_4^-$ forms which is transferred into the organic phase where the chlorination of the naphthalene occurs [8]. Butadiene is easily chlorinated in MeCN/CoCl$_2$ solution by electrolysis in a divided cell at a graphite anode set at 1.7 V vs Ag/AgCl. The 1,4-dichlorobutane product (*trans*-isomer) is obtained in 90% yield [9]. Chemically this chlorination requires high temperatures and chlorine gas. Anodic chlorination of methoxy- and ethoxybenzenes [10] affords preferentially the *p*-chlorinated products.

The electrolysis is carried out in DMF or dimethylacetamide, DMA, containing LiCl in a divided cell with a platinum anode at a potential 1.0–1.5 V vs SCE. At this potentials both Cl$^-$ and the organic substrate would be oxidized at the anode. Yields of about 90–95% are obtained. The p- to o-product ratio is 17:1 with DMF as solvent and 34:5 with DMA. (This difference may be attributed to different solvation effects.)

Iodination of aromatics

Iodination of aromatics can be accomplished in a rather special way [11]. First, a solution of iodine in trimethylorthoformate $CH(OCH_3)_3$ (TMOF) is anodically electrolyzed to obtain an active positive species, I^+, referred to as iodonium ion. The electrolytic oxidation is carried out in a divided cell in which the anolyte is a solution of iodine in TMOF/$LiClO_3 3H_2O$ and the catholyte the same colution without the iodine. Both electrodes are platinum. A current density of 0.1 A/cm^2 is sufficient to oxidize the iodine to I^+. The anolyte solution containing the iodonium ion is then mixed with a solution of the aromatic in TMOF and the iodination occurs chemically. This iodination is selective with regard to the electron-donating substituent X on the aromatic ring, as summarized below:

$$^1/_2 \; I_2 \xrightarrow[\text{anode}]{-e} I^+ \qquad \text{step 1}$$

	X = Cl :	no reaction
X = Cl, H, Me, MeO, *t*-Bu	X = H :	60%
	X = *t*-Bu :	97%, *p*-isomer : 100%

Iodinations via I^+ species are also possible by electrolysis in MeCN or CH_2Cl_2/TFA media [12], but they are not as selective as they are in the TMOF medium.

2.16 Hydroxylations

Direct anodic hydroxylations in aqueous or organic-aqueous media are difficult because water is usually oxidized at the anode at less positive potentials than the organic substrate. However, if the organic substrate is oxidized before the water, direct hydroxylation is possible by reaction of water with the produced organic cations. Hydroxylation of phenols takes place at platinum, PbO$_2$ and other electrodes, as illustrated below [1,2]:

Phenols

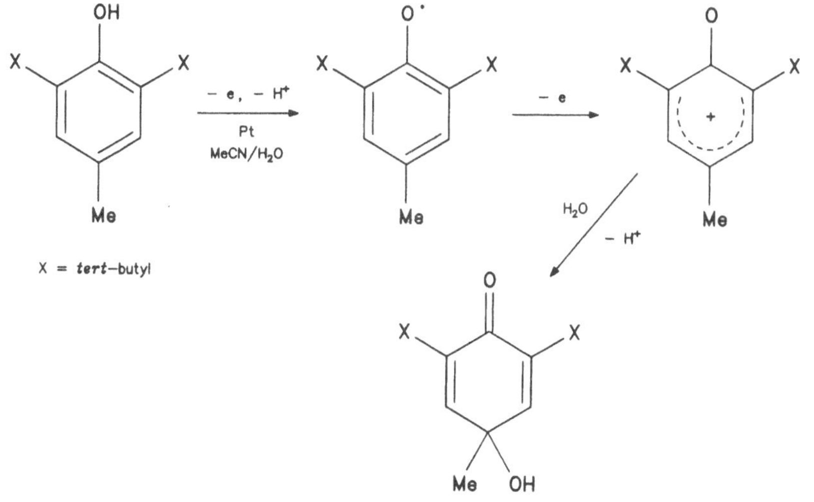

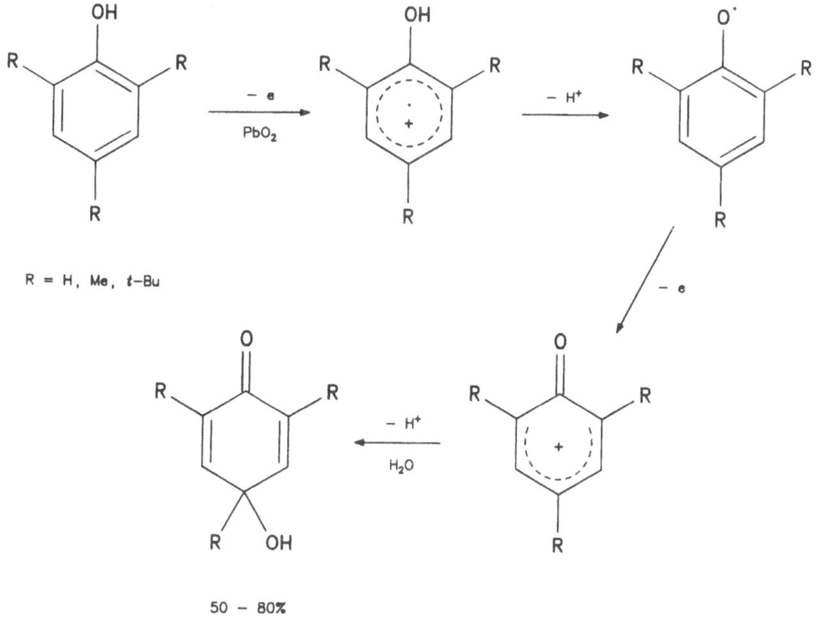

X = *tert*−butyl

R = H, Me, *t*−Bu

50 − 80%

Oxidations at the PbO_2 anode in acidic solutions may involve both electrocatalytic (via PbO_2) and direct electron transfer steps. The initial radical cation of the phenolic compound is most probably electrocatalytically formed while its further deelectronation leading to the oxonium ion may involve both indirect electrocatalytic and direct electron transfer steps. (Note: PbO_2 is not stable below 1.4 V vs SCE in 1 M H_2SO_4.) In the anodic oxidation of phenols in general PbO_2, C, Ni or Pt are suitable electrodes. Usually the PbO_2 is more effective. Benzene has been converted to p-benzoquinone at a PbO_2 electrode in sulfuric acid medium on a pilot-plant scale [3]. The reaction involves the hydrooxylated intermediates. Anthracene is transformed into anthraquinone and bianthrone, via hydroxylated intermediates by electrolysis in $MeCN/H_2O$ media [4,5].

Side chains of aromatics   Hydroxylations on side chains of aromatics are possible by electrolysis in $H_2O/MeCN/Et_4NBF_4$ solutions or in organic/H_2O emulsions [6,7]. Styrenes are thus dimerized upon hydroxylation, and by loss of water from the dihydroxy dimer afford the 2,5-diphenyltetrahydrofurans [8]:

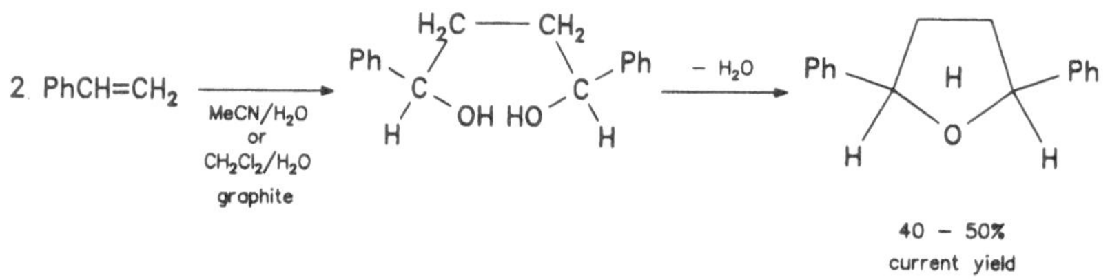

2.17 Alkoxylations

Anodic alkoxylations are reactions of a general type by which intermolecular and intramolecular C—O bonds are formed leading to a wide variety of useful synthetic intermediates. Methoxylations are the most **Methoxylations** common type of alkoxylations. They are easily performed in methanol solutions containing the organic substrate and KOH, NaOH or the alkali methoxide salt. At the anode both methanol and methoxide ion are subject to oxidation by loss of an electron. For synthetic work the oxidation of MeOH or MeO^- ion must not advance beyond the one electron oxidation stage, or the alcohol must not undergo oxidation before the organic substrate so that it can act as a nucleophile for the desired methoxylation. Methoxylation can proceed along two paths, as illustrated in Scheme 244:

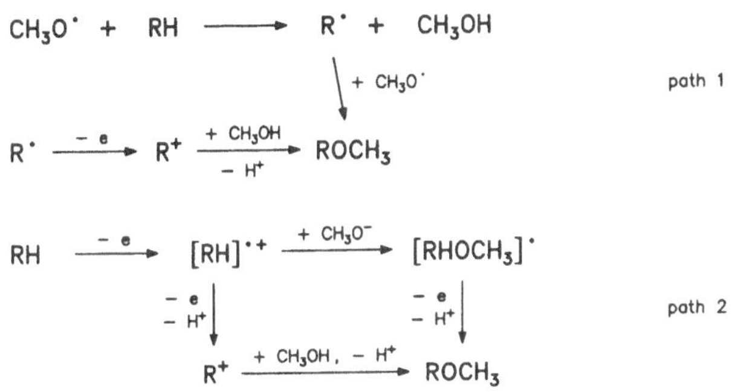

$$CH_3O^{\cdot} + RH \longrightarrow R^{\cdot} + CH_3OH$$

$$\downarrow + CH_3O^{\cdot} \qquad\qquad \text{path 1}$$

$$R^{\cdot} \xrightarrow{-e} R^+ \xrightarrow[-H^+]{+ CH_3OH} ROCH_3$$

$$RH \xrightarrow{-e} [RH]^{\cdot +} \xrightarrow{+ CH_3O^-} [RHOCH_3]^{\cdot}$$

$$-e \downarrow -H^+ \qquad\qquad -e \downarrow -H^+ \qquad\qquad \text{path 2}$$

$$R^+ \xrightarrow{+ CH_3OH, - H^+} ROCH_3$$

Scheme 2.44

Path 2 requires that the organic substrate RH be more easily oxidized at the anode than either MeOH or CH_3O^- ion. Both methanol and methoxide ion are very good nucleophiles – the ion being more powerful than the neutral methanol molecule, and either one attacks the cation radical $RH^{\cdot +}$ or the cation R^+, usually in an overall two-electron process.

A large variety of olefinic compounds [1–4] and aromatics [5–13] have been methoxylated. Olefins are methoxylated easily but often mixtures of products are obtained, as may be inferred from the following general reaction scheme:

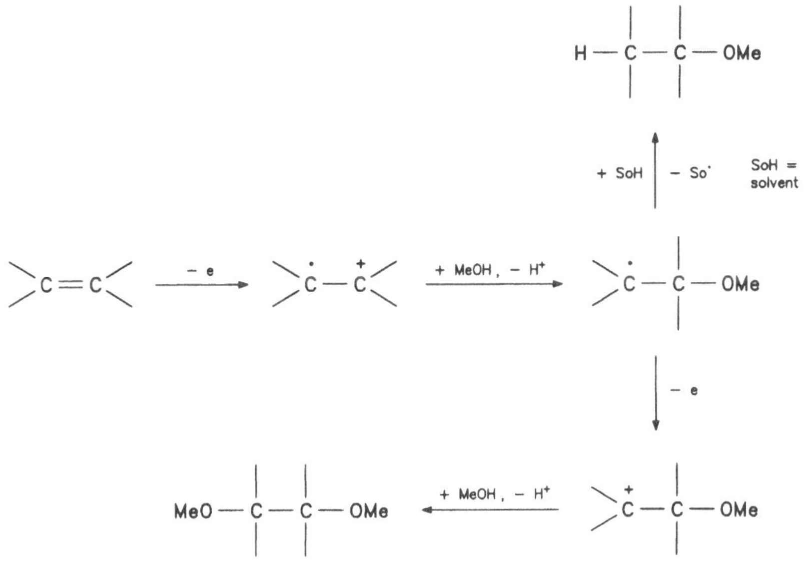

Methoxylation of olefinic Compounds and aromatics

$$H - \overset{|}{C} - \overset{|}{C} - OMe$$

$$+ SoH \uparrow - So^{\cdot} \qquad SoH = \text{solvent}$$

$$\overset{\diagup}{\underset{\diagdown}{}}C=C\overset{\diagup}{\underset{\diagdown}{}} \xrightarrow{-e} \overset{\cdot}{\underset{\diagdown}{}}C - \overset{+}{C}\overset{\diagup}{\underset{\diagdown}{}} \xrightarrow{+ MeOH, - H^+} \overset{\cdot}{\underset{\diagdown}{}}C - \overset{|}{C} - OMe$$

$$\downarrow -e$$

$$MeO - \overset{|}{C} - \overset{|}{C} - OMe \xleftarrow{+ MeOH, - H^+} \overset{+}{\underset{\diagdown}{}}C - \overset{|}{C} - OMe$$

Numerous intermolecular and intramolecular alkoxylations have been demonstrated on a laboratory scale of synthesis. Only selected typical examples will be considered for illustrative purposes, since the book's space does not allow even a limited literature survey of this topic. Kim and coworkers [5] studied the methoxylation of *p-tert*-butyltoluene as a model process in a bipolar cell with a packed bed electrode of graphite pellets. The side p-CH$_3$ chain was dimethoxylated in two steps in the chosen electrolysis medium MeOH/CH$_3$COOH/NaBF$_4$:

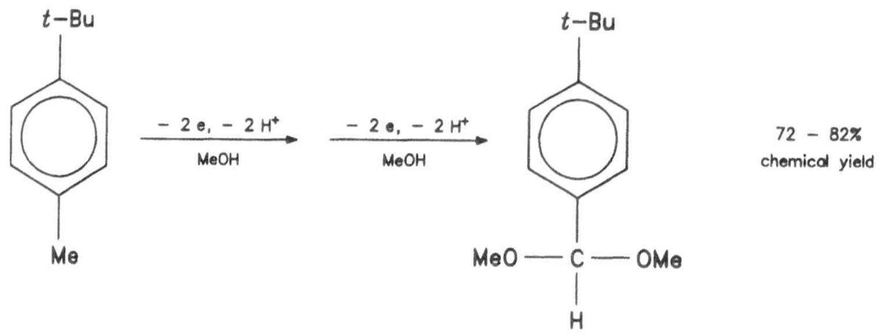

Some tar and oligomer formation were side products of this electrolysis. The electrolysis was carried out under constant current at 308 K. Cyclic voltammetry showed that the butyltoluene was oxidized at a less positive potential than the methanol, but under constant current conditions in the large scale process some methanol also was oxidized at the anode (because the anode potential had to be raised to keep the current constant). Arylcyclopropanes easily undergo dimethoxylation following opening of the cyclopropane ring, as illustrated in the following exampl [6,7]:

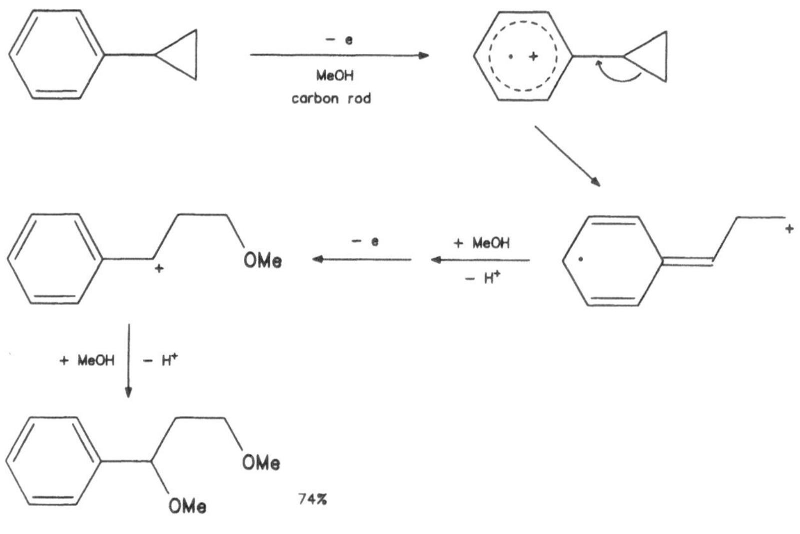

This electrolysis is performed in a three-neck flask equipped with carbon rod electrodes, anode and cathode, and a reflux condenser, with the substrate in methanol solution containing tetraethylammonium tosylate as electrolyte. The electrolyzed solution is diluted with water, and the products are extracted with ether and isolated by distillation in vacuo.

Swenton and coworkers [8] successfully employed electrochemical methoxylation in the synthesis of lithiated quinone ketals in connection with anthracycline antibiotic ring systems and the preparation of methoxyjuglone [9] (a natural product of biological importance), the latter from the starting compound 1,4,5-trimethoxynaphthalene. These reactions are synoptically depicted below:

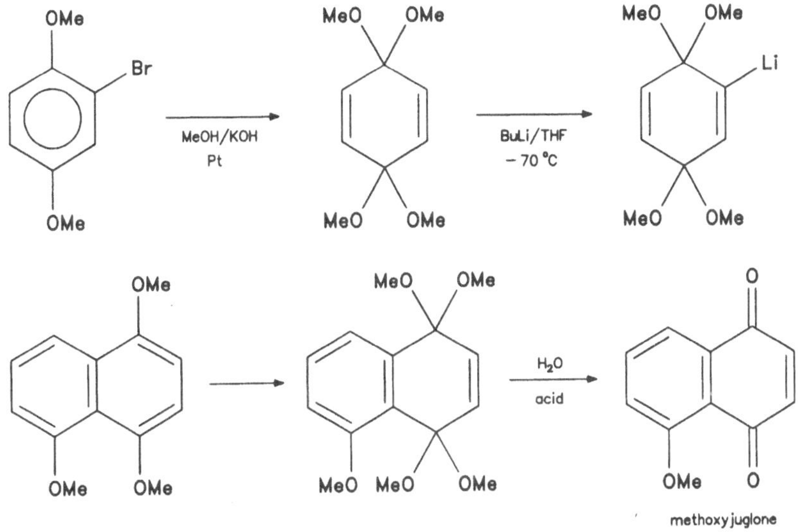

methoxyjuglone

Silyl-substituted 1,3-dienes are transformed to 1,1,4-trimethoxy derivatives [10] by anodic electrolysis on carbon anodes in MeOH/Et$_4$NOTs solutions in one-compartment cells [4]. The β-silyl group facilitates the deelectronation of the π bond and the C—Si bond splits easily. Methoxylations of piperidine and pyrrolidine-derived urethanes yield products that are intermediates for the synthesis of various alkaloids.[11,12] The amino acids lysine and ornithine were prepared via the methoxylated N-acylpiperidine [13]. Anodic methoxylation of acetals in MeOH/KOH solutions affords orthoesters most likely via the anodically formed CH$_3$O˙ radicals [14,15]. Intramolecular alkoxylation of some ethanolamines leads to the formation of five-membered rings, like that of 1,3-oxazolidine [16]:

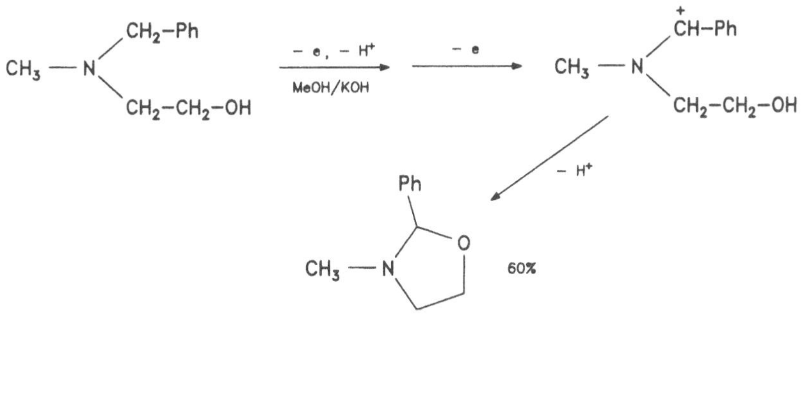

The methoxylation of N,N-dimethylformamide in MeOH/BF$_4^-$ solution occurs directly via the cation radical of the amide and also indirectly in MeOH/NH$_4$NO$_3$ solution by the action of NO$_3^-$ radicals produced at the anode [17]. It is interesting to note that in the electrolysis of methanol/dimethylformanide mixtures the oxidation of methanol is inhibited by the amide. The amide is thus oxidized at the anode, before the BF$_4^-$ ion, so that the direct methoxylation reaction becomes feasible. The reactions are summarized formally in Scheme 2.45:

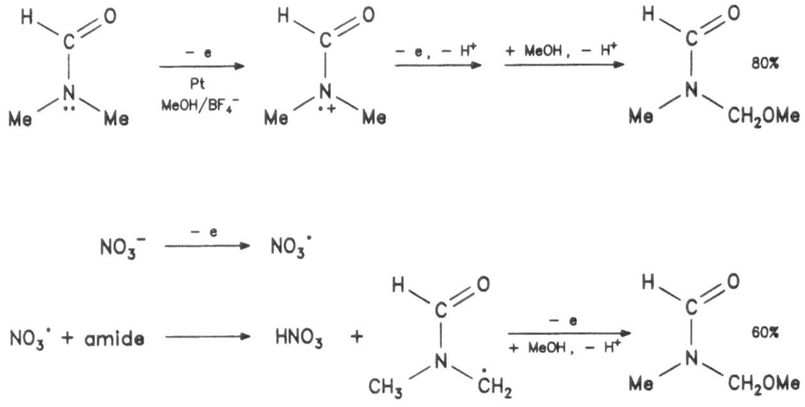

Scheme 2.45

> **Methoxylation of DMF (Scheme 2.45)**
>
> The methoxylation of the amide by direct electrolysis is conducted in a one-compartment cell (200 ml beaker) containing a solution of equal volumes of the amide and methanol (75 ml of each) and a few grams of Bu$_4$NBF$_4$ as electrolyte. Two platinum electrodes, placed two cm apart to minimize IR drop, serve as anode and cathode. A voltage is applied across the cell sufficient to allow a current of 2 A (1 A/cm^2) to pass through the solution which is vigorously stirred by a magnet bar. When about one half of the amide has been oxidized, the electrolysis is stopped and the products are recovered by distillation, and are identified by vapor phase chromatography (VPC) using authentic samples as standards (Adapted from Ref. 17, Copyright of American Chemical Society, 1972).

The methoxylation of methylbenzenes provides a route for the preparation of cyclohexa-1,4-dienes. Barba and coworkers [18] confirmed that an ECEC reaction sequence, Scheme 2.46, leads to methoxylated products either by addition to the ring or by substitution on the methyl chain.

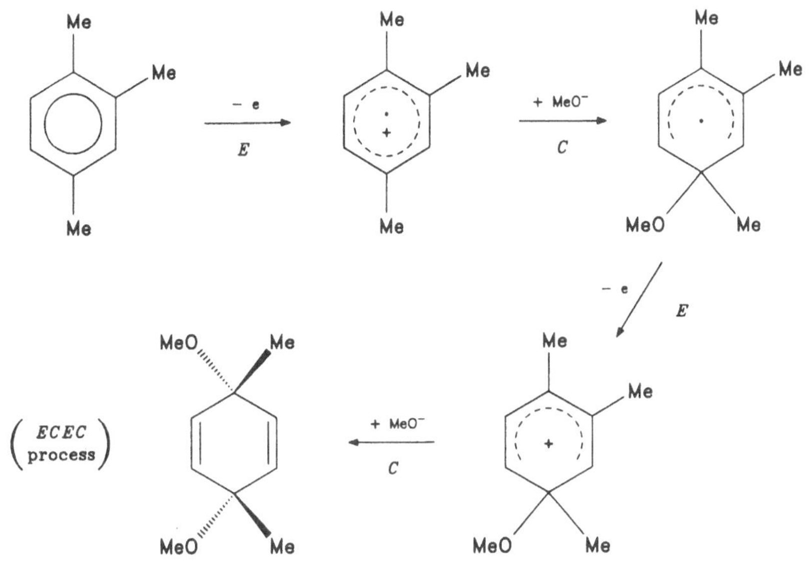

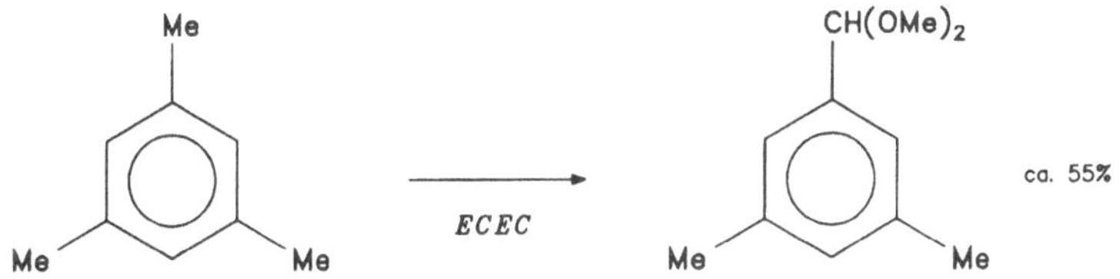

Methoxylation of Methylbenzenes, (Scheme 2.46)

Typically, the methoxylation is performed in MeOH/NaOMe solution in an undivided cell under a current of $50-60\,mA/cm^2$ at $25-30\,°C$ with carbon or platinum electrodes [21]. The sodium methoxide is prepared by addition of metallic sodium $0.2\,g$ to $70\,ml$ of methanol containing $1\,g$ of the methylbenzene substrate. The products are isolated by flash chromatography on a carbon-Celite column with hexane/ethyl acetate eluant, and identified by spectroscopic techniques.

Stereospecific methoxylations affording *cis* and *trans* products have been demonstrated with molecules such as acenaphthene [19] and *p*-xylene [20]. The *cis* and *trans* forms of the acenaphthene products were formed in ratios of 67/31 *cis/trans* at platinum and 36/64 at graphite electrodes apparently because of different orientational adsorption effects on the electrodes [19].

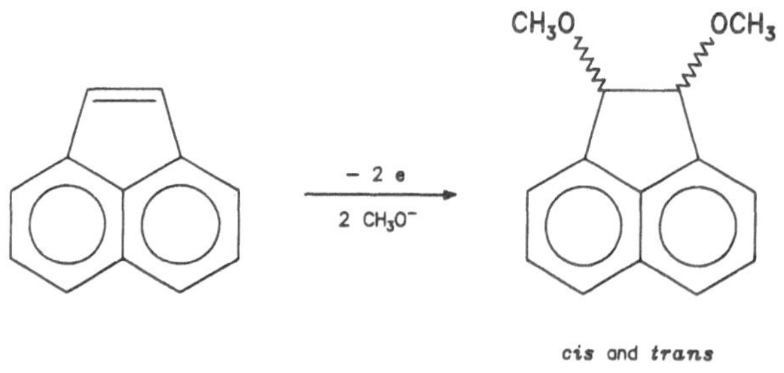

cis and *trans*

The methoxylation of alkylanisoles in MeOH/NaOMe solutions proceeds readily by a direct one-electron transfer to the anode as a first step to give the cation radical. This may be attacked by any of the available nucleophilic species, methanol, methoxy radical or methoxide ion. Thus the nuclear methoxylation of *p*-methylanisole leads to the 2,5-dienone by controlled hydrolysis of the ketal intermediate, most likely according to the following path [21]:

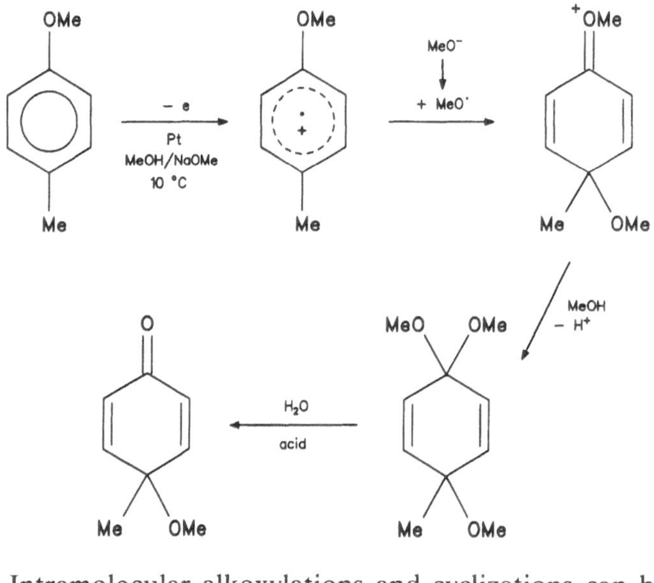

Intramolecular alkoxylations and cyclizations can be efficiently accomplished by anodic electrolysis in basic methanolic solution, as the following typical examples illustrate [22,23,24]:

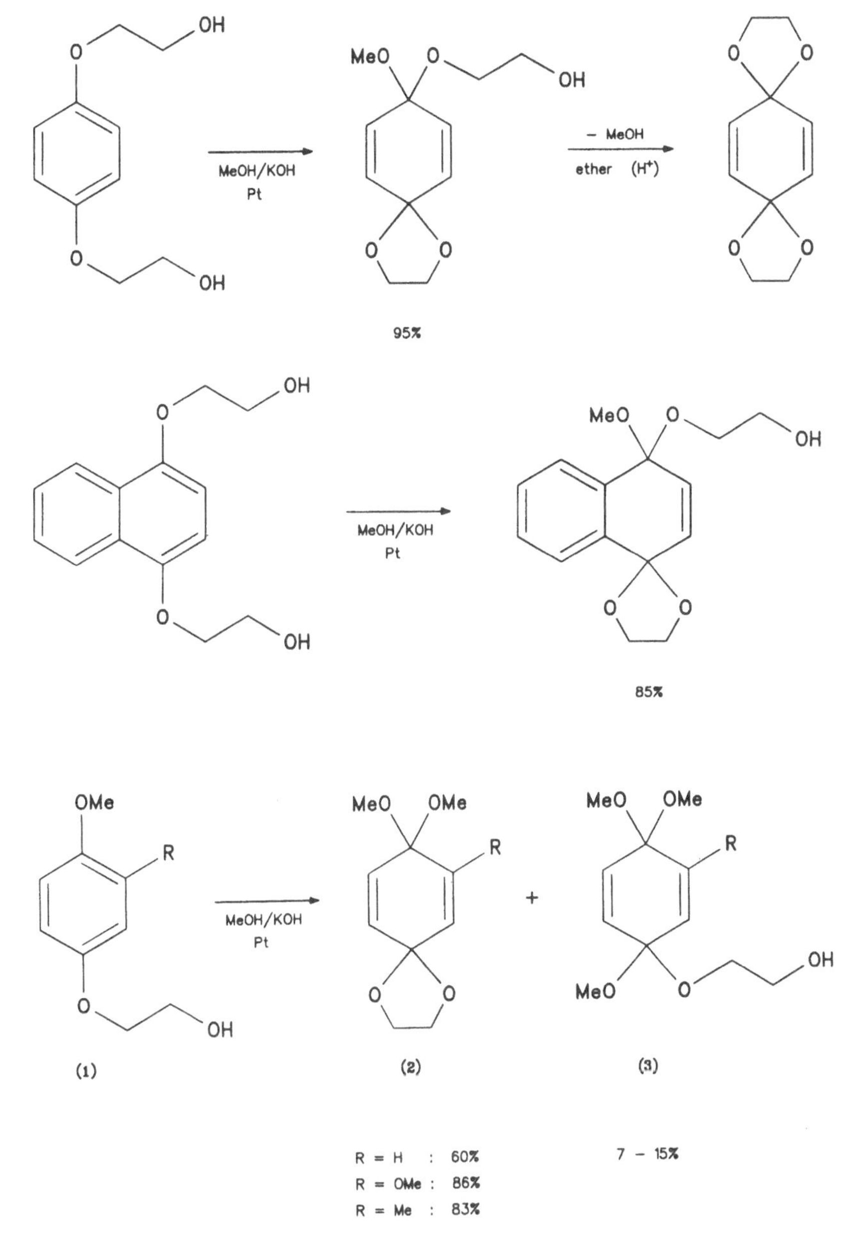

Scheme 2.46

Citronellol is transformed to the cyclic ether by anodic electrolysis in MeCN/Et$_4$NOTs solution [22]:

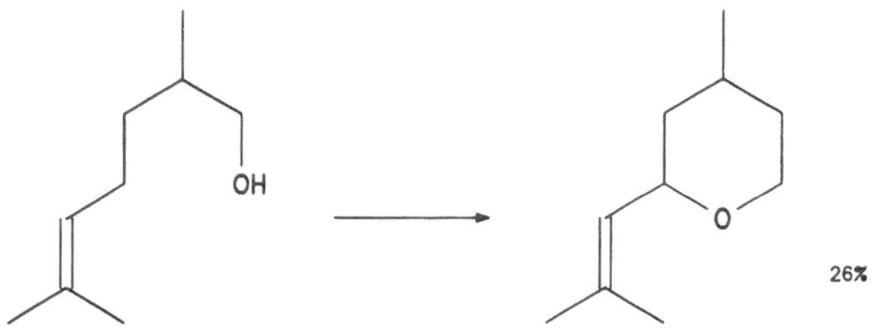

26%

rose oxide

Anodic Oxidation of Mixed Ethers of Hydroquinones

A typical experimental procedure concerning the aforementioned intramolecular alkoxylations is the procedure used by Swenton [23] for the oxidative transformation of compound (**1**) to (**2**) (see above formulas for **1** and **2** Scheme 2.46.

The electrolysis is carried out in an undivided cell equipped with platinum electrodes and a platinum wire to act as a reference electrode. A potentiostat is used as a dc power source. Three g of compound A is dissolved in 1% KOH methanolic solution (75% ml) and a current of 1 A is passed through the solution by properly adjusting the applied potential so as to keep the current constant as the electrolysis progresses. The potential region within which the desired oxidation occurs is determined from a i-E curve, as usual. The reaction is followed by periodic spectroscopic UV measurements at 288 nm, until the initial UV absorption falls to less than 5% of the initial value. The electrolyzed solution is then subjected to distillation in vacuo to remove the methanol, and to the residue is added 20 ml of water. From this aqueous solution the product is extracted with chloroform (3 × 20 ml) and the chloroform is evaporated. The colorless oily liquid that remains is chromatographed on a neutral alumina column to isolate the product B, which is a white crystalline solid. Identification of B is made by IR and NMR techniques (Adapted from Ref. 23, Copyright of the American Chemical Society, 1981).

2.18 Pyridinations

Pyridine is known to be a good nucleophile adding readily to various cation radicals and particularly to condensed aromatic compounds (anthracene, perylene, diphenylanthracene, and benzo(a)pyrene). The last one is known as a potent carcinogen, and has been studied in connection with in vivo bonding of hydrocarbons to deoxyribonucleic acid (DNA) [1].

The pyridination of diphenylanthracene (DPA) has been proposed to occur with the pyridine adding successively to the cation radical and to the dication in the fourth step [2]. Spectroelectrochemical investigations indicated that these steps are fast [3].

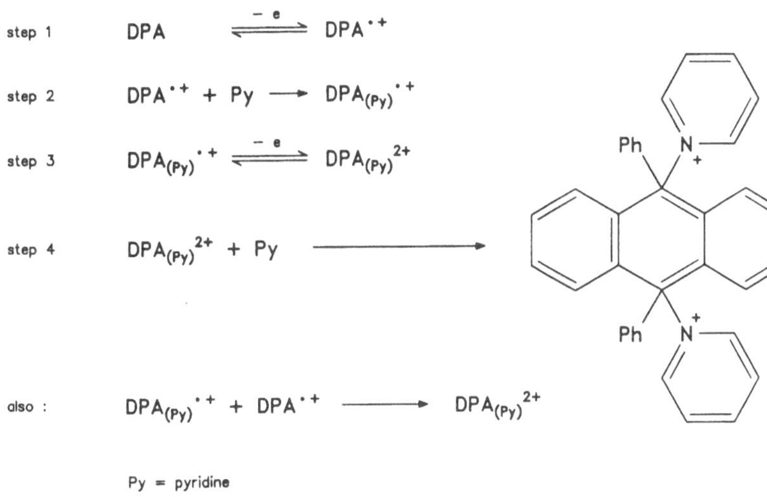

step 1 $DPA \rightleftharpoons DPA^{\cdot +}$ $\quad -e$

step 2 $DPA^{\cdot +} + Py \longrightarrow DPA_{(Py)}{}^{\cdot +}$

step 3 $DPA_{(Py)}{}^{\cdot +} \rightleftharpoons DPA_{(Py)}{}^{2+}$ $\quad -e$

step 4 $DPA_{(Py)}{}^{2+} + Py \longrightarrow$

also : $DPA_{(Py)}{}^{\cdot +} + DPA^{\cdot +} \longrightarrow DPA_{(Py)}{}^{2+}$

Py = pyridine

2.19 Nitrations

Aromatic nitrations are accomplished by anodic electrolysis in aqueous nitric acid and also in nitrite ion containing solutions [1–3]. The active nitronium ion, NO_2^+, generated at the anode, as shown below, reacts with the aromatic compound to yield the substituted nitro derivative. Nitrogen dioxide, $NO_2^{\cdot}$, is in equilibrium with N_2O_4.

$$2\ NO_2{}^{\cdot} \rightleftharpoons N_2O_4 \rightleftharpoons NO^+ + NO_3^-$$

$$NO_2{}^{\cdot} \xrightarrow[\text{anode}]{-e} NO_2{}^+$$

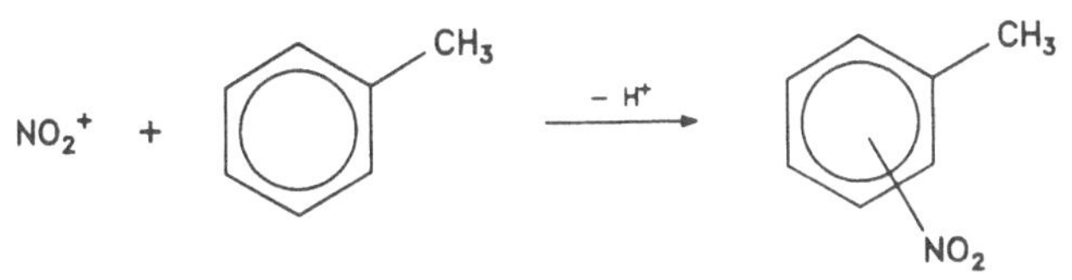

$$NO_2{}^+ + \text{(toluene)} \xrightarrow{-H^+} \text{(nitrotoluene)}$$

The electrolysis system $Bu_4N^+NO_3^-/McCN/N_2O_4$ can be used for nitrations. In such a medium the nitrogen dioxide is oxidized to NO_2^+ at 0.8 V as Ag/Ag^+. The possible existence and role of the positive ion $N_2NO_3^+$ near the anode might also be considered for certain nitrations as indicated below:

$$H_2O \xrightarrow{\;-\;e\;} 2\ H^+ + {}^1\!/_2\ O_2$$

$$NO_3^- + 2\ H^+ \longrightarrow [H_2NO_3]^+$$

$$[H_2NO_3]^+ + C_6H_5X \longrightarrow O_2NC_6H_4X + H_3O^+$$

$$X = alkyl, etc.$$

Nitrations of aromatics, alkenes, dienes, and others have been carried out in acetonitrile/BF_4^- and CH_2Cl_2/BF_4^- solutions at platinum electrodes using N_2O_4, (NO_2^+), as the nitrating agent [4]. Also, nitrations have been conducted in MeCN solutions containing a quaternary ammonium nitrite salt and styrene derivatives as substrates [5] in which the vinylic hydrogen was substituted by NO_2.

References

Section 2.1 Fundamental Nature

1. Yoshida K (1984) Electrooxidation in organic chemistry. Wiley-Interscience, New York
2. Bard AJ, Ledwith A, Shine H (1976) Adv Phys Org Chem 13:155
3. Baizer MM, Lund H (eds) (1983) Organic electrochemistry. Marcel-Dekker, New York
4. Kyriacou D (1981) Basics of electroorganic synthesis. Wiley-Interscience, New York
5. Kyriacou D, Jannakoudakis D (1986) Electrocatalysis for organic synthesis. Wiley-Interscience, New York
6. Fry AJ (1989) Synthetic organic electrochemistry. Wiley-Interscience, New York
7. Eberson L (1982) Adv Phys Org Chem 18:79
8. Eberson L (1987) Electron transfer reactions. Springer, Berlin Heidelberg New York
9. Ross SD, Finkelstein M, Rudd EJ (1975) Anodic oxidations. Academic press, New York
10. Uneama K, Steckhan E (eds) (1987) Topics in current chemistry, vol 142. Springer, Berlin Heidelberg New York
11. Weinberg NL, Weinberg HR (1968) Chem Rev 68:449

Section 2.2 Hydrocarbons

1a. Pletcher D, Fleischmann M, Korinek K (1971) J Electroanal Chem 33:478
1b. Clark DB, Fleischmann M, Pletcher D (1972) J Electroanal Chem 36:137
1c. Bertram J, Coleman JP, Fleischmann M, Pletcher D (1973) J Chem Soc Perkin, Trans 2:374
2. Coleman JP, and Pletcher D (1974) Tetrahedron Lett 147; Bertram J, Fleischmann M, Pletcher D (1978) Tetrahedron Lett 349

3. Kollmer H, Smith AO (1970) Chem Physics Lett 5:7
4. Olah GA (1967) J Am Chem Soc 89:2227
5. Tomat R, Rigo A (1986) J Appl Electrochem 16:8
6. Nyberg K (1971) Acta Chem Scand 25:534; Fleischmann M, Pletcher D (1969) Tetrahedron Lett 6225; Adams RN (1969) Accounts Chem Res 2:175; Brans K, Mann CK (1967) J Org Chem 32:1474
7. Dietrich M, Heinze J (1981) Angew Chem 24:508
8. Mann CK (1969) In: Bard AJ (ed) Electroanalytical chemistry, vol 3. Marcel-Dekker, New York
9. Steckhan E (1978) J Am Chem Soc 100:3526
10. Nyberg K (1970) Acta Chem Scand 24:1609
11. Becker JY, Miller LL, Boekelheide V, Morgan T (1976) Tetrahedron Lett 2939
12. Sato T, Kamada M (1977) J Chem Soc Perkin Trans 2:384
13. Sainsbury M, Schinazi RE (1972) J Chem Soc Chem Commun 718
14. Kotani E, Tekeuchi N, Tobinaga S (1975) J Chem Soc Chem Commun 550
15. Palmquist V, Nilsson A, Parker VD, Ronlan A (1976) J Am Chem Soc 98:2571
16. Sainsbury M, Wyatt J (1979) J Chem Soc Perkin Trans 1:108
17. Miller LL (1979) Pure Appl Chem 51:2125
18. Piersma BJ (1967) In: Gileadi E (ed) Electrosorption. Plenum Press, New York; Damhs H, Grear M (1963) J Electrochem Soc 110:1075; Bockris JOM (1966) Modern aspects of electrochemistry. Plenum Press, New York, chap 2
19. Bordaell FC, Cheng JP, Bausch MJ, Bases JE (1988) J Phys Org Chem 1:208
20. Yoshida K, Nagase S (1979) J Am Chem Soc 101:4268

Section 2.3 Hydroxy Compounds

1. Lund H (1971) In: Patai S (ed) Chemistry of the hydroxy group. Wiley-Interscience, London
2. Lund H (1957) Acta Chem Scand 11:491
3. Parker VD (1978) In: Bard AJ, Lund H (eds) Encyclopedia of electrochemistry. Marcel-Dekker, New York, vol VI, p 181
4. Fleischmann M, Korinek K, Pletcher D (1971) Electroanal Chem 31:39
5. Vertes G, Horanyi G, Nagg F (1972) Tetrahedron 28:377
6. Schäfer HJ, Kaulen J (1982) Tetrahedron 38:3299
7. Kaulen J, Schäfer HJ (1979) Synthesis 513
8. Lund T, Lund H (1984) In: Extended abstracts, vol 84. The Electrochemical Society Cincinnati, Ohio, Abst 330
9. Chum Li H, Ellis KJ, Islam N, Smith CZ, Sopher DW, Utley JHP (1984), Extended abstracts. The Electrochemical Society, Cincinnati, Ohio In: Abst 332
10. Chum Li H, Baizer MM (1985) The electrochemistry of biomass and derived materials. ACS Monograph
11. Bockris JOM, Piersma BJ, Gileadi E (1964) Electrochim Acta 9:1329
12. Pletcher D, Fleischmann M, Korinek K (1971) J Electroanal Chem 33:478
13. Hampton NA, Lee JB, MacDonald KI (1972) Electrochim Acta 19:921
14. Kyriacou D, Tougas TP (1987) J Org Chem 52:2318
15. McMillan JA (1962) Chem Rev 62:65
16. Kyriacou D, University of Massachusetts. Lowell (unpublished work)
17. Ref. 5, section 2.1
18. Kuguni Y, Nonaka T, Bochong Y, Watanaba N (1992) Electrochim Acta 37:353
19. Sundholm G (1971) J Electrochem Soc 31:265; (1971) Acta Chem Scand 25:3188
20. Belanger G (1976) J Electrochem Soc 123:818
21. Scholl PC, Lentsch SE, van de Mark MR (1976) Tetrahedron 32:303

Section 2.4 Oxide Covered Electrodes

1. Pohl JP, Picket H (1988) In: Trassati S (ed) Electrodes of conductive oxides. Elsevier, New York, p 192
2. Kyriacou D, Janhgen EGE, Tougas TP, University of Massachusetts, Lowell. Siver oxide anodes as biosensors. a. Debenedetto MJ (1991) PhD Thesis; b. DeMott J (1992) PhD Thesis
3. Stonehart P, Portandet EP (1968) Electrochim Acta 13:1805
4. Droog JMM, Huisman F (1980) J Electroanal Chem 15:211
5. Schäfer HJ, Kaulen J (1982) Tetrahedron 38:3299
6. Ref. 5, section 2.1
7. Fleischmann M, Korinek K, Pletcher D (1971) J Electroanal Chem 31:97; (1972) J Chem Soc Perkin Trans 2:1396
8. Fleischmann M, Pletcher D (1973) Adv Phys Org Chem 10:155; Pletcher D, Walsh F (1990) Chem Rev 90:837
9. Robertson PM (1980) J Electroanal Chem 111:97
10. Kyriacou D, Tougas TP (1987) J Org Chem 52:2318; Hampton NA, Lee JB, Morley JR, Scanton B (1969) Can J Chem 47:3789
11. Kyriacou D. University of Massachusetts, Lowell (unpublished work)
12. Bondarenko EG, Tomilov AP Zh (1969) Prikl Khim 42:2033
13. Campbell KD, Gulbankian AH, Edamura F, Kyriacou D (1985) US patent 4446440
14. Kolb DM (1978) Adv Electrochem Engn 11:125; Jütner K (1986) Electrochim Acta 31:8; Kokkinidis G, Theodoridou E, Jannakoudakis D, University of Thessaloniki, private communications; Adzic RR, Tripkovic AV, Markovic NM (1980) J Electroanal Chem 14:37; Kokkinidis G, Jannakoudakis D (1982) J Electroanal Chem 133:307
15. Ocon P, Beden B, Huser H, Lamy C (1987) Electrochim Acta 32:387

Section 2.5 Phenolic Compounds

1. Yoshida K (1984) Electrooxidation in organic synthesis. Wiley-Interscience, New York, p 229
2. Tablem A, Wang S, Swenton JS (1992) J Org Chem 57:2135
3. Nash RA, Skauen DM, Pardy WC (1958) J Am Pharm Assoc 47:443
4. Covitz FH (1980) French Patent 154350; Tabakovic I, Gruzic Z, Bejtovic Z (1983) J Heterocyc Chem 20:635
5. Svanholm V, Bechgaard K, Parker VD (1974) J Am Chem Soc 96:2409
6. Oyama N, Osaka T, Sazuki T (1987) J Electrochem Soc 134:3068
7. Torri S, Diminaka AL, Yoshikata A, Inokuchi T (1989) Tetrahedron Lett 30:2105
8. Smith C, Utley JHP, Petrescu M, Viertler H (1989) J Appl Electrochem 19:535
9. Lalvani SB, Rajagopal P (1992) J Electrochem Soc 139, No 1
10. Schutte JHP, du Plessis JAK, Lachman G, Vilgoen AM, Nelson AD (1989) J Chem Soc Chem Commun 1290

Section 2.6 Carbonyl Compounds

1. Weinberg NL, Weinberg HR (1968) Chem Rev 68:449
2. Fedorenko (1974) In: Tipson RS, Horton D (eds) Advances in carbohydrate chemistry and biochemistry. Academic Press, New York
3. Bockris JOM, Piersma BJ, Gileadi E (1964) Electrochim Acta 9:1329
4. Chiba T, Okimoto M, Nagai H, Takata Y (1982) Bull Chem Soc (Japan) 55:335
5. Koch VR, Miller LL (1982) J Electroanal Chem 43:318
6. Becker JY, Miller LL, Siegel TM (1975) J Am Chem Soc 97:849
7. Becker JY, Byrd RR, Miller LL, So YH (1975) J Am Chem Soc 97:853

8. Segura M, Barba F, Aldaz A (1986) Electrochim Acta 31:83
9. Ye S, Beck F (1991) Tetrahedron 47:5465
10. Ismail MT (1987) Bull Soc Chim (France) 438
11. Chiba T, Okimoto M (1992) J Org Chem 57:1375

Section 2.7 Carboxylic Acids

1. Kolbe H (1849) Liebigs Ann 69:257
2. Utley JHP (1974) In: Weinberg NL (ed) Techniques of electroorganic synthesis part I. Wiley-Interscience, New York, p 793
3. Weedon BCL (1960) Adv Org Chem 1:1
4. Finkelstein M, Petersen RC (1960) J Org Chem 25:136
5. Rand L, Mohar AF (1965) J Org Chem 30:388
6. Conway BE, Vigh AK (1967) Electrochim Acta 12:102
7. Vasiliev Yu B, Korsman ED, Freiden G (1982) Electrochim Acta 27:937
8. Einhorn J, Soulser JL, Bucheet C (1983) Can J Chem 61:584
9. Vasiliev Yu B, Budetzky US, Korsman EP (1982) Electrochim Acta 27:929
9a. Wladislaw B, Ayers AMJ (1962) J Org Chem 27:281; Coleman JP, Utley JHP, Weedon BCL (1964) J Chem Soc Chem Commun 438
10. Uneama K, Namba H (1988) J Org Chem 53:4598
11. Pletcher D, Smith CZ (1975) J Chem Soc Perkin Trans 1:948
12. Ogumi Z, Yamashita H, Nishio K, Takehara Z, Yoshizawa S (1983) Electrochim Acta 28:1687
13. Kraeutler B, Bard AJ (1978) J Am Chem Soc 99:77; (1978) J Am Chem Soc 100:2239

Section 2.8 Amines and Amides

1. Lord SS, Rojers LB (1954) Anal Chem 26:284
2. Nelson RF, Adams RN (1968) J Am Chem Soc 90:3925
3. Portis LC, Klug TJ, Mann CK (1974) J Org Chem 39:3488
4. Cauquis G, Fauvelot G, Rigani J (1968) Bull Soc Chim 12:4928
5. Seo ET, Nelson RF, Fritsch JM, Marcoux LS, Leedy DW, Adams RN (1966) J Am Chem Soc 88:3498
6. Greason SC, Weeder J, Nelson RF (1972) J Org Chem 37:4440
7. Barns KK, Mann CK (1967) J Org Chem 32:1474
8. Smith PJ, Mann CK (1969) J Org Chem 34:1821
9. Serve D (1976) Electrochim Acta 21:1171
10. Feldhues V, Schäfer HJ (1982) Synthesis 82:145
11. Serve D (1975) J Am Chem Soc 97:432
12. Wawzonek S, McIntyre T (1967) J Electrochem Soc 114:1025
13. Esteban PC, Leger JM, Lamy C, Genies E (1989) J Appl Electrochem 19:462
14. Hand RL, Nelson RF (1978) J Electrochem Soc 125:1059
15. Lowert BA, Marcoux Y, Bard AJ (1972) J Am Chem Soc 94:5538
16. Finkelstein M, Nyberg K, Ross SD, Servin R (1978) Acta Chem Scand B32:182
17. Shono T, Matsumura Y, Tsubata K (1981) J Am Chem Soc 103:1172
18. Hall LR, Iowamoto RT, Haluznik RP (1989) J Org Chem 54:2446
19. Rudd EJ, Finkelstein M, Ross SD, Nyberg K, Servin R (1976) Acta Chem Scand B30:640
20. Bolzan AE, Iwasita T (1988) Electrochim Acta 33:109
21. Degner D (1989) Dechema-Mon (Org Electrochem) 375
22. Swenton JS, Brian RB, Cung-Pin C, Chun Tzer-C (1989) J Org Chem 54:51

Section 2.9 Ethers and Esters

1. Ronlan A, Behgaard K, Parker VD (1973) Acta Chem Scand 27:2375
2. Hammerich O, Moe NS, Parker VD (1972) J Chem Soc Chem Commun 156
3. Svanholm U, Parker VD (1972) Tetrahedron Lett 471
4. Stewart RF, Miller LL (1980) J Am Chem Soc 102:4999
5. Venkachadan CS et al. (1991) Tetrahedron Lett 51:7592
6. Shono T, Matsumura Y (1969) J Am Chem Soc 91:2003
7. Weinberg NL, Beleau B (1973) Tetrahedron 29:279
8. Koch D, Schafer HJ, Steckhan E (1974) Chem Ber 107:3640
9. Callot HJ, Louati A, Gross M (1983) Bull Soc Chim II:317
10. Moeller KD, Tinao LV (1992) J Am Chem Soc 114:1033
11. Boyd JW, Schmaltz PW, Miller LL (1980) J Am Chem Soc 162:3856
12. Miller LL, Ramachandra V (1974) J Org Chem 39:369
13. Yadav AK, Misra SC (1986) Electrochim Acta 31:507

Section 2.10 Organic Halides

1. Miller LL, Hoffman AK (1967) J Am Chem Soc 89:593
2. Laurent A, Laurent E, Tardivel R (1973) Tetrahedron Lett 4861
3. Laurent E, Tardivel R (1976) Tetrahedron Lett 1779
4. Miller LL, Kuzawa E, Campbell CP (1970) J Am Chem Soc 92:2821
5. Becker JY (1971) J Org Chem 42:3997; Coch VR, Miller LL (1973) Tetrahedron Lett 693
6. German A, Brunel D, Moreau P (1989) J Fluorine Chem 43:249
7. Becker JY, Smart E, Fukunga T (1988) J Org Chem 53:5714
8. Hoffender H, Lorch HW, Wendt H (1975) J Electroanal Chem 66:183

Section 2.11 N-Heterocyclics

1. a. Kwee S, Lund H (1969) Acta Chem Scand B23:2121; b. Weinberg NL, Weinberg HR (1968) Chem Rev 68:449; c. Babbit JM et al. (1966) Chem Ind (London) 2127; d. Eisner V, Kuthan J (1972) Chem Rev 72; e. Kuthan J, Kurfurst A (1982) Ind Eng Chem Proc Res Div 21:191; f. Ludvic J et al. (1987) Electrochim Acta 32:1063; g. Goyal RN et al. (1987) Bull Soc Chim 78
2. Yoshida K (1979) J Am Chem Soc 101:2116; Yoshida K (1978) J Chem Soc Chem Commun 1108
3. Ref. 9, section 2.9
4. Yoshida K, Kitabayshi H (1987) Bull Chem Soc (Japan) 60:3093
5. Iwasaki T, Horikawa H, Matsumoti K, Mioshi M (1979) J Org Chem 44:1552; Nyberg K (1976) Synthesis, 545
6. Billon JP, Cauquis G, Combrisson J, Li AM (1960) Bull Soc Chim (France) 2062
7. Cauquis G, Falmy HM, Pierre G, Elnagdi MH (1970) J Heterocycl Chem 16:413
8. Zzochralska B, Elving P (1981) Electrochim Acta 26:1755
9. Lacoutre F, Barbey G (1983) Electrochim Acta 28:1617
10. Bellamy AJ, Innes DI (1982) J Chem Soc Perkin Trans 2:1599
11. Thaning M, Wistrand G (1989) Acta Chem Scand B43:290
12. Laupp G, Koleli F, Grunden E (1985) Auger Chem 24:864
13. Wrona MA, Dryhurst G (1988) J Org Chem 54:2718
14. Anne A, Moiroux J (1988) J Org Chem 53:2816
15. Cupta PN, Rather GM (1988) J Ind Chem Soc LXV 633
16. Pragst E, Bock C (1975) J Electrochem Soc 61:47

Section 2.12 Organosulfur Compounds

1. Cottrell PT, Mann CK (1969) J Electrochem Soc 116:1499
2. Shono T, Matsumara Y, Mizozuch M, Hayashi J (1979) Tetrahedron Lett 3861
3. Jansson R (1984) Chem Eng News 62:43
4. Cuttler LH (1979) AIChE Symposium Ser 185 75:103
5. Kimura M, Matsuhara S, Sawaki Y (1984) J Chem Soc Chem Commun 1619
6. Porter DN, Utley JHP (1978) J Chem Soc Chem Commun 255
7. Saeva FD, Morgan BP, Fisher MW, Haley NF (1984) J Org Chem 49:390
8. Glass RS, Petson A, Wilson G (1986) J Org Chem 51:4337
9. Uneama K, Torii S (1972) J Org Chem 37:367
10. Silber JJ, Shine HJ (1971) J Org Chem 36:2923
11. Kim K, Hull VJ, Shine HJ (1974) J Org Chem 39:2534
12. Parker VD, Eberson LJ (1970) J Am Chem Soc 92:7488
13. Ref. 1, section 2.1
14. Hammerich O, Moe NS, Parker VD (1972) J Chem Soc Chem Commun 156

Section 2.13 Anodic Cyanations

1. Nilsson S (1969) Dis Faraday Soc 45:242
2. Miller LL (1979) Pure App Chem 51:2125
3. Koyama K, Susuki T, Tsutsumi S (1967) Tetrahedron 23:2675
4. Andreades S, Zahnow EW (1969) J Am Chem Soc 91:4181
5. Parker VD, Bargert BE (1965) Tetrahedron Lett 4065
6. Nilsson S (1973) Acta Chem Scand 27:329
7. Susuki T, Koyama K, Omori A, Tsutsumi S (1968) Bull Chem Soc (Japan) 41:2663
8. Yoshida K, Nagase S (1979) J Am Chem Soc 101:4268
9. Papouchado L, Adams RN, Felberg WS (1969) J Electroanal Chem 21:408
10. Cauquis G, Serve D (1979) Bull Soc Chim (France) 2:145
11. Shine JH, Ristagno CV (1972) J Org Chem 37:3424
12. Nyberg K (1976) Acta Chem Scand 30:2499
13. March J (1977) Advanced organic chemistry. McGraw-Hill International Student Edition Kagakusha, Ltd. p 503
14. Ref. 20, section 2.2
15. Eberson L, Parker VD (1969) J Chem Soc Chem Commun 340
16. Inoue T, Tsutsumi S (1965) Bull Chem Soc (Japan) 38:661
17. Nyberg K (1971) Acta Chem Scand 33:2499
18. Yoshida K (1979) J Am Chem Soc 99:6111
19. Chiba T, Takata Y (1977) J Org Chem 42:2973
20. Yoshida K, Fueno T (1970) J Chem Soc Chem Commun 711
21. Eberson L, Helgée B (1978) Acta Chem Scand B32:313
22. Eberson L, Radner F (1992) Acta Chem Scand 46:312

Section 2.14 Acetoxylations and Acetamidations

1. Eberson L (1967) J Am Chem Soc 89:4669
2. Eberson L, Oberrauch E (1981) Acta Chem Scand B35:193
3. Eberson L, Nyberg K (1966) J Am Chem Soc 88:1685
4. Nyberg K (1970) Acta Chem Scand B24:1609
5. Bum Z, Cedheim L, Eberson L (1977) Acta Chem Scand B31:662
6. Ross SD, Finkelstein M, Petersen RC (1970) J Org Chem 35:781
7. Ross SD, Finkelstein M, Petersen RC (1964) J Am Chem Soc 86:4139

8. So YH, Becker JY, Miller LL (1975) J Chem Soc Chem Common 262
9. Shono T, Matsumura Y, Onomura O, Ozaki M, Kazanawa T (1987) J Org Chem 52:536
10. Hembrok A, Schäfer HJ (1985) Angew Chem Int Ed Eng 24:1655
11. Clark DB, Fleischmann M, Pletcher D (1973) J Chem Soc Perkin Trans 2:1578
12. Eberson L, Nyberg K (1966) Tetrahedron Lett 2389
13. Baizer MM, Wagenknecht JH, Monsanto Chemical Co., USA
13a. Nyberg K (1969) J Chem Soc Chem Commun 774
14. Hammerich D, Parker VD (1974) J Chem Soc Chem Commun 245
15. Ref. 11, section 2.14
16. Laurent A, Laurent F, Locher P (1975) Electrochim Acta 20:857
17. Faita G, Fleischmann M, Pletcher D (1970) J Electroanal Chem 25:455
18. Fleischmann M, Pletcher D (1972) Chem Eng Techn 44:187
19. Eberson L, Olefsson B (1969) Acta Chem Scand B23:2355

Section 2.15 Anodic Halogenations

1. Meyers RA (ed) (1992) Encyclopedia of physical science and technology. Academic Press, New York, p 576
2. Childs WV (1982) In: Weinberg NL (ed) Techniques of electroorganic synthesis, vol 5. Wiley-Interscience, New York
3. O'Sullivan O, Klink EJP, Liu CC, Yeager EB (1985) J Electrochem Soc 132:2424
4. Drakesmith FG, Hughes DA (1986) J Fluorine Chem 32:103
5. Matsuda Y, Nishiki T, Sakota N, Nakagua A (1984) Electrochim Acta 29:35
6. Casalbore G, Mastragostino M, Valchev S (1975) J Electroanal Chem 61:33
7. Shono T, Matsumura Y, Onomura O, Ozaki M, Kanazawa T (1987) J Org Chem 52:536
8. Ibvisagic Z, Pletcher D, Brooks WN, Healy KP (1985) J Appl Electrochem 15:719
9. Takasy Y, Matsuda Y, Shinigu A, Morita M, Saito M (1991) Chemistry Lett. (Japan) 1685
10. Matsuda Y, Hayashi H (1981) Chemistry Lett (Japan) 661
11. Shono T, Matsumura Y, Kalosh S, Ikeda K, Kamota T (1989) Tetrahedron Lett 30:1649
12. Miller LL, Watkins BF (1976) J Am Chem Soc 98:1515; Lines R, Parker VD (1980) Acta Chem Scand B34:47

Section 2.16 Hydroxylations

1. Ronlan A (1971) J Chem Soc Chem Commun 1643
2. Nilsson A, Ronlan A, Parker VD (1973) J Chem Soc Perkin Trans 1:2337
3. Fremery M, Höver H, Schwartzlose G (1974) Chem Ing Tech 46:635
4. Majeski EJ, Stuart JD, Obnesoge WE (1968) J Am Chem Soc 90:633
5. Parker VD (1970) Acta Chem Scand B24:2757
6. Sternerup H (1974) Acta Chem Scand B28:579
7. Ponsold K, Kash H (1979) Tetrahedron Lett 4463
8. Steckhan E, Schäfer HJ (1974) Angew Chem 13:472

Section 2.17 Alkoxylations

1. Inoue T et al. (1963) Tetrahedron Lett 1409
2. Couture R, Beleaw B (1972) Can J Chem 50:3424
3. Brettle R, Sutton JR (1975) J Chem Soc Perkin Trans 1:1947
4. Engles R, Schäfer HJ, Steckhan E (1977) Justus Liebigs Ann Chem 204

5. Kim HJ, Kusabak K, Hokazono S, Morooka S, Kato S (1987) J Appl Electrochem 17:1213
6. Shono T, Matsumura Y (1970) J Org Chem 35:4157
7. Brettle R, Sutton JR (175) J Chem Soc Perkin Trans 1:1955
8. Manning MJ, Raynolds PW, Swenton JS (1976) J Am Chem Soc 98:5008
9. Jackson DK, Swenton JS (1977) Synth Commun 7:333
10. Yoshida J, Murata T, Isoe S (1987) Tetrahedron Lett 28:211
11. Shono T, Matsumura Y, Tsubaka K (1981) J Am Chem Soc 103:1172
12. Palmquist V, Nilsson A, Ronlan A, Parker VD (1979) J Org Chem 44:196
13. Warning K, Mitzlaff M, Jensen H (1978) Justus Liebigs Ann Chem Ann Chem 1707
14. Shono T, Matsumura Y (1969) J Am Chem Soc 91:2803
15. Goosens HJM, Top AWH (1978) Synthesis 283
16. Weinberg NL, Brown EA (1966) J Org Chem 31:4054
17. Rudd EJ, Finkelstein M, Ross DS (1972) J Org Chem 37:1763
18. Barba I, Chincilla, R, Gomez C (1991) J Org Chem 56, 3673
19. Weinberg NL, Weinberg HR (168) Chem Rev 471
20. Barba I, Alonso F (1989) J Org Chem 54:4365
21. Nilsson A, Palmquist V, Peterson T, Ronlan A (1978) J Chem Soc Perkin Trans 1:708
22. Shono T, Ikeda A, Kimura T (1971) Tetrahedron Lett 3599; Margaretha P, Tissol P (1975) Helv Chim Acta 58:933
23. Dolson MG, Swenton JS (1981) J Org Chem 46:177
24. Dolson MG, Swenton JS (1981) J Am Chem Soc 103:2361

Section 2.18 Pyridinations

1. Parker VD, Eberson L (1970) Acta Chim Scand B24:3542; Blackman GM, Will JP (1974) J Chem Soc Chem Commun 67; Yoshida K, Nagasi S (1979) J Am Chem Soc 101:4268
2. Manning G, Parker VD, Adams RN (1969) J Am Chem Soc 91:7584
3. Blount HN, Kuwana T (1973) J Electroanal Chem 27:271

Section 2.19 Nitrations

1. Achord JM, Hussey C (1981) J Electrochem Soc 128:2556
2. Kargin Yu M (1987) J Gen Chem (USSR) 56:1863
3. Bloom AJ, Fleischmann M, Mellor JM (1987) Electrochim Acta 32:785
4. Eberson L, Radner F (1986) Acta Chem Scand B40:71
5. Laurent A, Laurent E, Locher P (1975) Electrochim Acta 20:857

3 Cathodic Organic Reactions

3.1 Phenomenological Nature

Anion radical formation

Cathodic or electroreductive organic reactions proceed by means of the electrolytically generated primary intermediate products, anion radicals, $R^{\overline{\cdot}}$, radicals, $R^{\cdot}$, and anions, R^{-}. Anion radicals result from either direct electron transfer of an electron from the electrode to an organic neutral molecule or indirectly by the action of a mediator. Organic anion radicals may act as bases and as reducing agents, depending on the environment in which they are generated. Because of the presence (or potential presence) of the extremely electrophilic proton in most of the commonly used electrolysis media the study of anion radicals per se is much more difficult than the study of the anodically produced cation radicals.

Birch reduction. Grignard reaction

Two typical examples of one-electron transfer reactions in conventional chemical organic synthesis are the *Birch reduction* and the *Grignard reaction*, as indicated by the examples below:

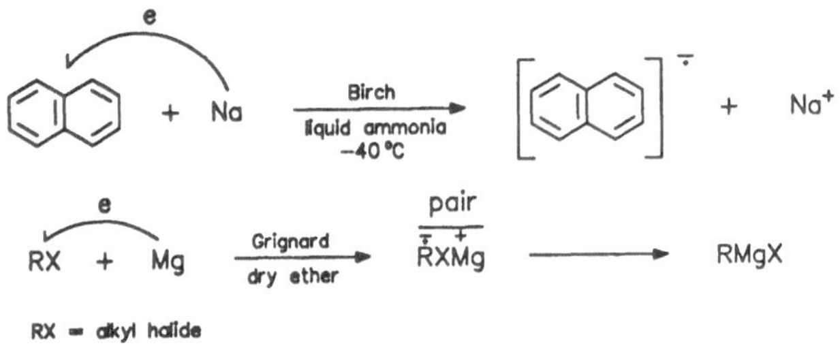

Direct Electron transfer reactions

In electrolytic cells the ET reactions would be effected by direct ET from the cathode to the organic molecule, as depicted below:

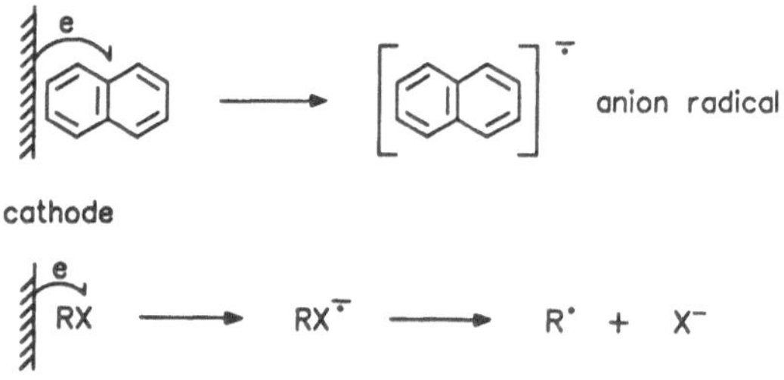

3.2 Carbon-Hydrogen Bond Formations. Unsaturated Systems

Hydrogenation of unsaturated carbon-carbon bonds can be electrochemically effected directly or indirectly, as is symbolically indicated below:

direct
electroreduction

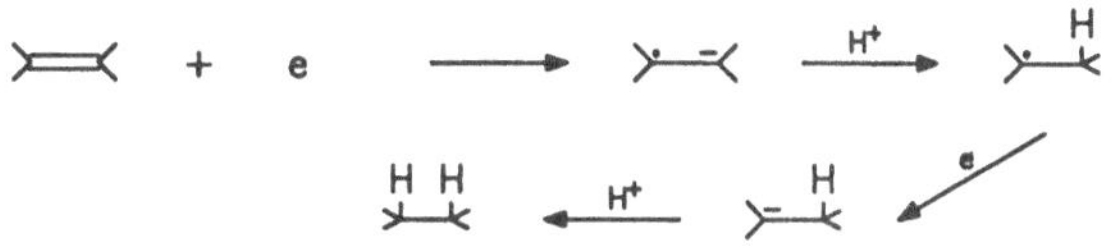

indirect,
electrocatalytic reduction
(black Pt, Raney Nickel surfaces)

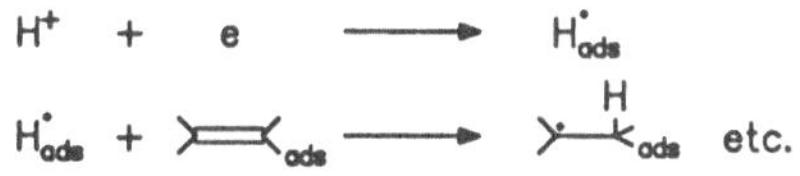

Conjugated and polycyclic aromatic systems are generally much more easily reduced than are nonconjugated or monocyclic systems. Ethylene has not yet been directly electroreduced, whereas butadiene is reduced (−2.5 V vs SCE). Benzene requires severe Birch-type reduction conditions but naphthalene and anthracene are readily reduced at the cathode [1].

Activated olefins

Activated olefins are easily electroreduced, and under controlled conditions yield the corresponding hydrodimers in high yields. For example, acrylonitrile, upon direct one-electron per molecule reduction, gives the useful material adiponitrile in high yield [2,3], Scheme 3.1.

$$H_2C{=}CHCN \;+\; e \;\xrightarrow[Q_4N^+]{}\; H_2\dot{C}-\bar{C}HCN$$

$$2\;H_2\dot{C}-\bar{C}HCN \;\longrightarrow\; NC-\bar{C}H-CH_2CH_2-\bar{C}H-CN$$

$$\downarrow 2H^+$$

$$NC-(CH_2)_4-CN$$

Scheme 3.1

Baizer process

This electrohydrodimerization reaction forms the basis of the largest known electroorganic industrial process in the world today [3] (Baizer process, Monsanto Co., USA).

Cathodic reductions of unsaturated systems in protonic media tend to produce hydrogenated products. Hydrodimers would be produced if proton availability is limited near the electrode surface where the primary radical anions are generated. In the case of the electrohydrodimerization of acrylonitrile, the presence of the quaternary ammonium ion, Scheme 3.1, apparently modifies the structure of the electrical double layer so that the availability of protons from H_2O within this double layer region is considerably limited. This double layer structure effect allows the monoanions to couple rather than hydrogenate, thus affording predominantly adiponitrile instead of propionitrile. Such synthetically very important *double layer effects* on the course of electrochemical reactions have been referred to as "electrocatalysis of the second kind" [4]. Intramolecular reductive cyclization of activated olefins is exemplified below [5]:

Double layer effects

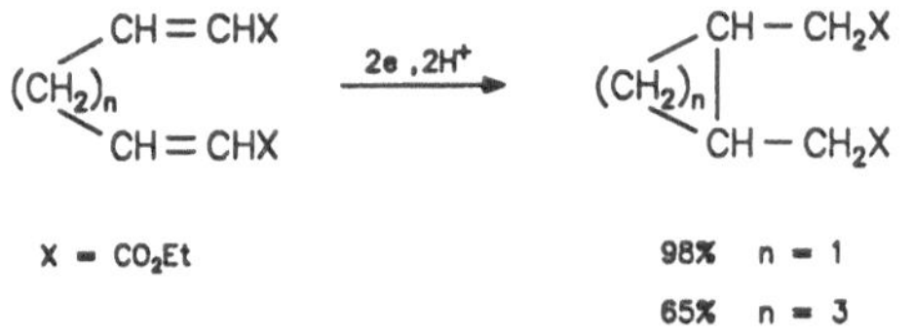

$$(CH_2)_n \begin{matrix} CH=CHX \\ \\ CH=CHX \end{matrix} \xrightarrow{2e,2H^+} (CH_2)_n \begin{matrix} CH-CH_2X \\ | \\ CH-CH_2X \end{matrix}$$

X = CO$_2$Et

98%   n = 1
65%   n = 3

Good proton donors

Carbon-hydrogen bond formations, or protonations, proceed readily in media that are good proton donors. In the absence of water, good proton sources may be acidic substances such as benzoic acid, phenol, and the so-called carbon acids (e.g. diethyl malonate). Some compounds may act as self-protonating sources as it happens in the cathodic reduction of indenes to indans [6]:

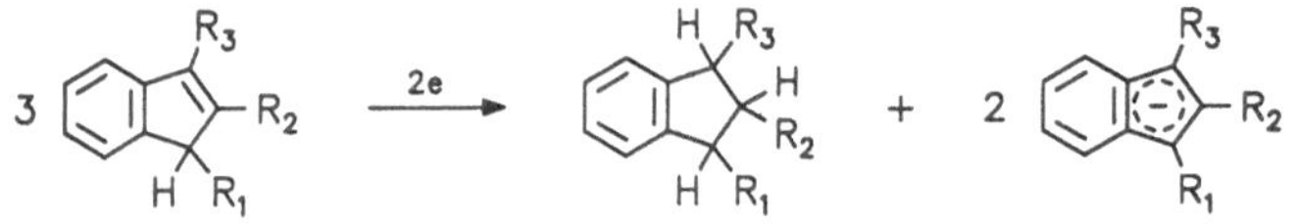

It would be interesting to note that electrolysis in DMF and in H_2O gave different isomers as major products: the *cis* isomer in the former and the *trans* isomer in the latter medium.

Ketones can be electrocatalytically converted to hydrocarbons and alcohols at platinized platinum electrodes:

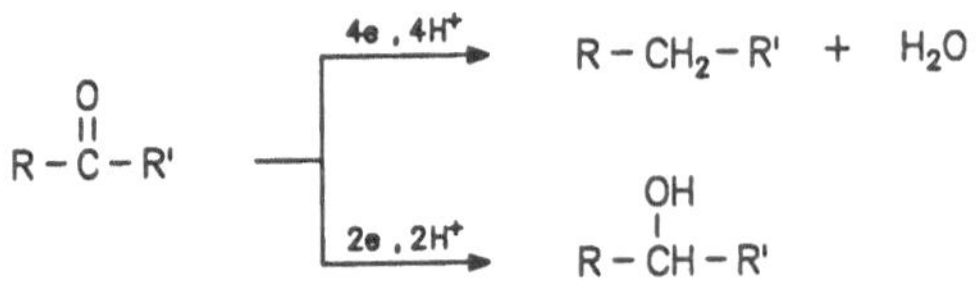

Catalytic hydrogenation

Pletcher studied this type of reduction in relation to medium composition and observed that the hydrocarbon yield of acetophenones increased as the ethanol content in the ethanol-water medium increased [7]. Catalytic hydrogenations are favored at low overpotential electrodes (Pt, Ni, Cu, Fe) whereas direct electroreductions and dimerization would be most efficient at high-overpotential electrodes (Hg, Pb, Cd). Hydrogen over-potential values vary from $0.0\,V$ for platinized platinum to $1.2\,V$ for Cd relative to the normal hydrogen electrode [8]. High- and low-hydrogen overvoltage electrodes have been used on an industrial scale [9]. Aqueous media are preferred if the organic substrate is soluble in water. Pyridine and tetrahydrocarbazole are hydrogenated at lead cathodes in aqueous sulfuric acid solution (Hoechst, BASF, Germany) and adiponitrile is reduced to 1,6-hexadiamine at nickel-copper electrode in aqueous hydrochloric acid solution. In protogenic media such as methanol-sodium hydrooxide, hydrogenations are possible at electrocatalytic electrodes, such as electrodes having surfaces of black platinum, Raney nickel, or graphite modified with platinum. Pletcher and coworkers studied the use of high surface electrocatalytic electrodes (Ni, Pt, Pd, Re, Rh) for the hydrogenation of various functional groups [10]. Excellent efficiencies were obtained at these electrodes with substrates, such as the following:

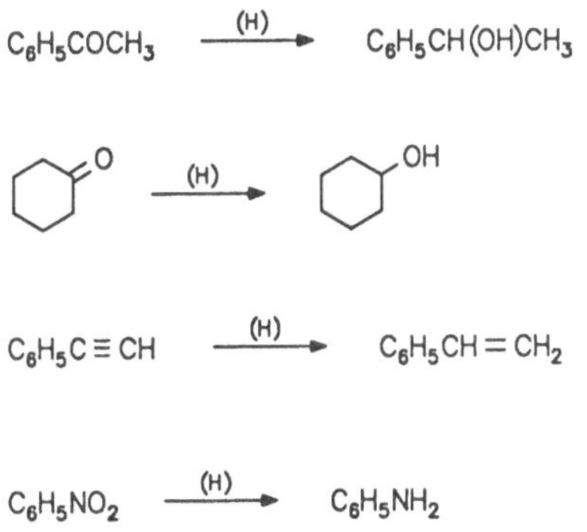

Direct electro-reductions and dimerization

Electrocatalytic electrodes

$$C_6H_5COCH_3 \xrightarrow{(H)} C_6H_5CH(OH)CH_3$$

cyclohexanone $\xrightarrow{(H)}$ cyclohexanol

$$C_6H_5C\equiv CH \xrightarrow{(H)} C_6H_5CH=CH_2$$

$$C_6H_5NO_2 \xrightarrow{(H)} C_6H_5NH_2$$

These hydrogenations were performed in MeOH/NaOH media under constant current of about $50\,mA/cm^2$. Electrocatalytic electrode surfaces can be prepared in situ by electrodeposition of the catalytic metal ion [10].

3.3 Carbonyl Compounds

3.3.1 General Reduction Mechanism

The carbonyl function is a good and versatile electrophore. In principle, a carbonyl group can be reduced by one- two- and four-electron processes giving respectively a diol, a monohydroxy alcohol and even a hydrocarbon.

Zuman proposed [1] general overall reduction mechanisms, in acidic and neutral solutions. They are summarized in Scheme 3.2

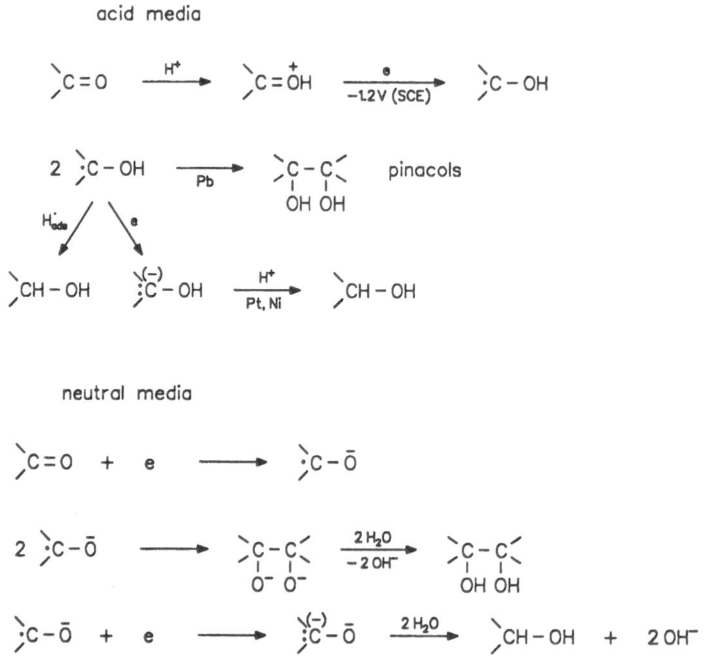

acid media

$$\>C=O \xrightarrow{H^+} \>C=\overset{+}{O}H \xrightarrow[-1.2V\,(SCE)]{e} \>C-OH$$

$$2\>\overset{.}{C}-OH \xrightarrow{Pb} \>\underset{OH}{\overset{|}{C}}-\underset{OH}{\overset{|}{C}}\< \quad pinacols$$

$$\overset{H_{ads}^{.}}{\swarrow} \quad \overset{e}{\searrow}$$

$$\>CH-OH \qquad \>\overset{(-)}{C}-OH \xrightarrow[Pt,\,Ni]{H^+} \>CH-OH$$

neutral media

$$\>C=O + e \longrightarrow \>\overset{.}{C}-\bar{O}$$

$$2\>\overset{.}{C}-\bar{O} \longrightarrow \>\underset{O^-}{\overset{|}{C}}-\underset{O^-}{\overset{|}{C}}\< \xrightarrow[-2\,OH^-]{2\,H_2O} \>\underset{OH}{\overset{|}{C}}-\underset{OH}{\overset{|}{C}}\<$$

$$\>\overset{.}{C}-\bar{O} + e \longrightarrow \>\overset{(-)}{\overset{.}{C}}-\bar{O} \xrightarrow{2\,H_2O} \>CH-OH + 2\,OH^-$$

Scheme 3.2

Acetone is reduced to alcohol, pinacol, and propane (at Cd) depending on the cathode material; pinacol formation is favored at lead cathodes [2].

Electrohydrogenations have been studied using nickel, copper and tin cathodes. Copper cathodes modified with electrodeposited nickel were found to be most effective [3]. Electrocatalytic hydrogenations were performed using slurry-type electrodes [4].

3.3.2 Reactions via the Carbonyl Electrophore

The carbonyl group is both a good electrophore and an activating group. It can therefore participate in a great variety of intermolecular and intramolecular electrochemical bond formations. Hydrogens on α-carbons to carbonyl are easily abstracted by electrogenerated bases; and hydroxyl groups on such carbons are cathodically replaced by hydrogen.

Little and coworkers [5] reported an electroreductive intramolecular cyclization involving ketonic or aldehydic groups and α,β-unsaturated ester subunits in the molecular structure. The electrolysis is carried out in a divided cell in acetonitrile-water solution using a mercury cathode. Thus the construction of ring systems was effected, as shown in Scheme 3.3. It was proposed that the unsaturated ester subunit was the electrophore.

Construction of ring systems

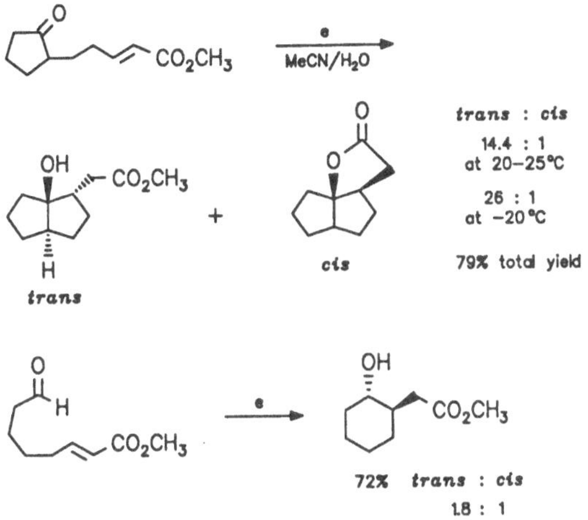

Scheme 3.3

Electroreductive couplings of carbonyls with methylalkyl chloride (MACl) are facilitated by the presence of Ni(II) bipyridine (Nibpd) complexes [6]. Mixed couplings of ketones with activated olefins afford lactones via cyclization of the intermediate hydroxy products [7]. Also couplings of adehydes with acrylic acid yield lactones [8]. These couplings are shown in Scheme 3.4

Electro-reductive couplings

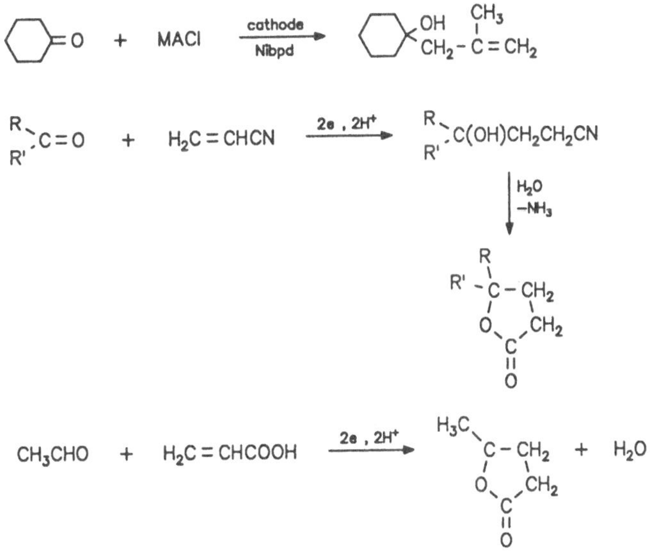

Scheme 3.4

Electrohydro-dimerizations

Electrohydrodimerizations of cyclohexanones give the glycol and the hydroxy ketone products, the glycol product being predominant as the water content in the mixed MeCN/H$_2$O solvent increases [9].

Electrocatalytic reductions

Pletcher and Lain [10] demonstrated that electrocatalytic reductions of carbonyl compounds are possible at graphite cathodes modified by electrodeposited Ni^{2+} ions. Thus benzaldehyde and acetophenone were reduced to their corresponding alcohols and hydrocarbons (mixed products) by electrolysis in acidic ethanol-water solutions containing 10^{-2} M Ni^{2+} ion. Some hydrogen also is evolved in these electrolyses.

Synthesis of biochemically interesting compounds

Electrochemical studies of carbonyls has been pursued in connection with the synthesis of biochemically interesting compounds. For example, the synthesis of 2-pyridyl substituted 1,3-indandiones was studied in basic, acidic and aprotic media [11]. In all these media the indandione moiety is reduced at less negative potentials than the pyridinium part of the molecule, Scheme 3.5. This is an example of the difficulty in predicting the actual electrophore in a complex molecular structure.

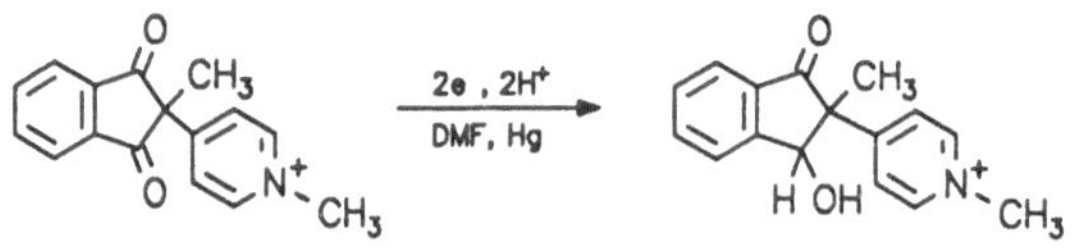

Scheme 3.5

Reduction of triacetone-amine

The electroreduction of triacetoneamine (1) was studied by Smirnov et al. [12] at a zinc electrode. The products were the monoalcohol (2), the dimer (3) and the tetramethylpiperidine (4).

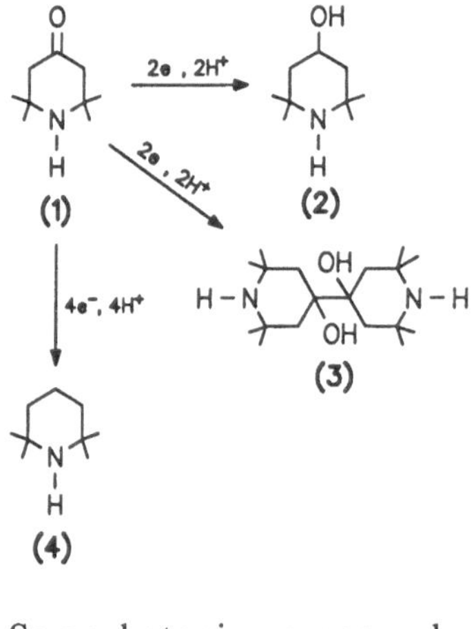

Some ketonic compounds may be reduced at the cathode in organic solvents that contain phenol as a proton donor. The following example is interesting. The primary radical anion is protonated and can be reoxidized at the anode to give the starting compound [13].

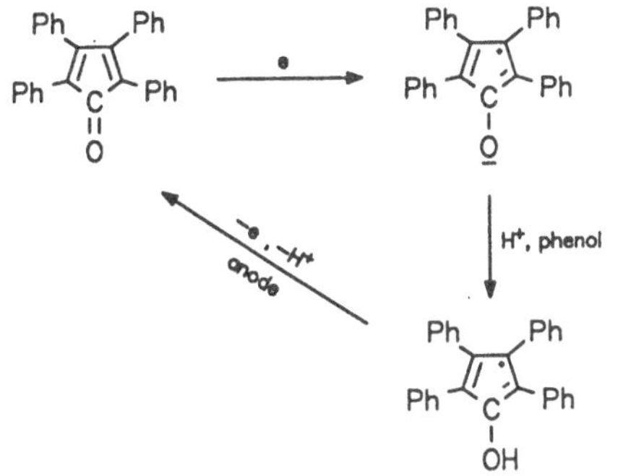

Steroidal carbonyls are electroreduced to give their methylene analogs in fairly good yields. If the electrolysis is performed in deuteriosulfuric acid at lead cathodes the oxygen of the carbonyl is replaced by deuterium [14].

Electrolytic reduction of aldehydes affords hydrodimerization products by a one-electron reduction process. Formaldehyde is thus dimerized to glycol [15]. A pilot-plant process was developed for the production of 4,4'-dihydrobenzoin from *p*-hydroxybenzaldehyde [16]. The electrolysis was performed in an aqueous 2 M NaOH solution at a mercury cathode (mercury gave more reproducible yields than a cadmium electrode in this case). **Reduction of aldehydes**

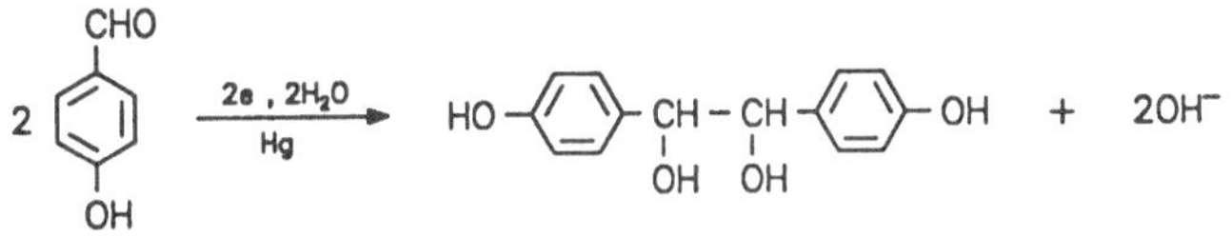

In principle, aldehydic carbonyls may be hydrogenated all the way to give hydrocarbons and water:

Acrolein and crotonaldehyde have been reduced to mixtures of saturated and unsaturated hydrocarbons [17]. In the presence of primary amines aldehydes undergo amination. Ketones also react similarly [18].

Recently, Shono reported an electroreductive cyclization method using γ and δ cyanoketones [19]. By this method five-membered rings can be constructed such as those of the natural products dihydrojasmone and methyldihydrojasmonate. The reaction is initiated by direct electronation of the carbonyl group. This step is followed by intramolecular bond formation with the cyano group and further reduction to the α-hydroximine, and then to the final cyclized products. The reaction is summarized in Scheme 3.6 **Cyclization of cyanoketones**

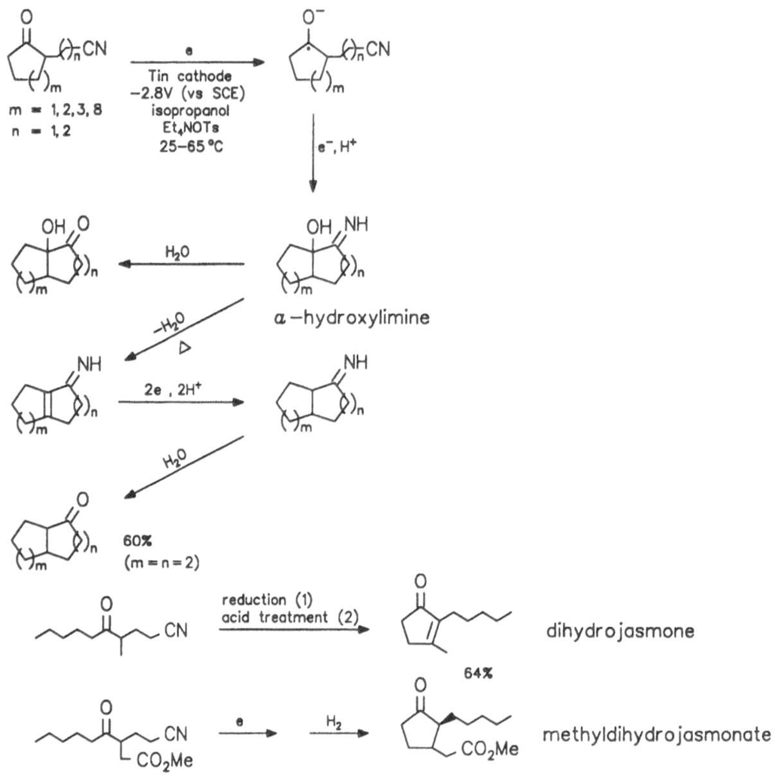

Scheme 3.6

Some ketonic structures such as that of phenalin-1-one react with acetic anhydride following electroreduction in DMF/Bu$_4$NClO$_4$ medium [20]. Mono- and diacetoxy derivatives are obtained.

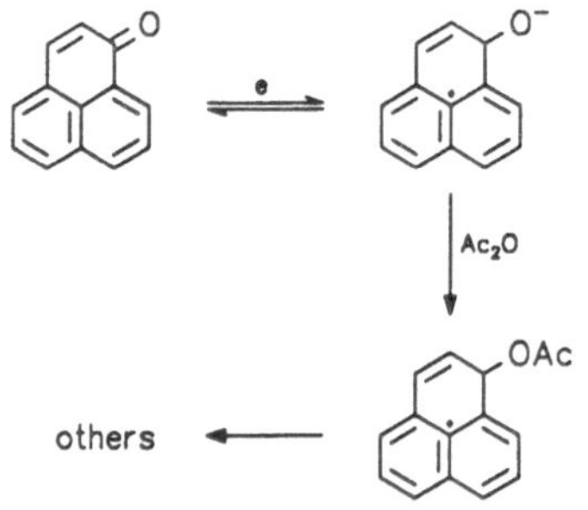

Electroreductive splitting of C—H bonds in ketones

It was mentioned earlier that hydrogens on α-carbons relative to a carbonyl group are acidic. In this connection we note that Bukhtiarov and coworkers [21] reported that aliphatic ketones undergo electro-reductive splitting of C—H bonds at transition metals as cathodes (Pt, Ni, Fe, Co) in aprotic solvents containing quaternary ammonium salts as electrolytes. Hydrogen is produced at the cathode, while the anions may react in various ways depending on the composition of the electrolysis medium:

$$R - \overset{\overset{O}{\|}}{C} - CH_3 \quad \xrightarrow{\bullet} \quad \left[R - \overset{\overset{O}{\|}}{C} - CH_2 \right]^{-} \quad + \quad H^{\cdot}$$

An interesting electroreduction of α-hydroxyaldehydes was recently reported [22]. This reduction is selective at carbon electrodes coated with films of dimethyldioctadecyl ammonium salts. The films apparently provide a hydrophobic surface on the electrode, and this surface allows reductive selective replacement of the hydroxyl group by hydrogen. The electrolysis is carried out in buffered (pH 7) aqueous solution at 20 °C, in a divided cell and at a cathode potential of −1.5 V versus the SCE. This is also an example of the activating influence of the carbonyl group on the C—OH bond α to the carbonyl.

Electro-reduction of α-hydroxy-aldehydes

$$H_3C - \overset{\overset{CH_3}{|}}{\underset{\underset{OH}{|}}{C}} - \overset{\overset{}{}}{\underset{\underset{O}{\|}}{C}} - H \quad + \quad 2e \quad + \quad 2H^+ \quad \longrightarrow \quad H_3C - \overset{\overset{CH_3}{|}}{\underset{\underset{H}{|}}{C}} - \overset{}{\underset{\underset{O}{\|}}{C}} - H$$

100% selectivity

3.4 Carboxylic Acids and Esters

Direct reduction of the COOH group has not generally been feasible at the cathode thus far. Only in salicylic [1] acid and some other hydroxybenzoic acids has the carboxyl group been reduced [2].

OH / COOH $\xrightarrow{2e, 2H^+}$ OH / CHO + H_2O (ref. 3)

80%

The reduction can advance to the alcohol stage at lead cathodes in sulfuric acid solutions. Alkire [3] observed that imposition of an alternating current component on the direct current component facilitated these reductions as regards both yields and reaction rates. In the case of salicylic acid the apparent rest potential shifted to a less negative value and the yield increased from 30 to 80% when the ac component was applied to dc component during the electrolysis [3].

Effect of an alternating current component

The electroreduction of a carboxylic acid to the corresponding alcohol is formally expressed as a two-step process [4]:

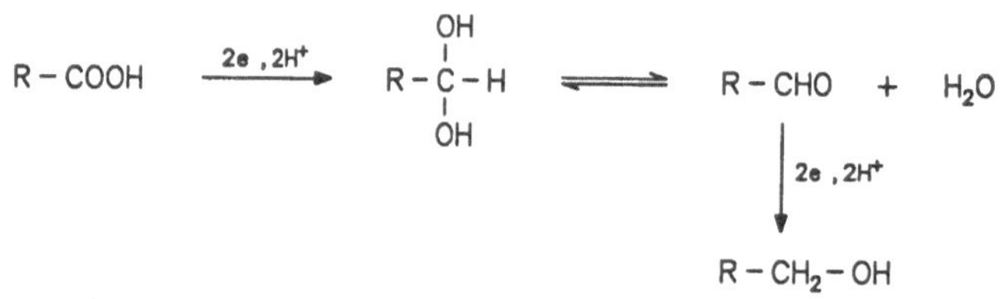

$$R-COOH \xrightarrow{2e^-,2H^+} R-\underset{\underset{OH}{|}}{\overset{\overset{OH}{|}}{C}}-H \rightleftharpoons R-CHO + H_2O$$

$$\downarrow 2e^-,2H^+$$

$$R-CH_2-OH$$

Because the aldehyde is more easily reduced than the acid, the reduction most often advances past the aldehyde stage. It is possible sometimes to arrest the reduction at the aldehyde stage. Heterocyclic and α,β-unsaturated carboxylic acids may be reduced to their aldehydes [5,6]. It has been claimed (patent) that the benzyl alcohol was obtained from benzoic acid in 83% yield and 85% current efficiency when 2-ethyl-1-hexanol was added to the electrolysis medium [7].

3.4.1 Preparation of Metal Carboxylates and Metal Alkoxides

Paired electrosynthesis

Aliphatic acids are not reduced to alcohols by the cathodic method but they yield carboxylate ions as a result of their proton reduction. When transition metals are used as anodes a *paired electrosynthesis* is achieved whereby transition metal carboxylates are produced, the electrochemical cell processes occurring according to the general Scheme 3.7

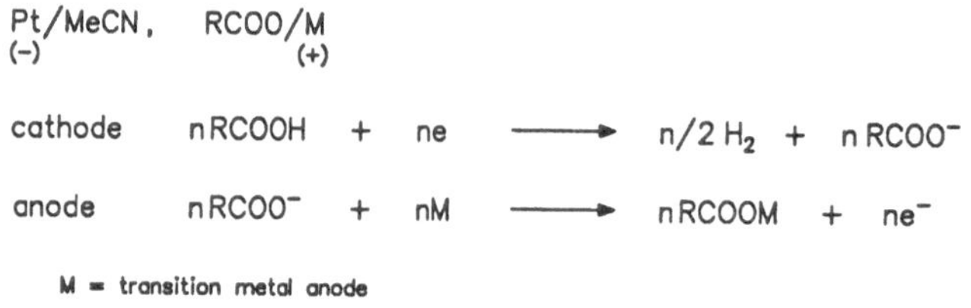

Pt/MeCN,   RCOO/M
(−)   (+)

cathode $n\,RCOOH + ne \longrightarrow n/2\,H_2 + n\,RCOO^-$

anode $n\,RCOO^- + nM \longrightarrow n\,RCOOM + ne^-$

M = transition metal anode

Scheme 3.7

Carboxylates of Cr, Mn, Ni, Co, Cu and Fe were prepared by electrolysis in organic solvents containing various carboxylic acids including long chain fatty acids [9]. Small amounts of Et_4NClO_4 may be added to enhance the conductivity of the medium.

Consumable anode methodology

Alkoxides also can be prepared by this methodology. Tungsten alkoxides were prepared using a tungsten plate as anode in absolute methanol or ethanol [10]. The *consumable anode* methodology has been considered for various applications [11,12].

Direct electrochemical synthesis of metal alkoxides by anode dissolution of the appropriate metals can be effected in absolute alcohols mixed with polar solvents so as to enhance the conductivity of the medium. Undivided cells are used. The alkoxides of Sc, Ti, Zn, Nb, Ta and Ge have thus been prepared. Suitable electrolytes are the salts Bu_4NBr, Bu_4NBF_4, Et_4NBr,

Et₄NCl, NH₄Cl and NaBr. Solid alkoxides are purified by recrystallization from alcohol. Liquid alkoxides are isolated by distillation under vacuum at room temperature [11,12]. An advantage of anodic metal dissolution in nonaqueous media is the possible formation of lower oxidation states of metallic ions. Thus novel metal complexes with organic or inorganic species may be produced.

3.4.2 Reduction of Esters

Esters are in general reduced with great difficulty at the cathode. In hydrous media α,β-unsaturated esters usually afford hydrodimers as major products via electronation of the unsaturated bonds [13]. The proton availability must play an important role in the overall reduction. In acetonitrile or dimethylformamide cyclizations take place rather than dimerizations, the latter being favored in protogenic media such as methanol [12,14]:

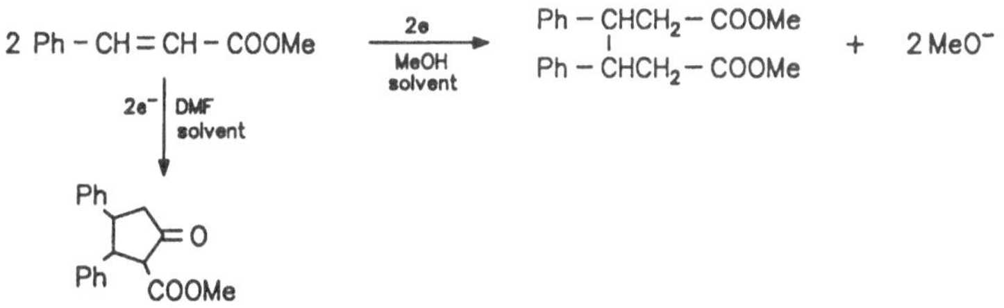

Cyclizations and dimerizations

Electrochemical reduction of certain esters in the presence of amines results in the formation of amidated esters. This reaction is difficult by chemical methods [15]:

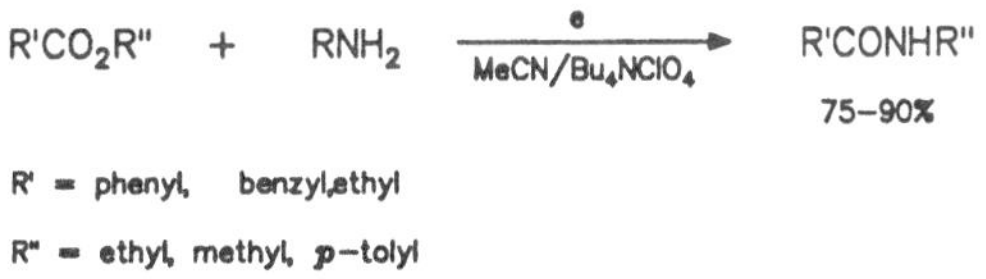

Amidated esters

The electroreductive cyclization of the olefinic ester (**1**) containing the terminal aldehedic function afforded the *trans* monocyclic ester (**2**) in good yield [16]. **Note:** This interesting reaction occurs best in dry acetonitrile with diethyl malonate at a proton source.

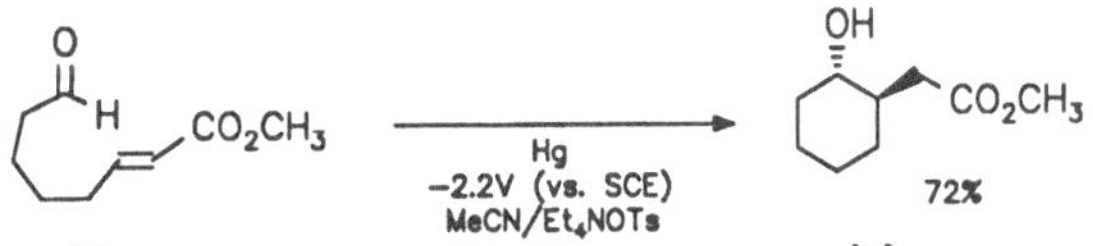

3.5 Organohalides

3.5.1 General Electroreduction Mechanism

The electrochemical reduction of organohalides has been very widely investigated since the first studies of Stackelberg and Stracke [1] in 1949. It is now generally accepted [2] that the reduction of a carbon-halogen (C—X) bond takes place according to the general mechanism shown in Scheme 3.8.

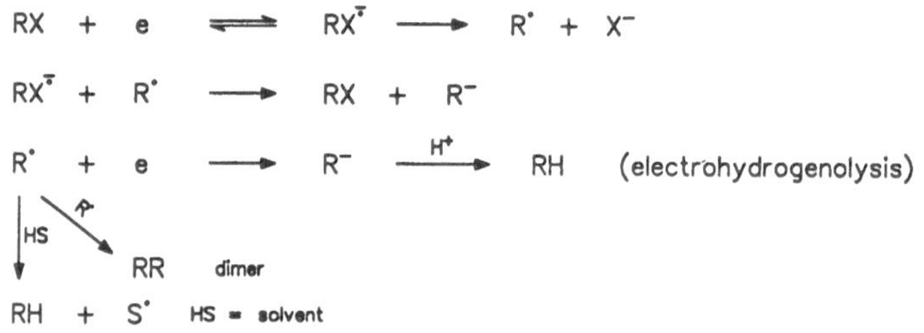

$$RX + e \rightleftharpoons RX^{\bar{\cdot}} \longrightarrow R^{\cdot} + X^-$$

$$RX^{\bar{\cdot}} + R^{\cdot} \longrightarrow RX + R^-$$

$$R^{\cdot} + e \longrightarrow R^- \xrightarrow{H^+} RH \quad \text{(electrohydrogenolysis)}$$

$$RR \quad \text{dimer}$$

$$RH + S^{\cdot} \quad HS = \text{solvent}$$

Scheme 3.8

Conceivably, a C—X bond may be cleaved by a concerted mechanism without the intermediate generation of the aion radical RX⁻.

3.5.2 Products from Aromatic and Aliphatic Halides

The cathodic reduction of aromatic halides in general yields anion radicals as primary products which are rapidly decomposed into radicals and halide ions [3]. The aromatic fragments, Ar⁻, are very often more electrophilic than the anion radical ArX⁻ and are immediately electronated to yield the anion Ar⁻ which reacts with an electrophile to produce the substituted product, as for example [3a], in the very selective electrohydrogenolysis of pentachloropyridine:

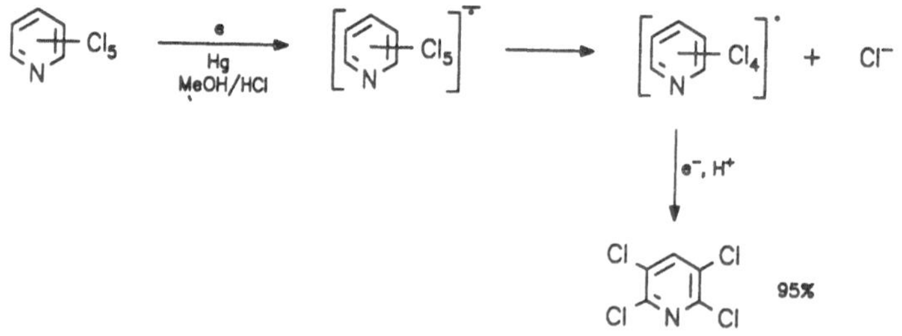

Relative ease of reduction Aliphatic short chain halides are reduced at lower potentials than do aromatic and vinyl halides. With vinyl halides stereoisomers are formed [4],

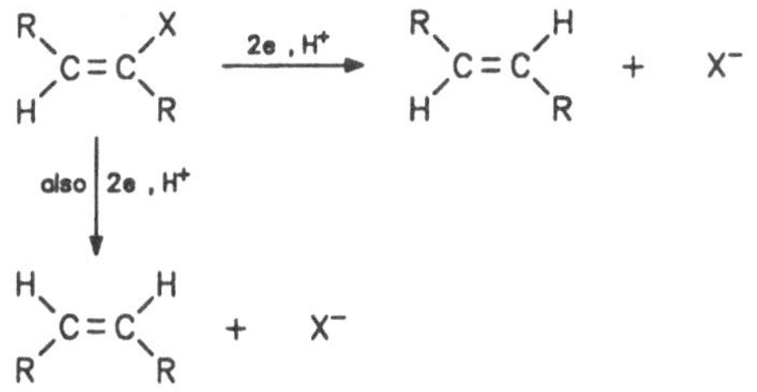

Electroreductions, as in the cases shown above, are believed to be direct electronations of the substrate, with an electron being added to the lowest vacant orbital followed by expulsion of a halide ion from the radical anion [5]. A second electron is added to the neutral species and the anion is protonated. (Protonation may also occur before halide expulsion from the radical anion.) Geminal and vicinal halides are reduced more easily than the monohalides.

Direct electronations

The reduction potential of carbon-halogen bonds increases in the order C—I, C—Br, C—Cl, C—F. In aprotic media one-electron reductions are favored, leading to couplings and a variety of other products. Very reactive intermediates can be formed in aprotic solvents: benzynes, carbenes [6–8], and polymerizable species [9]. Long chain alkyl chlorides and fluorides are very difficult (or practically impossible) to reduce at the cathode. Bromides and iodides, however, are reduced at the cathode even when their alkyl chains are long as in fatty acids. Carbon tetrachloride affords dichlorocarbene by reduction in aprotic solvents. The carbene may be trapped by an olefin [10]:

Benzynes carbenes

Dichloro-carbene

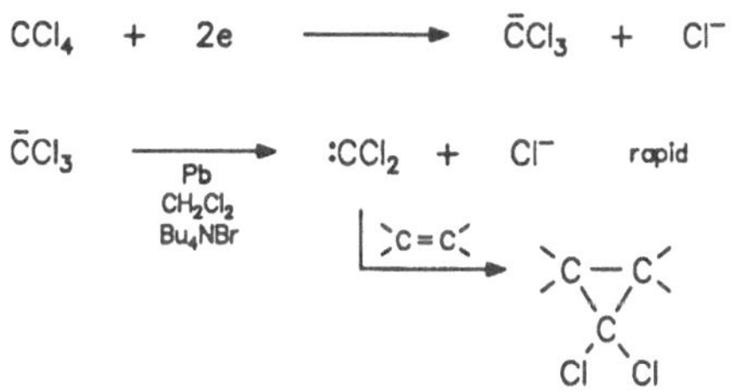

At lead cathodes organolead compounds may be produced by reaction of the generated radicals with the metallic electrode [11]

Organolead compounds

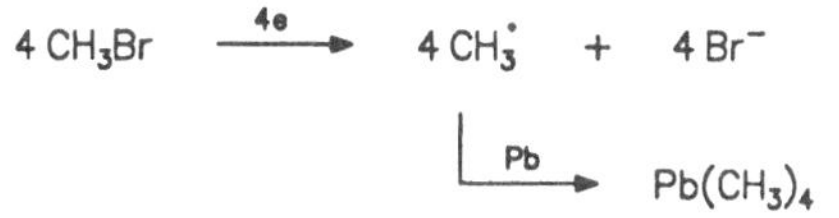

Certain organodihalides, as in the following case, are intermolecularly cyclized by reduction in aprotic media [12]:

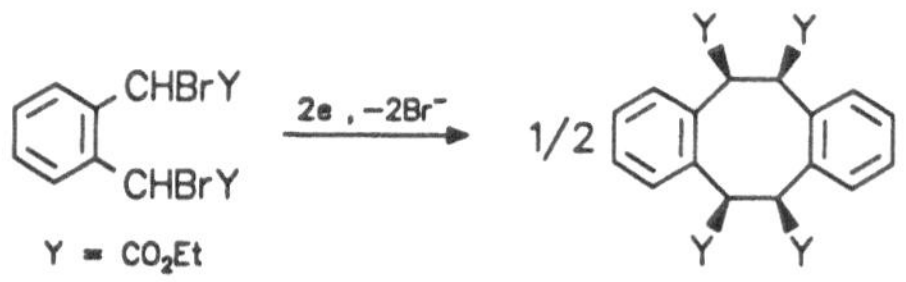

Cathodic dehalogenation, strained structures

Highly strained hydrocarbons can be prepared by cathodic dehalogenation of certain polyhalogenides [13–16]. Examples of strained structures are shown below:

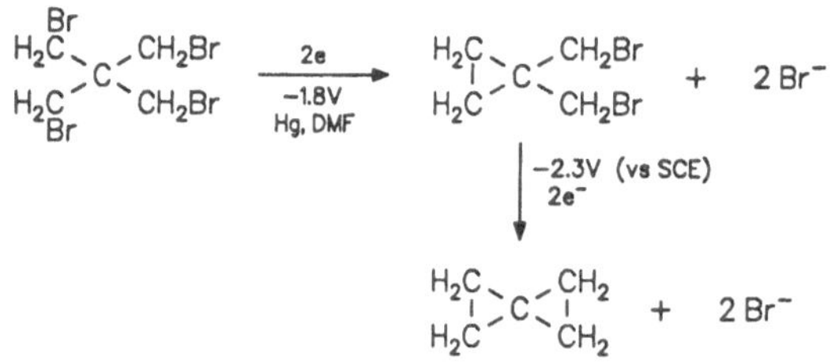

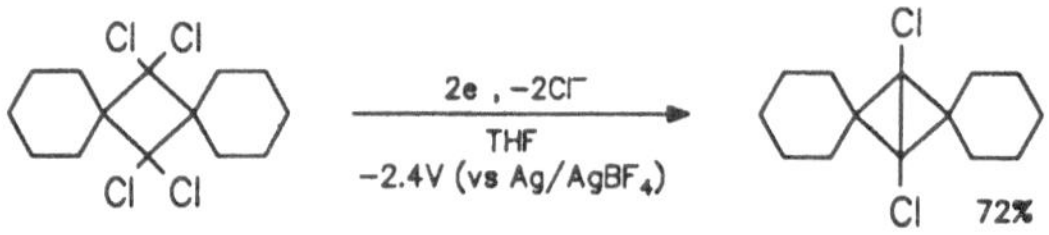

Cyclopropanations

Dehalogenation of certain dihalides may easily lead to cyclopropanations. An example is the dehalogenation of α,α' dihaloketones [17] at mercury cathode in DMF solution. The cyclization process is considered occurring as follows:

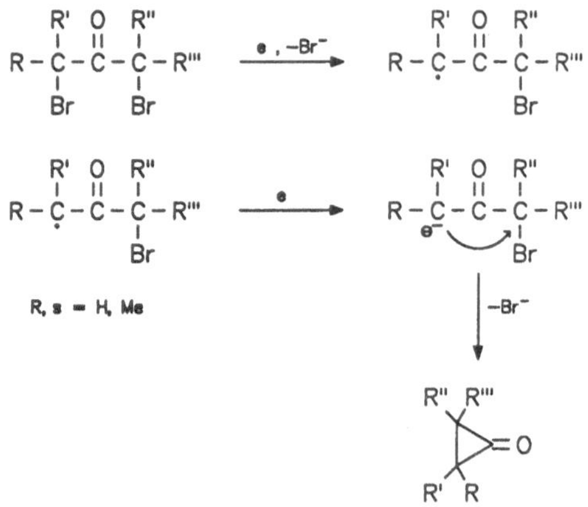

3.5.2.1 Sacrificial Anode Methodology Using Organohalides

In this section we consider the potential usefulness of *sacrificial* metallic anodes for organic synthesis with organic halides as cathodic reactants. Sacrificial anodes are usually electrodes of the metals zinc, magnesium and aluminum. These electrodes dissolve upon anodic polarization to give the corresponding metal ions. To be useful for synthetic application they must dissolve before any other species surrounding the electrode is oxidized – either because of thermodynamic or kinetic reasons, so that the main anodic half-reaction is the dissolution of the electrode, e.g.

$$\underset{\text{anode}}{\underset{\text{////////}}{Zn}} \longrightarrow Zn^{2+} + 2e$$

The practical merit of this type of electrolysis is that it makes possible the use of undivided cells – a significant advantage as regards cell construction and minimum potential drop in the solution (especially when the solvent is organic). Also, the metal ions generated in situ are sometimes essential for the occurence of the desired reaction.   **Undivided cells**

Sacrificial zinc anodes were employed by Sibille and coworkers in electrochemical cyclopropanations [18]. Alkenes and *gem*-dihaloalkanes are the starting materials, which in the presence of zinc ions generated in situ react as shown in Scheme 3.9. It was proposed that an organozinc carbenoid forms as an intermediate for this cyclopropanation as it happens in the chemical cyclopropanation methods.   **Zinc anodes**

$$CH_2Br_2 + RCH=CHR' + 2e \longrightarrow \underset{CH_2}{RCH - CHR'} + 2Br^-$$

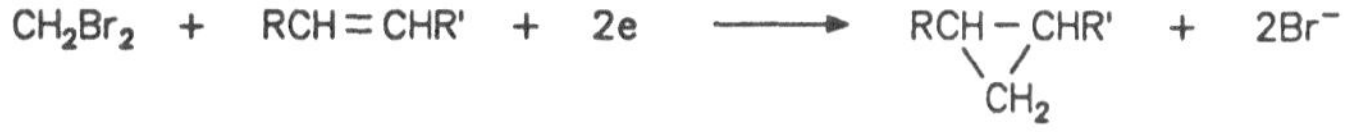

Scheme 3.9

General Cyclopropanation Procedure. Scheme 3.9 Alkenes and gem-dibromoalkanes or the diodo analogs are electrolyzed in a one-compart-ment cell fitted with a carbon fiber cathode and a sacrifical zinc anode. A 90/10 CH_2Cl_2/DMF solvent mixture with Bu_4NBr as electrolyte containing the alkene and an excess of the dihalo compound is electrolyzed at 40 °C under constant current (200 mA) and an argon atmosphere. The products are recovered by extraction with ether or pentane and are purified by column chromatography (silica gel-pentane, ether). Some typical alkenes and dihalides and respective cyclopropane product yields are the following:

$CH_3CH=CHCH_2OH$ +	CH_2Br_2	%yield	(56)
$CH_3CH=CHCH_2OH$	CH_2I_2		(47)
$PhCH=CHCH_2OH$	CH_2Br_2		(59)
cyclooctene	CH_2Br_2		(75)
$PhCH=CH_2$	CH_2I_2		(33)
gereniol	CH_2Br_2		(70)

Organo metallic chemistry

CCl_3^-/Zn^{2+}

Electrosynthetic methods using sacrificial anodes should be of interest in organo metallic chemistry. The following coupling reaction is especially interesting since the CCl_3^- anion produced under other conditions decomposes repidly to yield the carbene CCl_2. The in situ generated zinc ions are shown in the reaction scheme to point out that they are necessary for this coupling reaction [19].

$$Zn \longrightarrow Zn^{2+} + 2e \quad \text{anode}$$

$$CCl_4 + 2e \longrightarrow CCl_3^- + Cl^- \quad \text{cathode}$$

$$C_6H_5Br + CCl_3^- \xrightarrow{(Zn^+)} C_6H_5CCl_3 + Br^-$$

Aluminum or magnesium anodes

Organic halides electroreduced in the presence of carbonyls and with aluminum or magnesium sacrificial anodes afford alcohols [20]. The cell reaction is shown symbolically below with magnesium as anode.

$$Mg \longrightarrow Mg^{2+} + 2e^- \quad \text{anode}$$

$$RX + 2e^- \longrightarrow R^- + X^- \quad \text{cathode}$$

$$R^- + ^{\backslash}C=O \longrightarrow R-\overset{|}{\underset{|}{C}}-O^- \xrightarrow{H^+} R-\overset{|}{\underset{|}{C}}-OH$$

Couplings, carboxylations

Alicyclic rings

A sacrificial aluminum anode was used for the electroreductive coupling of α-halogenoesters with aryl or vinyl halides. Moderate to good yields of β,γ-unsaturated esters were obtained [21]. A good review on the use of sacrificial anodes in synthetic electrochemistry specifically for processes involving carbon dioxide has been recently presented by Silvestri and coworkers [22]. Reductive electrocarboxylations and the reduction of CO_2 alone are discussed. Alicyclic three-, five- and six-membered ring products were formed by electroreductive coupling of dimethyl maleate with alkyl dihalides in undivided cells fitted with sacrificial aluminum anodes to provide the apparently needed Al^{3+} ions [22].

$$\underset{X}{\overset{X}{(CH_2)_n}} + \overset{}{\underset{}{>=<}}_Z \xrightarrow[\substack{DMF/Bu_4NBF_4 \\ Al^{3+}}]{2e} (CH_2)_n \overset{}{\underset{Z}{\rceil}} + 2X^-$$

$$20-70\%$$

X = Br, Cl   n = 1–4   Z = electron withdrawing group

Electrolysis of bromotrifluoromethane in DMF solution in a one-compartment cell fitted with a sacrificial zinc anode converts the bromo compound to trifluoromethyl alcohol and some organozinc compounds (CF_3ZnBr, $(CF_3)_2Zn$). Addition of tetramethylene diamine to the electrolysis medium seem to promote alcohol formation [22,23].

3.5.3 Various Reductive Dehalogenations

Organohaloanions produced cathodically may enter into various reactions in the presence of suitable electrophiles. Lund reported the following example [24]:

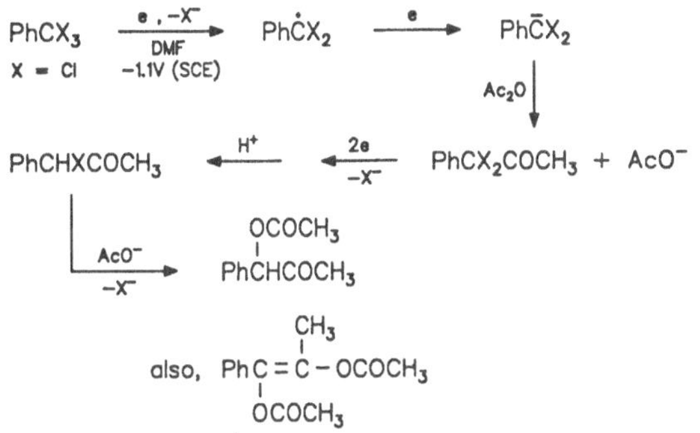

Haloanions

$$PhCX_3 \xrightarrow[\substack{DMF \\ -1.1V \, (SCE)}]{e \, , \, -X^-} PhC\cdot X_2 \xrightarrow{e} Ph\bar{C}X_2$$
$$X = Cl$$

with $Ph\bar{C}X_2 \xrightarrow{Ac_2O}$ (downward)

$$PhCHXCOCH_3 \xleftarrow{H^+} \xleftarrow[-X^-]{2e} PhCX_2COCH_3 + AcO^-$$

$$PhCHXCOCH_3 \xrightarrow[-X^-]{AcO^-} \underset{\substack{| \\ OCOCH_3}}{PhCHCOCH_3}$$

also, $\underset{\substack{| \\ OCOCH_3}}{PhC} = \overset{\substack{CH_3 \\ |}}{C} - OCOCH_3$

The electroreductive coupling depicted above occurs in DMF/Bu$_4$BF$_4$/Ac$_2$O medium. If the electrolyte is NaClO$_4$ or NH$_4$ClO$_4$ different products are obtained: 1-acetoxy-1-phenyl-2-propanone and the 1-chloro-analog. Such electrolyte effects on the nature of the products may be attributed to differences in the structure of the electrical double layer [25,26].

Electro-reduction coupling

Electroreductive couplings of halohydrins are feasible if the hydroxy function is protected (esterified). Cipris [27] thus obtained 1,4-butanediol from bromoethyl acetate by a one-electron reductive process.

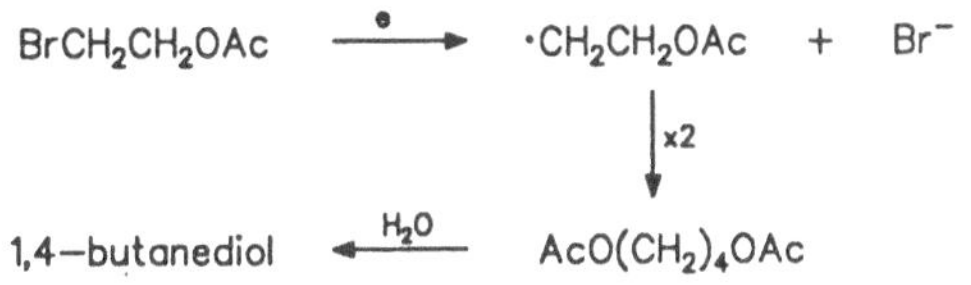

Couplings of halohydrius

$$BrCH_2CH_2OAc \xrightarrow{e} \cdot CH_2CH_2OAc + Br^-$$

with $\downarrow x2$

$$1,4\text{-butanediol} \xleftarrow{H_2O} AcO(CH_2)_4OAc$$

Reductive dehalogenation of 3-bromo lactams in the presence of acetic anhydride gives only the *trans*-acetyl derivative [28].

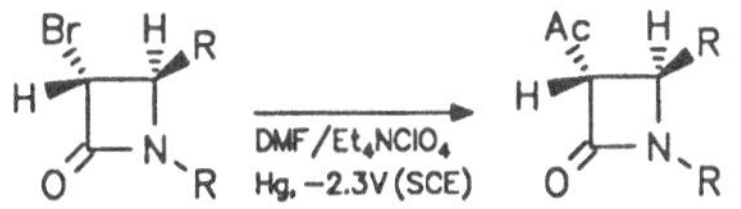

Reductive dehalogenation

Synthetically useful groups can be produced by partial reductive dehalogenation, such as the following [29]:

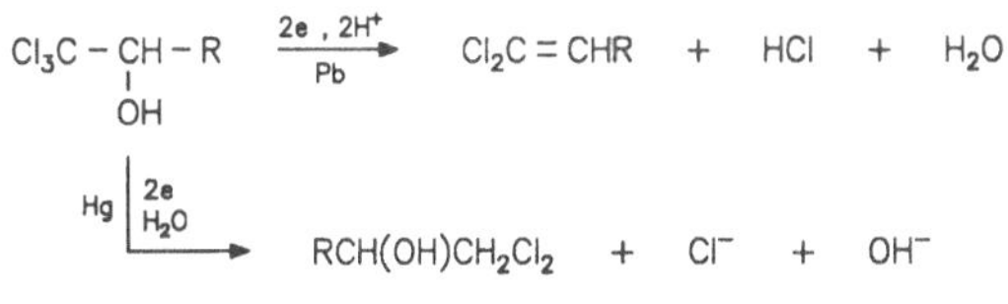

$$Cl_3C - \underset{\substack{| \\ OH}}{CH} - R \xrightarrow[Pb]{2e \, , \, 2H^+} Cl_2C = CHR + HCl + H_2O$$

$$\underset{\substack{Hg \Big| \substack{2e \\ H_2O}}}{} \longrightarrow RCH(OH)CH_2Cl_2 + Cl^- + OH^-$$

Splitting
C—H bonds   In certain halogen containing molecular structures it is possible to split a C—H bond instead of a carbon-halogen bond if the electrolysis is performed in aprotic media and the cathode is a transition metal (Fe, Ni, Pt). This is possible because the C—H bonds in these molecules (carbon acids) are relatively weak and the transition metal surfaces are good catalysts for hydrogen formation [30–32]. An example is the following [30]:

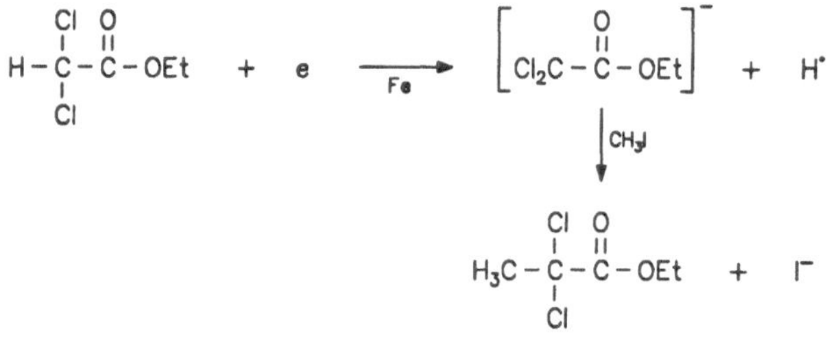

Benzyne   Reductive dehalogenations of aromatic dihalides aprotic solutions may generate intermediates like the benzyne or anions which enter into reactions as exemplified in Scheme 3.10 [33,34].

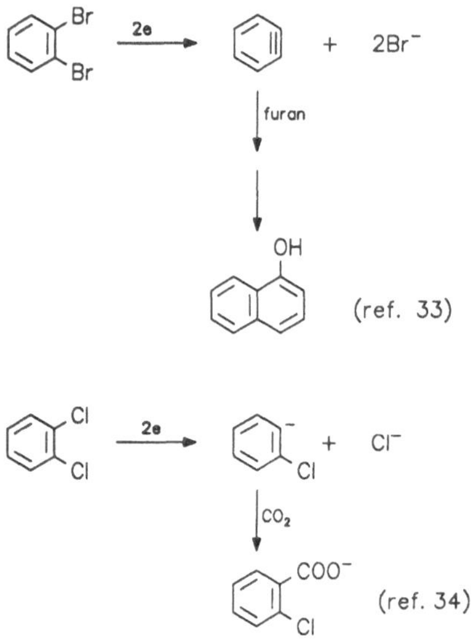

Scheme 3.10

3.5.4 Electrohydrogenolysis of Polyhalogenated Compounds

Electrohydrogenolysis of polyhalogenated pyridines (agricultural chemicals) was investigated from an industrial perspective [35]. Pentachloropyridine and tetrachloropicolinic acid were electroreduced selectively to symmetrical tetrachloropyridine and 3,6-dichloropicolinic acid, respectively. The latter was produced on a commercial scale (under the trade name Lontrel, Dow Chemical Co., California). Lontrel is one of the most valuable (biodegradable) plant growth regulators and was until recently manufactured in an unpartitioned "frame and plate" cell with a *spongy* silver cathode and stainless steel anode in 10% aqueous NaOH solutions. The overall cell reaction is shown in Scheme 3.11. This electrochemical process has been viewed as a typical large scale heterogeneous organic electrocatalysis process [36].

3,6-Dichloro-picolinic acid – "Lontrel"

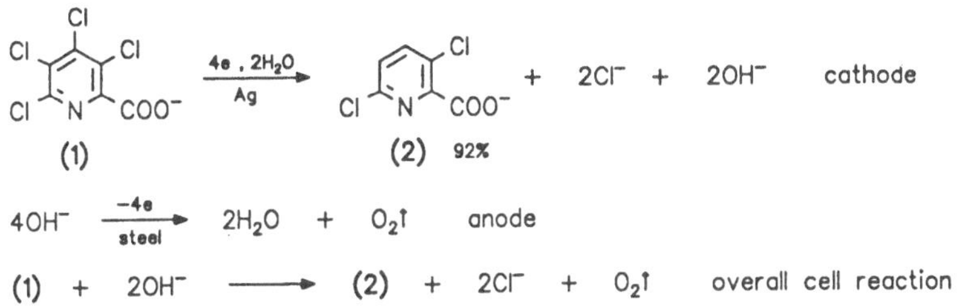

Scheme 3.11

Preparation of 3,6-Dichloropicolinic Acid by Electrohydrogenolysis of Tetrachloropicolinic Acid (Scheme 3.11)

This very selective electrohydrogenolysis (that is, hydrogenolysis of C—Cl bonds by electrolysis) is achieved in a simple undivided cell in aqueous alkaline solution. The electrolysis cell is a 400-ml glass beaker fitted with a 20-mesh silver screen cathode and a stainless steel rod anode positioned concentrically by means of a neoprene rubber disc as shown in the cell schematic, Fig. 3.1. The solution to be electrolyzed consists of 150 ml distilled water, 12 g reagent grade sodium hydroxide and 10 g of tetrachloropicolinic aicd (added in steps). The cell is placed in a water bath thermostated at 25–35 °C. The solution (suspension) is vigorously stirred during electrolysis.

a. *Activation of the silver cathode*: Before adding the organic acid to the solution in the cell the silver electrode is *anodized* by applying to it a positive potential starting from zero and gradually raising it to $+0.6\,V$ (vs SCE). The potential is held at 0.6 V for about 3 minutes until practically all the electrode surface becomes covered with silver oxides and the current begins falling towards a minimum value. The polarity is then reversed and the silver electrode potential is set at $-1.3\,V$, which is the optimum potential for the desired reaction. The surface silver oxides are reduced and the current falls to a steady value (ca 0.3 A) when all the oxides are apparently reduced and the

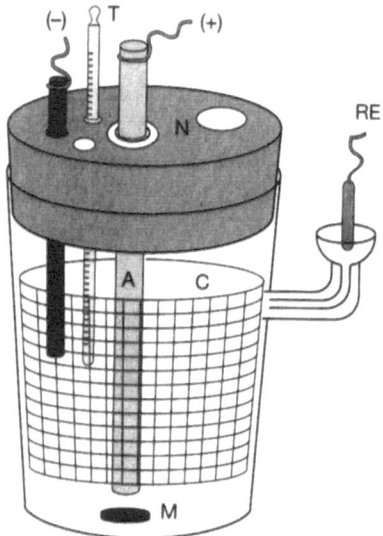

Fig. 3.1. Schematic of electrolysis cell, for the electrohydrogenolysis of tetrachloro-picolinic acid
A, stainless steel rod anode
C, cathode, silver screen
N, neoprene rubber stopper with holes for gas exit
RE, reference electrode, SCE
M, magnet bar
T, thermometer

silver surface assumes a "spongy" structure (believed to be the active form).

b. *Electrolysis*: The electrolysis is now carried out at -1.3 V potentiostatically, the organic acid being added to the solution in the cell in several steps: About 2 g of the organic acid is macerated with a small portion of the alkaline solution from the cell and the mixture is added to the solution in the cell. The current rapidly rises from 0.3 to 3 A as a result of the electroreduction of the organic acid (the sodium salt in solution). The whole amount (10 g) of the organic acid is thus added to the cell in steps over a period of about 8 hours. When all the acid is reacted the current falls to 0.3 A and the electrolysis is ended. The electrolyzed solution is acidified with 25 ml of conc. hydrochloric acid, whereupon copious precipitation (white crystals) of the product, 3,6-dichloropicolinic acid, occurs. The product is recovered by filtration, and after washing with dilute hydrochloric acid and some distilled water, it is dried (100 °C oven) (7 g, 98% purity, 90% yield). Identification is made by comparing IR, NMR and chromatographic data with those of an authentic sample. The impurities (2%) were trichloropicolinic acid and a trace of monochloropicolinic acid.

It was found that periodic activation in situ of the cathode by anodization (every 3 hours) was very beneficial. Also, best reproducibility was obtained by cleaning the used silver cathode after every run with 1:1 conc. HCl/H_2O. This procedure was found to remove traces of electrodeposited metals, Fe, Cu, Ni, Pb which were present in the materials used, and which adversely affected the electrode's activity. The anodization procedures (as was determined by X-ray fluorescence analysis) removed most metallic impurities except copper, which apparently had the worst effect on the effectiveness of the silver cathode.

In situ reactivation of cathode

It would be reasonable to postulate that two forces are largely responsible for this very selective electrohydrogenolysis: a) *specific adsorption* forces on the spongy silver and b) orientational electrostatic forces favoring electron transfer to 4 and 5 positions of the ring.

It should be noted that practically no reduction was possible on a shiny silver surface; and the reduction was not selective at other electrodes (Hg, Cu, C).

3.5.5 Electrochemically Induced Nucleophilic S$_{RN}$1 Substitutions

Aryl and Heteroaryl halides have been used as substrates for some electrochemically induced nucleophilic substitutions [37]. In conventional organic reactions movement of electrons in general occurs as a movement of an electron pair or a movement of the bond by an S_N^2 process (polar process):

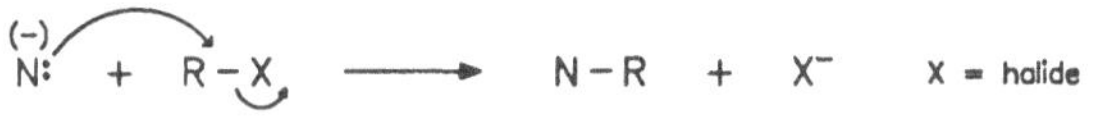

$$N: + R-X \longrightarrow N-R + X^- \quad x = halide$$

For radical producing chemical or electrodic processes one-electron transfers take place from the nucleophile, N^-, or from the electrode to the organic substrate $R—X$ as depicted below:

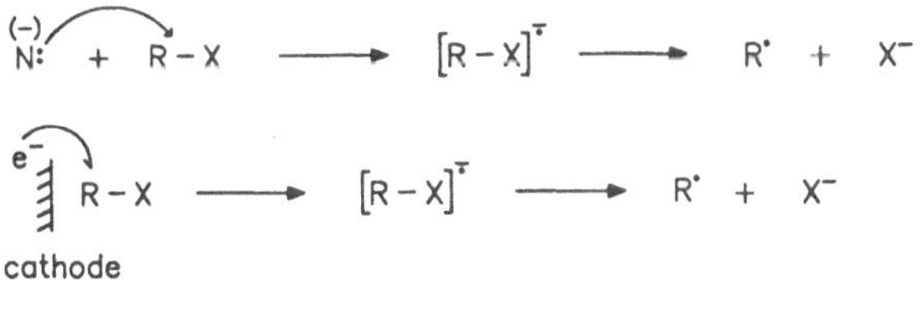

Electron transfer

$$N: + R-X \longrightarrow [R-X]^{\overline{\cdot}} \longrightarrow R^{\cdot} + X^-$$

$$R-X \longrightarrow [R-X]^{\overline{\cdot}} \longrightarrow R^{\cdot} + X^-$$

cathode

Cathodic reductions of aromatic halides, Ar X generally afford as primary products anion radicals:

Aromatic halides → anion radicals

$$ArX + e \rightleftharpoons Ar^{\overline{\cdot}} \quad E^{\circ} \text{ characteristic potential}$$

The anion radical is usually an unstable species and decomposes rapidly to an organic radical and the halide ion,

Anion radicals → organic radical

$$ArX^{\overline{\cdot}} \longrightarrow Ar^{\cdot} + X^-$$

The fragment Ar· is more electrophilic than the $ArX^{\bar{\cdot}}$ species and the carbon which was bound with the halide X tends to combine with a nucleophile and also to abstract a hydrogen atom from the environment. Thus competitive chemical reactions would occur. For our present purposes consider the reaction sequence below:

1. $ArX + e^- \rightleftharpoons ArX^{\bar{\cdot}}$ E_1°
2. $ArX^{\bar{\cdot}} \longrightarrow Ar^· + X^-$
3. $Ar^· + Y^- \longrightarrow ArY^{\bar{\cdot}}$
4. $ArY^{\bar{\cdot}} - e^- \rightleftharpoons ArY$ E_2°
5. $ArY^{\bar{\cdot}} + ArX \rightleftharpoons ArY + ArX^{\bar{\cdot}}$

Substitution and catalytic cycle

The electron electrocatalyst

If $E_2^\circ < E_1^\circ$, reaction 5 will be thermodynamically favored, and X will be substituted by Y, thus yielding the new product ArY and also the anion radical $ArX^{\bar{\cdot}}$ to repeat the catalytic cycle (chain reaction). The electron "shuttles" between the electrode and the solution so that in principle there should be no *net* charge passed through the cell. The electron itself is the electrocatalyst. The process is designated as an $S_{RN}1$ process (or radical nucleophilic unimolecular). By analogy to proton transfer catalysis $H^+C\ H^+$ it is also designated as an E C E process. In Alder's view [38] "there is no logical reason why electron transfer catalysis should not be as general as acid base (proton transfer) catalysis." A typical example of $S_{RN}1$-type synthesis is the aromatic substitution reported by Saveant [39], Scheme 3.11a. The monoanion (1) derived from a β-cyanocarbonyl compound reacts with the benzoyl substituted aromatic bromohalide (2) to give the substituted products (3). The electrolysis is carried out in DMF at the potential of the first voltammetric wave, the height of which decreases upon addition of the nucleophile (1), as expected.

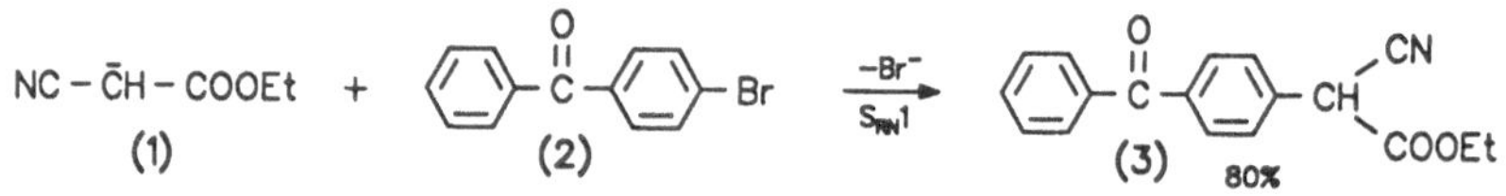

Scheme 3.11a

A rather unique $S_{RN}1$-type electrosynthesis was recently reported by Degrand concerning the direct preparation of diaryl diselenides and ditellurides [40]. The overall reaction is shown below:

$$2ArX + E_2^{2-} \xrightarrow{S_{RN}1} ArEEAr + 2X^-$$

ArX = arylhalide
E = Se or Te

Cathodic polarization of Se or Te electrodes generates the E_2^{2-} dianions so that the reactions (1) and (2) become possible;

$$2E + 2e \longrightarrow E_2^{2-} \qquad (1)$$
$$ArX + e \rightleftharpoons ArX^{\bar{\cdot}} \longrightarrow Ar^· + X^- \qquad (2)$$

Thus the dichalcogenides of the 4-bromobenzophenone are obtained:

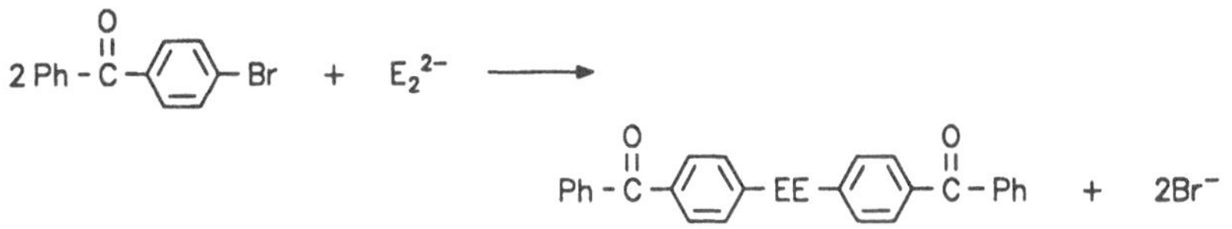

$$2\,Ph-\overset{\overset{\displaystyle O}{\|}}{C}-\!\!\left\langle\bigcirc\right\rangle\!\!-Br \quad + \quad E_2^{2-} \quad \longrightarrow$$

$$Ph-\overset{\overset{\displaystyle O}{\|}}{C}-\!\!\left\langle\bigcirc\right\rangle\!\!-EE-\!\!\left\langle\bigcirc\right\rangle\!\!-\overset{\overset{\displaystyle O}{\|}}{C}-Ph \quad + \quad 2Br^-$$

This electrolysis is performed in an H-type cell in acetonitrile or dimethyl-sulfoxide containing a quaternary ammonium salt as electrolyte, and at a cathode potential of -1.6 to $-1.7\,V$ versus SCE. Simultaneous generation of E_2^{2-} anions and reduction of 4-bromobenzophenone take place. The Se and Te were contained in the form of pellets in a tea-bag type electrode.

3.5.6 Organic Fluorides

Perfluorocyclopentene and perfluorocyclobutene afford polymers by electrolysis in aprotic solvents containing Et_4NBF_4 as electrolyte [41]. The polymers form as lustrous blue-black needles attached to the electrode (platinum) surface. They are electronic conductors. No polymers are formed if the electrolyte is lithium perchlorate.

3.6 Nitrocompounds

The nitro group is one of the most versatile electrophores in organic electrosynthesis [1,2]. Nitroalkanes are reduced in both hydrous and anhydrous media, giving hydroxylamines and amines. In basic solutions the reduction is difficult because the *aci* anion form of the nitroalkane is resistant to electroreduction.

Electroreduction of nitrobenzene occurs at copper and nickel electrodes in neutral and basic hydrous methanolic solutions [3]. Phenylhydroxylamine and aniline are produced depending on the potential of the electrode. Electrocatalytic hydrogenations at Devarda copper and Raney nickel electrodes takes place by reaction of adsorbed species as depicted in Scheme 3.12

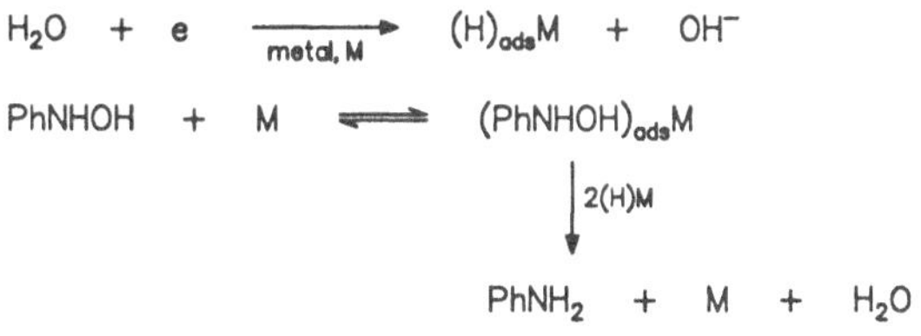

Copper and nickel electrodes

$$H_2O \; + \; e \; \xrightarrow[\text{metal, M}]{} \; (H)_{ads}M \; + \; OH^-$$

$$PhNHOH \; + \; M \; \rightleftharpoons \; (PhNHOH)_{ads}M$$

$$\Big\downarrow 2(H)M$$

$$PhNH_2 \; + \; M \; + \; H_2O$$

Scheme 3.12

Complete reduction of nitrobenzene to aniline requires six electrons [4], Scheme 3.12a

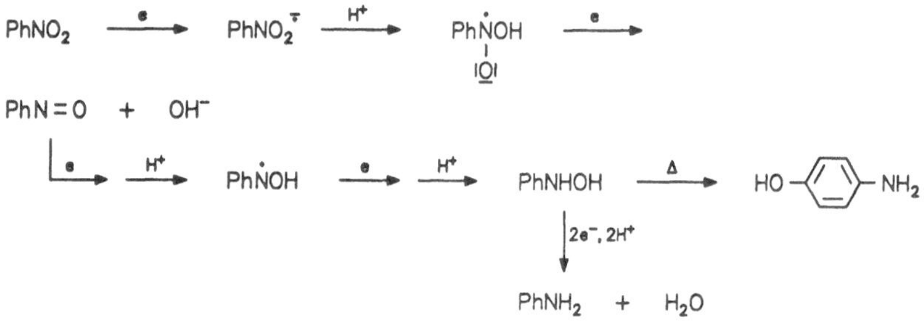

$$PhNO_2 \xrightarrow{\text{e}} PhNO_2^{\overline{\bullet}} \xrightarrow{H^+} Ph\overset{\bullet}{N}OH \xrightarrow{\text{e}}$$
$$\underset{|\underline{O}|}{|}$$

$$PhN=O + OH^-$$

$$\overset{|}{\underset{\downarrow \text{e}}{}} \xrightarrow{H^+} Ph\overset{\bullet}{N}OH \xrightarrow{\text{e}} \xrightarrow{H^+} PhNHOH \xrightarrow{\Delta} HO-\langle\text{ring}\rangle-NH_2$$

$$\Big\downarrow 2e^-, 2H^+$$

$$PhNH_2 + H_2O$$

Scheme 3.12a

In strongly acidic hot solutions it is possible to reduce nitrobenzene to *p*-aminophenol by a four-electron process. Electroreduction in aprotic solutions generates free radicals and nitrite ions [5]. Nitrobenzene was recently reduced in a solid polymer electrode cell using a Nafion membrane *composite electrode* [6] prepared with deposited platinum, nickel and copper. Incidentally, the electrolytic reduction of nitrobenzene was first studied by Haber to demonstrate the role of the electrode potential [7].

Composite electrode)

Ceramic electrode

Reduction of nitrobenzene at a ceramic electrode developed by Beck and Gabriel affords preferably aniline [8]. This electrode was prepared by heating a titanium sheet to 450 °C while covered with titanium acetylacetonate. The reduction is depicted below (heterogeneous electrocatalysis):

$$Ti(OH)_4 + e + H^+ \longrightarrow Ti(OH)_3 + H_2O$$

$$6\,Ti(OH)_3 + C_6H_5NO_2 + 4\,H_2O \longrightarrow 6\,Ti(OH)_4 + C_6H_5NH_2$$

TiO$_2$/Ti electrode

Reduction of nitrobenzene at a TiO$_2$/Ti electrode at higher temperatures was said to favor formation of p-aminophenol [9]. Copper metal generated at -0.95 V (vs SCE) is so active as to reduce nitrophenol chemically, being in the process oxidized to copper oxide and copper hydroxide [10]. Because several paths are available in the electrochemical reduction of nitroaromatics all the parameters, potential, solution pH and even the degree of stirring must be considered [11].

3.6.1 *N*-Heterocyclizations

A variety of heterocyclizations is possible under electroreductive conditions with nitrocompounds as substrates. For example, the reduction of nitrocompounds possessing a keto group affords *N*-heterocyclic compounds via the intermediate hydroxylamines [12].

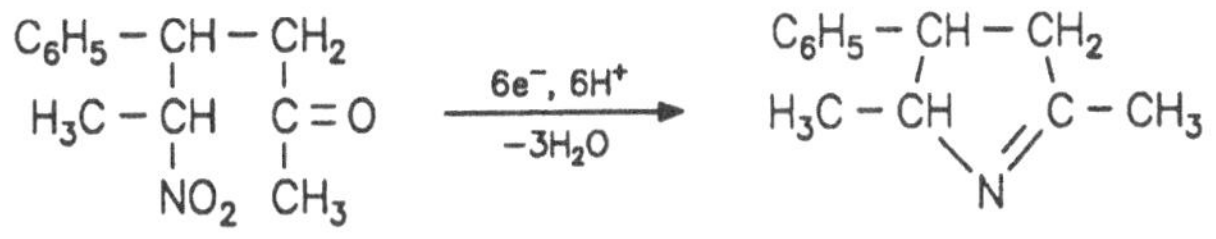

$$C_6H_5-CH-CH_2 \atop H_3C-CH \quad C=O \atop NO_2 \quad CH_3 \quad \xrightarrow[-3H_2O]{6e^-,\,6H^+} \quad C_6H_5-CH-CH_2 \atop H_3C-CH \quad C-CH_3 \atop N$$

Controlled potential electrolysis of *o*-nitrophenylazophenylnitromethane (1) led to the 1,2,4-triazine compound (4) in high yield despite the very complex reduction process [13], Scheme 3.13.

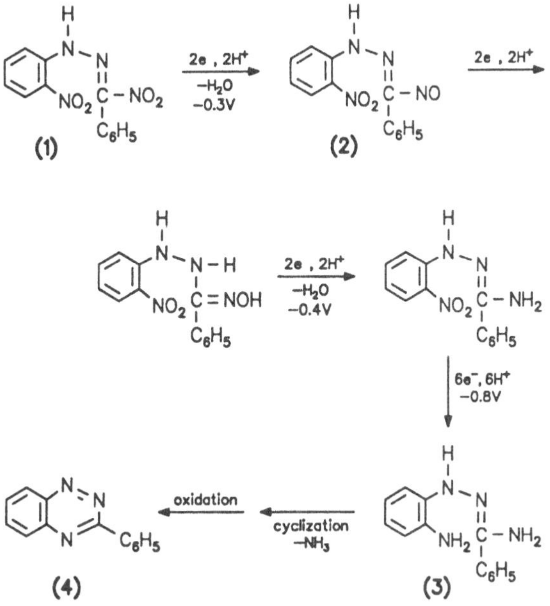

Scheme 3.13

Quinoline-N-oxides and *N*-hydroxyquinoliones can be prepared in good yields by controlled potential electrolysis [14], Scheme 3.14.

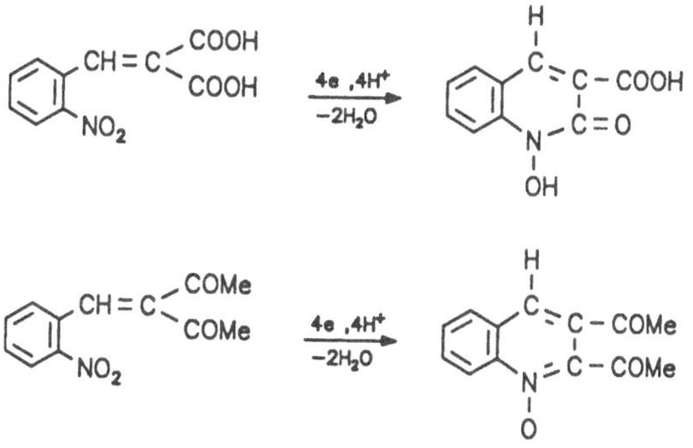

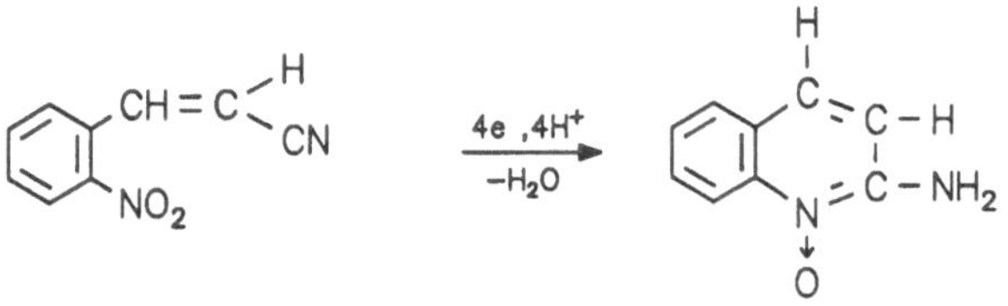

Scheme 3.14

Wagenknecht and Johnson [15], synthesized more than eighty new products according to the overall reaction scheme below:

$$PhNO_2 \ + \ 4RX \ \xrightarrow[DMF/Bu_4NBr]{4e} \ Ph-\underset{OR}{\underset{|}{N}}-R \ + \ R_2O \ + \ 4X^-$$

In aprotic solvents in the presence of acetic anhydride reductive acylations of aliphatic and aromatic nitro compounds take place [16]. (Nitroso compounds also afford similar products.) The reactions are shown in Scheme 3.15.

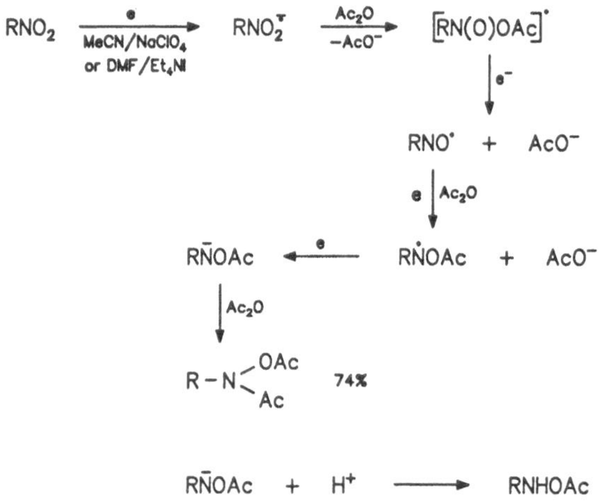

Scheme 3.15

These electrolyses are carried out in divided cells using mercury cathodes at −0.3 to −0.6 V (vs Ag/AgBr) and platinum anodes. Generally, isolated product yields are in the range of 40–70%.

Paired electrosynthesis   An efficient paired electrosynthesis was recently developed by Cault and Moinet [17] for the preparation of N-sulfonylbenzisoxazolones from 2-nitrobenzoic and sulfinic acids. The electrolysis is carried out in aqueous phosphate buffered solutions in a cell fitted with porous electrodes. The cell reaction is represented in Scheme 3.16.

$$ArNO_2 \ + \ 4e \ + \ 4H^+ \ \longrightarrow \ ArNHOH \ + \ H_2O \qquad \text{at cathode}$$

$$ArNHOH \ \longrightarrow \ ArNO \ + \ 2e \ + \ 2H^+ \qquad \text{at anode}$$

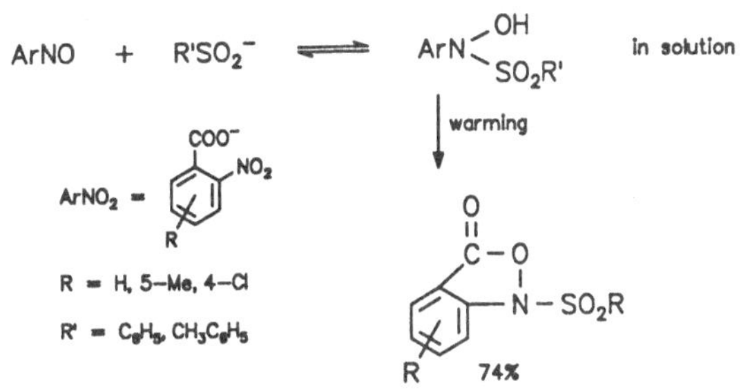

Scheme 3.16

Reduction of *tert*-nitroalkanes [18] and aliphatic or aromatic nitroolefins [19] in acidic alcoholic solutions are interesting since they give products useful in organic synthesis:

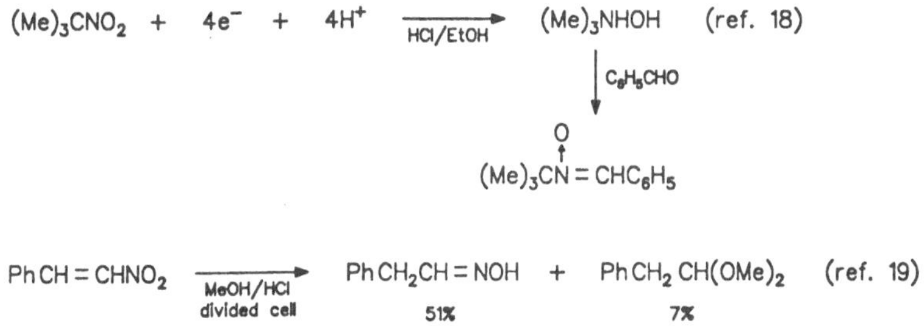

$$(Me)_3CNO_2 \ + \ 4e^- \ + \ 4H^+ \xrightarrow{\text{HCl/EtOH}} \ (Me)_3NHOH \quad (\text{ref. 18})$$

$$\downarrow C_6H_5CHO$$

$$(Me)_3C\overset{O}{\overset{\uparrow}{N}}=CHC_6H_5$$

$$PhCH=CHNO_2 \xrightarrow[\text{divided cell}]{\text{MeOH/HCl}} \ PhCH_2CH=NOH \ + \ PhCH_2CH(OMe)_2 \quad (\text{ref. 19})$$
$$\phantom{PhCH=CHNO_2 \xrightarrow[\text{divided cell}]{\text{MeOH/HCl}} \ } 51\% 7\%$$

3.6.1 Dinitro Compounds

Vicinal nitro compounds are converted to olefinic derivatives by electro-reductive elimination of NO_2^- ions [20]. Evans and Bower [21] studied a series of vicinal dinitro compounds and proposed a mechanism including expulsion of nitrite anion, as is formally depicted in Scheme 3.16a.

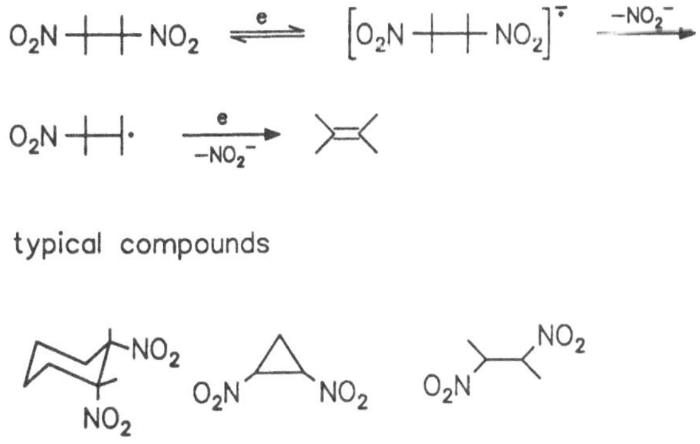

typical compounds

Scheme 3.16a

3.6.2 Reduction on Transition Metal Electrodes

Aliphatic nitro compounds exhibit properties of carbon acids. They are readily reduced on transition metal electrodes by splitting C—H bonds and producing hydrogen and nitroanions at the cathode [22].

$$RR'CHNO_2 \xrightarrow{\ \ e\ \ } [RR'CNO_2]^- + H^\cdot$$

R, R' = H, Me, Et

Organometallic complexes can be produced if the anode is a metal that is anodically dissolved and its cation able to complex with the organic nitroanion formed at the cathode in an undivided cell.

The Nitro group can be reduced by metal powders generated in situ by electrolysis of sulfate salts of metals (Sn, Zn, Fe, Cu). Tin and zinc were found to be most effective powders for these reductions. The electrolysis was performed in benzene-water emulsions (two-phase solvent system) with a carbon tube anode and a rotating carbon or aluminum cylinder as cathode [23].

3.7 *N*-Heterocyclics and Other Heteroatom Structures

The cathodic electrochemistry of *N*-heterocyclics has been intensely studied and continues to be of great interest because of the biological significance of these compounds [1].

3.7.1 Pyridines, Purines, and Similar Compounds

Pyridine and quinoline have been electroreduced to piperidine and tetra-hydroquinoline, respectively, on an industrial scale. At mercury and lead cathodes purines [2], pyrimidines and pyridopyrazines [3] have been reduced to give hydroderivatives and other products. Pyridine is reduced at $-3.1\,V$ (vs Ag/Ag$^+$) in aprotic solutions at a mercury cathode. In the presence of EtOC(O)Cl the product *N*-(ethoxycarbonyl)pyridine is formed [4]. Reduction of pyridine at lead cathode in aqueous sulfuric acid and in the presence of zinc salts gives α,α'-bipiperidyl [5]. Phthalazine reduction in alkaline medium gives the hydro dimer and the hydrogenated product [6].

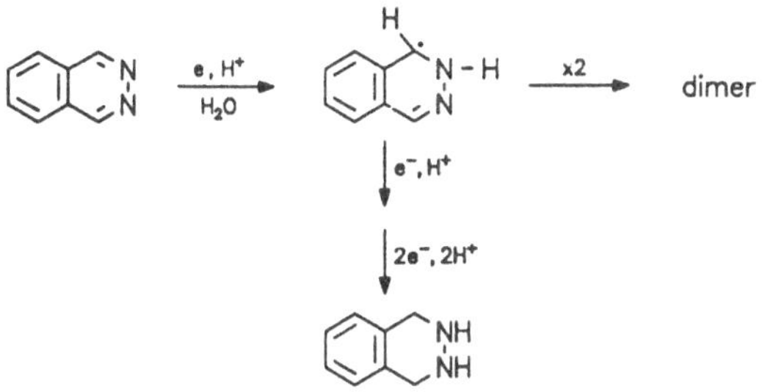

Purine, upon reduction and hydrolysis of the product undergoes opening of the six-membered ring. Cinnolines under go ring contraction with expulsion of ammonia [6].

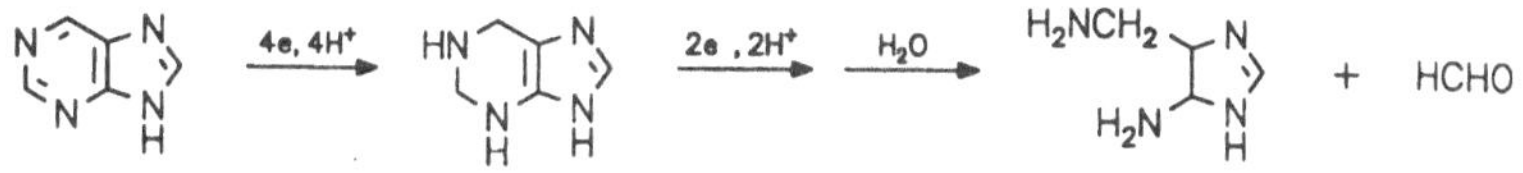

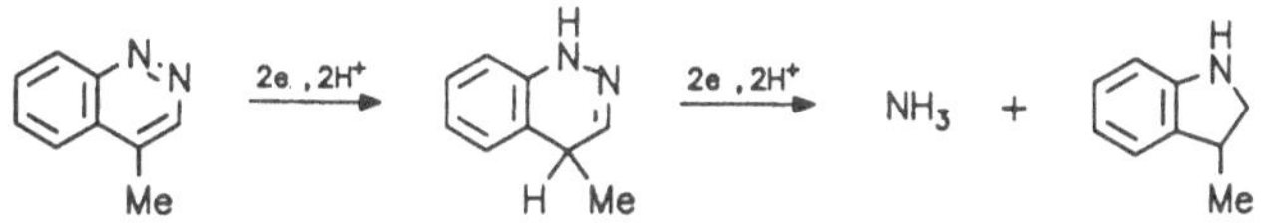

Compounds containing the pyrazine ring undergo hydrogenation of the pyrazine nitrogens [7] in a stepwise reduction process:

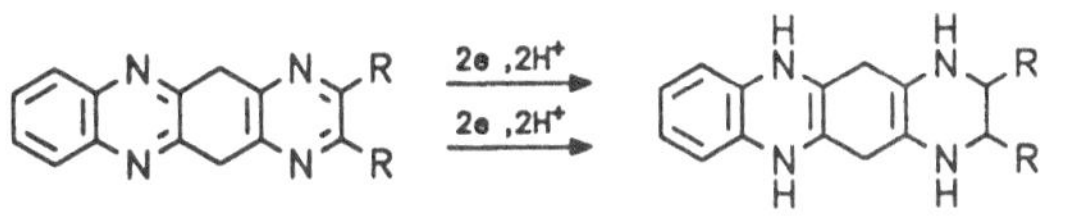

R = *p*-OCH₃Ph

Pyrimidines are reduced by a complex four-electron process to give contracted ring products [8] as shown in Scheme 3.17

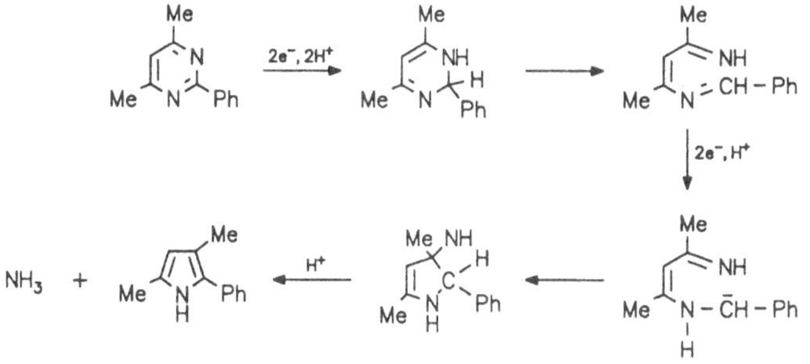

Scheme 3.17

Triazoles are converted to 1,2,4-triazines in two steps [9]: first, electro-reduction of (1) to obtain the expanded ring product (2), and second, chemical oxidation of (2) to obtain the triazino product (3), as shown in Scheme 3.18.

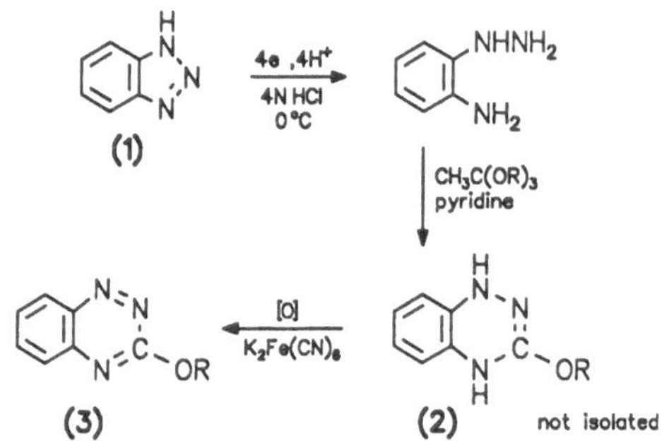

Scheme 3.18

The incorporation of the COOMe group to phenazine has been effected by cathodic reduction of phenazine in acetonitrile containing chlorofor-mate and tetraethylammonium perchlorate as supporting electrolyte [10], Scheme 3.19.

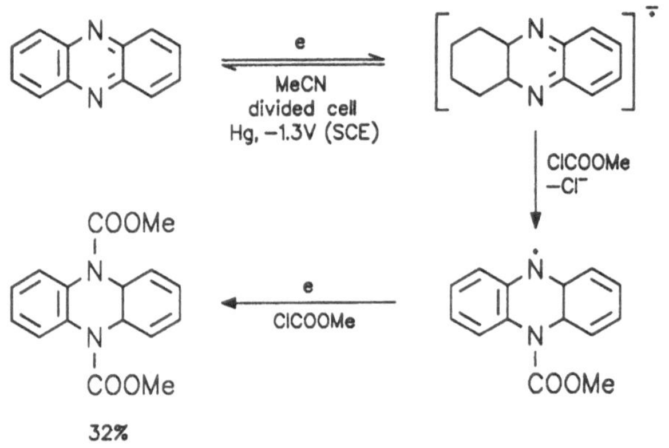

Scheme 3.19

Preparation of β-enaminoketones is accomplished by cathodic electrolysis of *N*-alkylisoxazolium tetrafluoroboranes in aqueous solutions at mercury electrodes [11]. Yields of 90–95% are obtained. The following plausible reaction mechanism was proposed, Scheme 3.20.

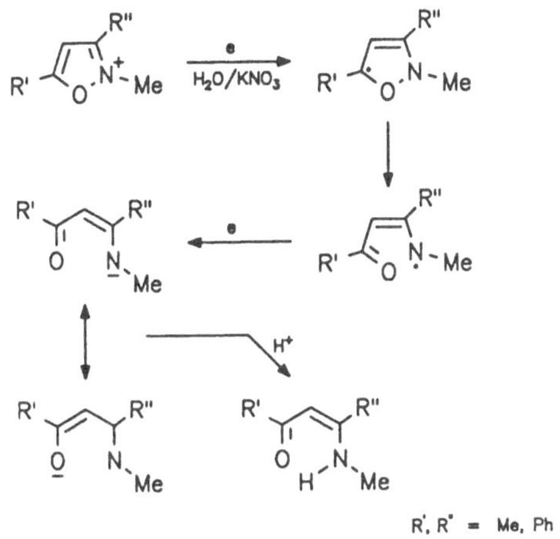

Scheme 3.20

3.7.2 Pyridinium and Similar Salts

Simple pyridinium salts are cathodically reduced by one- or two-electron processes, giving dimers or products from the cleavage of $\overset{+}{N}$—R bonds:

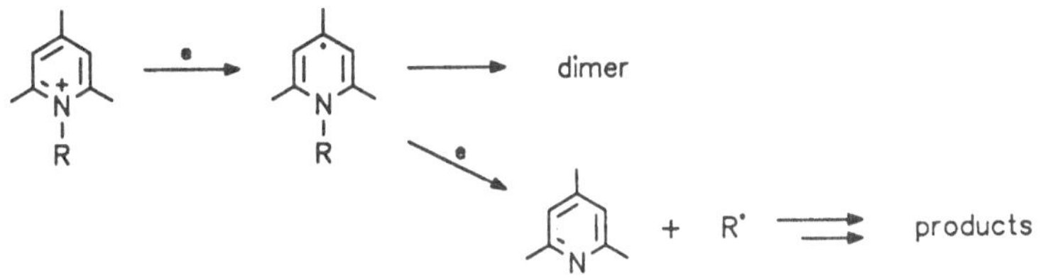

Dimers and cleavage products

Pragst and coworkers reduced a series of 1-substituted 2,4,6-trimethyl-phridinium ions and obtained the 3,10-diaza-tricyclo 5.3.1.1 dodeca-4,8-dienes [12].

Weber and Osteryoung [13] studied the reduction of 4-cyano-1-methyl-pyridinium ion in acetonitrile. Two reduction waves were observed. Dimerization products were obtained when controlled potential electrolysis was performed at potentials corresponding to either of the two polarographic waves. Thus the conclusion was drawn that reduction products were dependent on the time scale of the experimental techniques used. The process was explained as follows:

Polarography of pyridiniumion

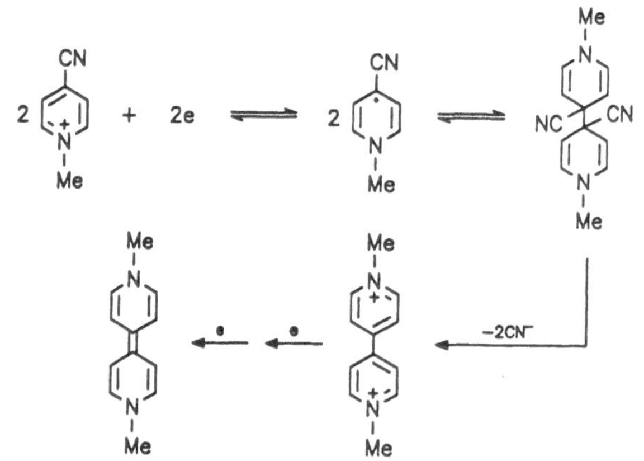

Stable radicals and carbanions

Heteroazolium salts with ring heteroatoms of O, Si and Se are reduced at mercury in dimethylformamide reversibly, affording stable radicals which at more negative potentials give their carbanions [14].

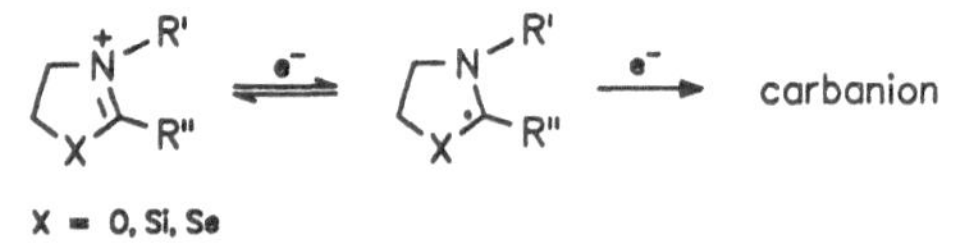

Electrochromic applications

The electroreduction of *n*-heptyl viologen (heptyl V) was studied with regard to electrochromic applications [15]. The one-electron transfer is rapid and reversible between the colorless and colored forms:

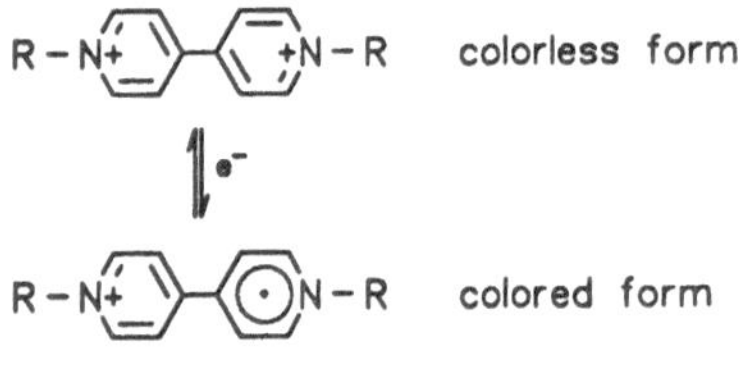

3.7.3 *N*-Heterocyclics with Carbonyl Groups

Phthalazinones, pyridazinones pyridones and pyrimidones undergo complex electroreductive transformations and reactions with various substances present in the electrolysis media. Phthalazinones can be reduced without reducing the carbonyl group under potentiostatic electrolysis conditions [16].

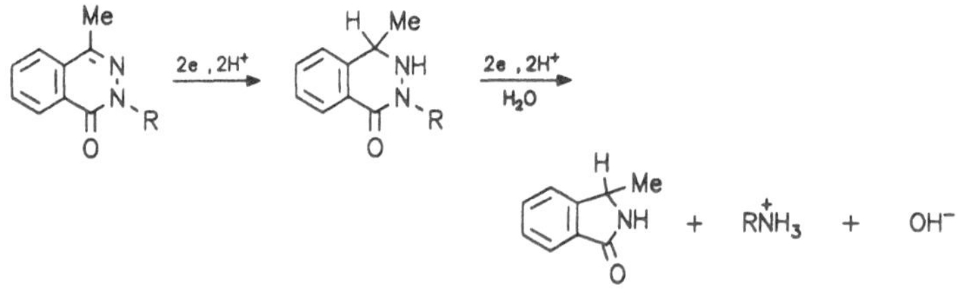

Phthalazinones

An efficient intramolecular C—N bond formation is the cathodic conversion of 1(2*H*)-phthalazinone, (**1**), to the phthalazino(2,3-b) pathalazine-5, 12(7*H*,14*H*)-dione [17] (**3**). This reduction could not be effected satisfactorily by chemical methods for the intended large scale production process. The electrochemical method was found to be very effective. It will be discussed in detail as a typical electroorganic synthesis example (for possible laboratory practice). The desired overall cyclization reaction is shown in Scheme 3.21.

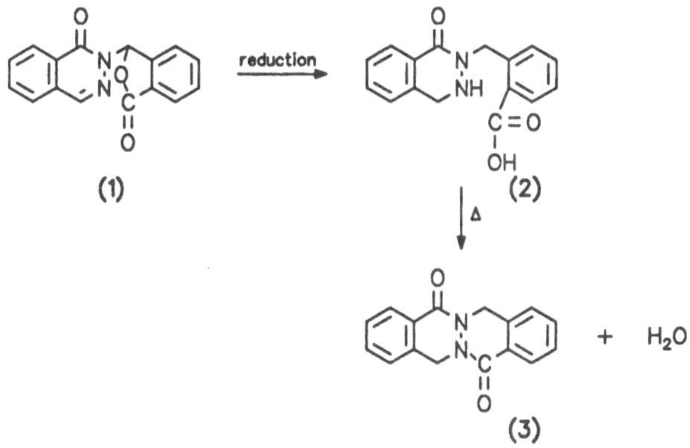

C—N bond formation

Scheme 3.21

a. *Working hypothesis*: It was assumed that if compound (**1**) could be reduced to give the intermediate comound (**2**) the desired dione compound (**3**) would be produced by elimination of water from the intermediate (**2**). It is reasonable to expect, therefore, that electrolytic reduction of (**1**) to (**2**) at 90–95 °C would give directly, in one step, the desired product (**3**). A voltammetric wave with a tin wire electrode showed that reduction took place at about −1.3 to −1.5 V versus the SCE, very close to the potential for hydrogen evolution. Because compound (**1**) is only slightly soluble in water a suitable aqueous-organic solvent system would be required for preparative electrolysis. After several trials a 1/1 mixture of water/sulfolane was found to be most suitable. Maximum yields were obtained by electrolysis in a divided cell at pH 5.8 to 6.2 and at temperatures of 90–95 °C.

b. *Electrolysis Cell. Procedure*: The cell is a 250-ml glass beaker at the center of which is positioned a porous ceramic cup to contain the anolyte, 20 ml of dilute hydrochloric acid, and the anode, a graphite rod, placed at the center of the cup. The catholyte is 50 ml of a 1/1 water/sulfolane mixture containing 5 g of the substrate (**1**), 4 g of sodium chloride and 4 g of sodium acetate trihydrate. A few drops of conc. hydrochloric acid is added to the catholyte to adjust the pH to 5.8–6.0. This pH value is periodically adjusted during the electrolysis by addition of a few drops

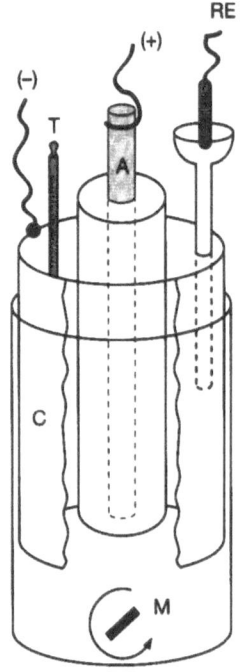

Fig. 3.2. Schematic of electrolysis cell
C, cylindrical tin foil cathode
A, graphite rod anode into ceramic cup anode compartment
RE, reference electrode with Luggin capillary
T, thermometer
M, magnet stirring bar

of hydrochloric acid. The reference electrode is an ordinary saturated calomel electrode (SCE). The cell is placed in a mineral oil bath thermostated at 90–95 °C, and the electrodes are connected to the potentiostat. The cell assembly is schematically shown in Fig. 3.2. The cathode potential is set at −1.35 V and a current of 1.5 A is passed for about 5 h, while the catholyte is magnetically stirred. The catholyte is then withdrawn and diluted with 500 ml of warm water (60 °C) whereupon copious crystallization of the product occurs. The solids are filtered, washed and dried in vacuo at 60 °C, giving 3.9 g of product (**3**) of 98% purity (82% yield). Identification of the product is made by spectroscopic and chromatographic analysis and comparison of data with those of an authentic sample.

Electroreductive ring contractions of *N*-heterocyclics with more than one nitrogen in the ring are useful reactions for the formation of five-membered rings. Hydrogenation of ring nitrogens and elimination of ammonia affords contracted ring products. Thus the indalozinones [18] are obtained from benzo-1,2,3-triazine-4(3*H*)-ones as shown in Scheme 3.22.

Contracted rings

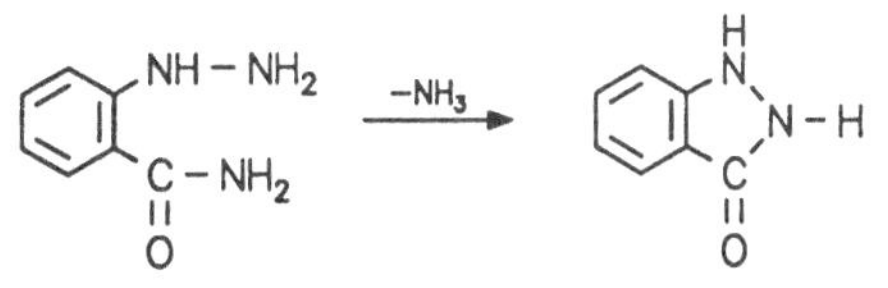

Scheme 3.22

3.8 Organosulfur Compounds

The reductive electrochemistry of organosulfur compounds has been, and continues to be, widely investigated [1]. Elemental sulfur itself is electro-reduced and may be used as such for certain synthetic applications [2]. Simonet prepared thiophenes by an interesting procedure using an intimate mixture of sulfur and carbon powder as cathode, and suggested that electrodically formed sulfur anion radicals attacked the acetylenic sub-strate $RC{\equiv}CR'$ so that the thiophene product was formed as indicated in the following scheme:

Thiophenes

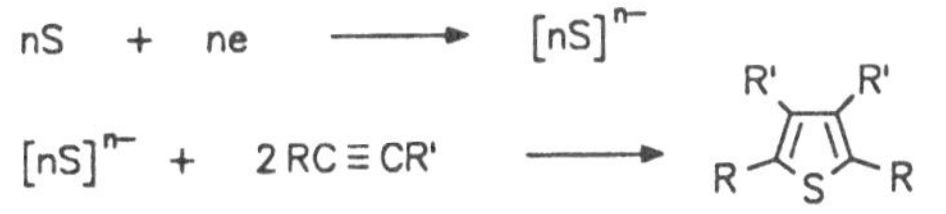

In general, reduction of single C—S bonds is very difficult, while double bonds are reduced relatively easily, as may be seen from polarographic observations [3]. (With mercury electrodes certain sulfur compounds readily form organomercury compounds.)

Organo-mercury compounds

Carbon disulfide is readily reduced at the cathode in aprotic solvents. The reduction product can enter into reactions with alkyl halides as indicated below [4]:

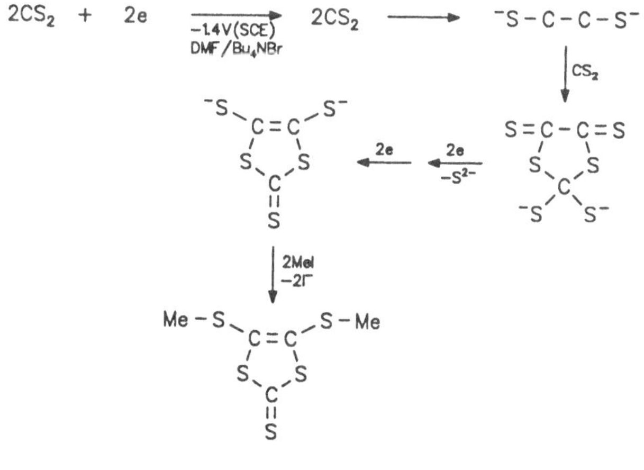

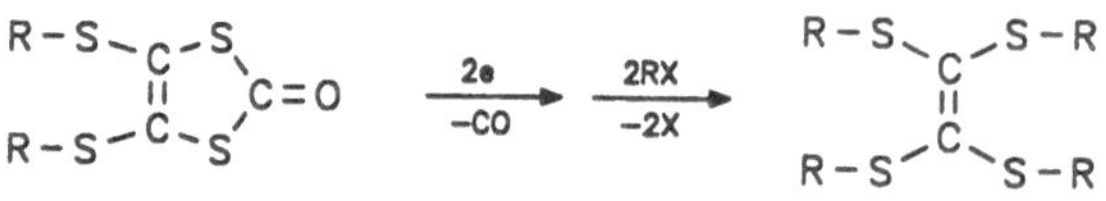

Typical organosulfur compounds that are directly reduced at the cathode include thiosulfonic acids and esters, sulfonyl chlorides, thiocarbonates, disulfides, sulfoxides and sulfones, thioamides and thiols [5,6].

Thioacetals and monothioacetals   Thioacetals are prepared by reduction of dithioesters in the presence of dimethyl sulfate in methanol solution, and monothioacetals by reduction of thioketones in the presence of alkylating agents as indicated below [8]:

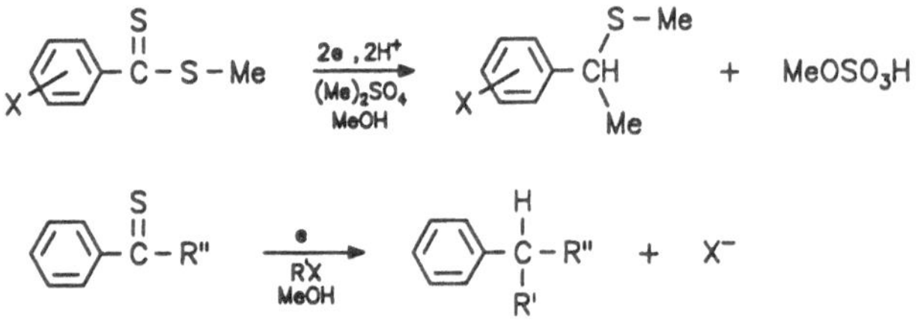

X = H, 3–Cl, 4–Cl, 2,3,4–OMe
R' = Me, Et etc; R" = t–Bu, Ph, Me

Tetrathioethylenes and thioanisole   Electroreductions of trithiocarbonates and related compounds in the presence of alkyl halides yield mixtures of E- and Z-tetrathioethylenes ($E/Z \approx 10/1$) and thioanisole [9].

$$2(ArS)_2C=S \quad + \quad 4RX \quad + \quad 4e \quad \longrightarrow$$

$$ArS(RS)C=C(SR)SAr \quad + \quad 2ArSR \quad + \quad 4X^-$$

R = Me

Sulfinic acids   Electroreduction of disulfides, RSSR, in the presence of oxygen in organic solutions affords sulfinic acids [10,11] via reaction of the produced RS$\cdot$ radicals with ion O_2^-, Scheme 3.23.

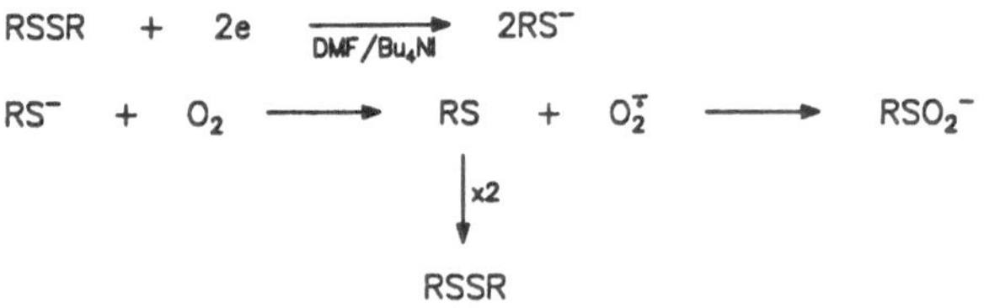

Scheme 3.23

The electrogenerated RS$^-$ anion acts as a one-electron transfer agent to generate the superoxide ion which reacts with the RS$\cdot$ radical to give the sulfinic acid anion, RSO_2^-.

Coupled products   Coupled products are obtained by reduction of sulfenyl or sulfonyl compounds via the ArS$\cdot$ radicals [12]:

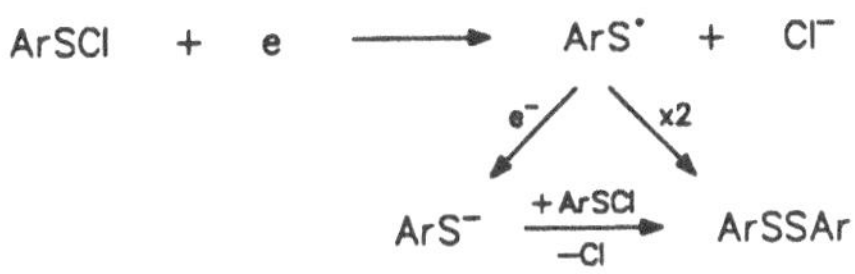

$$\text{ArSCl} \quad + \quad e \quad \longrightarrow \quad \text{ArS}^{\bullet} \quad + \quad \text{Cl}^-$$

$$\text{ArS}^- \xrightarrow[-\text{Cl}]{+\text{ArSCl}} \text{ArSSAr}$$

Sulfenyl halogen bonds are more easily reduced (cleaved) than sulfonyl or S—S bonds. Electrochemical alkylation of dithiolethiones is easier than the chemical alkylation. The following is an example [13]:

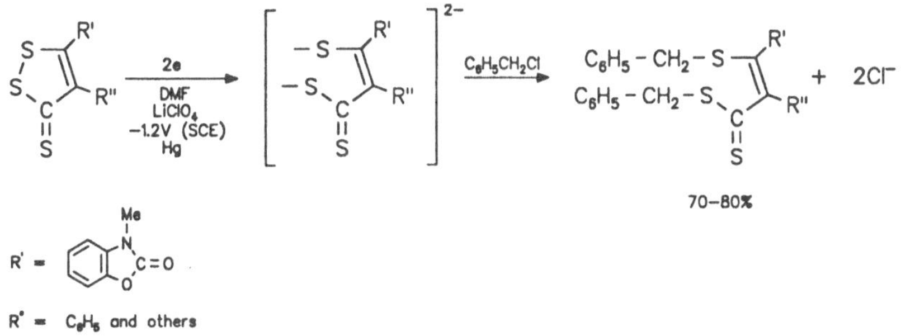

Alkylation

R' = [structure with Me, N, C=O]

R' = C_6H_5 and others

These alkylations may be effected in situ or ex situ depending on the relative reduction potentials of the alkylating agent and the sulfur compound [14]. Thiols undergo cleavage of the SH bond to give thiolate anions, RS^-, and mixed coupled products [15]:

Thiolate anions

$$\text{RSH} \xrightarrow[\text{Ni, Fe}]{\bullet} \text{RS}^- + \text{H}^{\bullet}$$

$$\text{RS}^- + \text{PhCH}_2\text{Cl} \longrightarrow \text{RSCH}_2\text{Ph} + \text{Cl}^-$$

Splittings of S—H bonds occur on transition metal electrodes. Activated thiols such as $EtOCOCH_2SH$ (pK 7.8) yield hydrogen at the cathode most likely by direct reduction of dissociated protons and can do so on other than transition metal electrodes [15].

Splitting of S—H bonds

Sulfoxides and sulfones are reduced to sulfides. Sulfones may be reduced to sulfinic acids and mercaptans [5,16]. Simonet reported an "unexpected" dimerization of phenyl vinyl sulfone in slightly protic media [16]:

Sulfides, sulfinic acids, and mercaptans

$$3 \text{ PhSO}_2\text{CH} = \text{CH}_2 \xrightarrow{\text{reduction}} \text{PhSO}_2\text{CH}_2\text{CH}_3$$

and $\text{PhSO}_2\text{CH}_2\text{CH} = \text{CHCH}_2\text{SO}_2\text{Ph}$

dimer

Dimethyl sulfoxide is reduced at the cathode by cleavage of the C—S bond, indicating the semipolar nature of the S—O bond [17]:

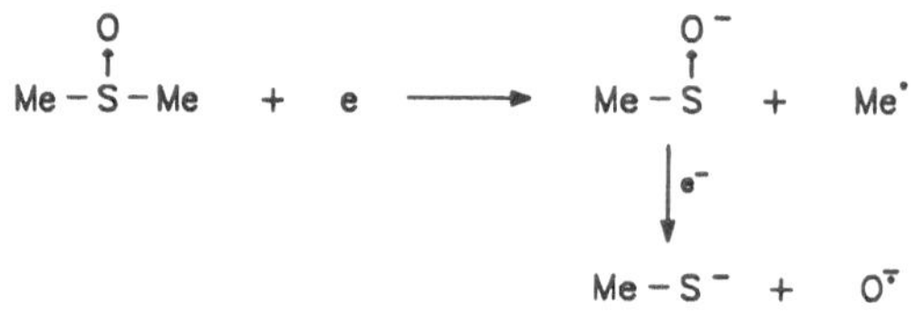

$$Me-\overset{\overset{O}{\uparrow}}{S}-Me \quad + \quad e \quad \longrightarrow \quad Me-\overset{\overset{O^-}{\uparrow}}{S} \quad + \quad Me^{\cdot}$$

$$\downarrow e^-$$

$$Me-S^- \quad + \quad O^{\bar{\cdot}}$$

**Composite
C—S
electrode**
Recently an interesting composite carbon-sulfur electrode was invented by Simonet and coworkers [18]. This electrode acts as a source of polysulfides with nucleophilic properties. Upon cathodic polarization the sulfur dissolves and chemical reactions can take place in the catholyte compartment in an electrolysis cell, as illustrated below:

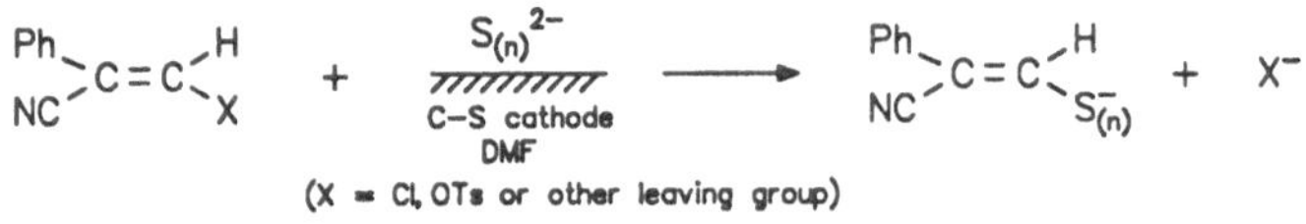

$$\overset{Ph}{\underset{NC}{>}}C=C\overset{H}{\underset{X}{<}} \quad + \quad \overset{S_{(n)}^{2-}}{\underset{\underset{DMF}{C-S\ cathode}}{\overline{/////////}}} \quad \longrightarrow \quad \overset{Ph}{\underset{NC}{>}}C=C\overset{H}{\underset{S_{(n)}^-}{<}} \quad + \quad X^-$$

$$(X = Cl, OTs \text{ or other leaving group})$$

3.9 Some Special Cleavages of C—O Bonds in Hydroxy Compounds

There is a distinct interest in the cleavage of hydroxy groups in organic substances [1]. Only a few C—OH bonds can be transformed to C—H bonds by direct electrohydrogenolysis. Alcohol derivatives provide an indirect way for converting C—OH bonds to C—H bonds, if the hydroxy group can be replaced by a good leaving group prior to electrolysis, such as a bromo or iodo group:

$$ROH \quad \longrightarrow \quad RBr \quad \overset{2e,H^+}{\longrightarrow} \quad RH \quad + \quad Br^-$$

In the reduction of phenols the intermediate diethylphosphates, and in certain cases the mesylates, may be used as leaving groups [1b,d]:

$$ArOH \quad \longrightarrow \quad ArOPO(OC_2H_5)_2 \quad \overset{2e,2H^+}{\longrightarrow} \quad ArH \quad + \quad HOP(OC_2H_5)_2$$

$$ROH \quad \longrightarrow \quad R(OSO_2)CH_3 \quad mesylate$$

$$\downarrow 2e,2H^+$$

$$RH \quad + \quad HSO_3CH_3$$

**Tosylates
mesylates**
In this connection we note that tosylates are electroreduced by cleavage of their O—S bonds (hence they are used to protect hydroxy groups). In contrast, mesylates split their C—O bonds preferentially, giving the hydrocarbon products [1c]. Propylene carbonate and other diol carbonate derivatives are reduced at the dcathode only with great difficulty [2] (very negative potentials are required). Cyclic sulfate derivatives of diols yield

the corresponding alkenes along with other products, the alkenes being produced by a one-electron process as indicated below [3,4]:

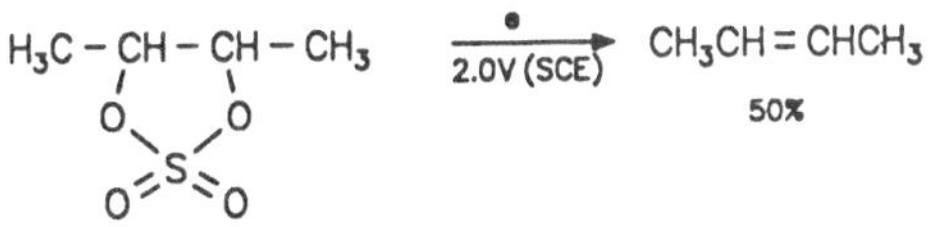

Alkenes

The electrochemical reduction (deoxygenations) of hydroxy compounds via esterification as indicated above is effective but it requires very negative potentials. The reduction of esters of oxalic, squaric and oxamic acids is an easier operation for practical purposes. Utley and coworkers [5] studied such cathodic cleavages. For benzylic esters a one-electron reduction process was indicated as follows:

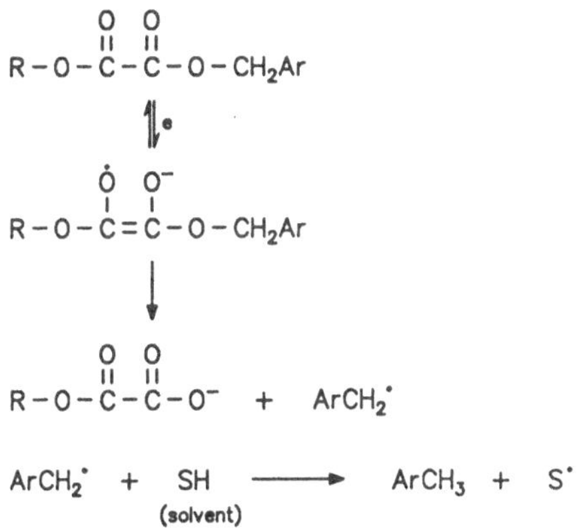

Benzylic esters

An example of the electroreduction of oxalate esters is given in Scheme 3.24. The oxalate esters are formed in situ by transesterification with diethyl oxalate.

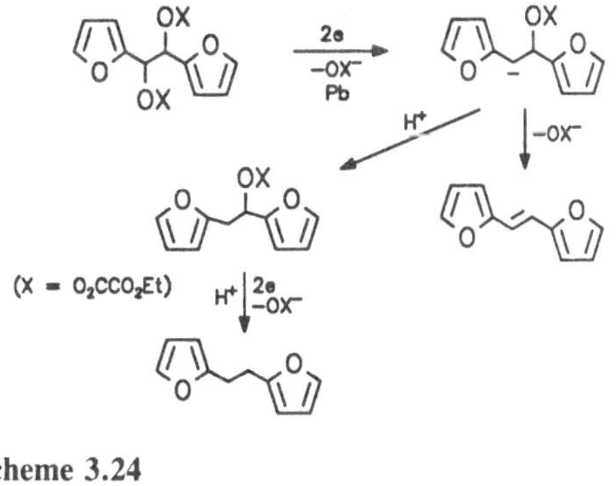

Oxalate esters

Scheme 3.24

Furfuryl alcohols and hydroxyfuroin are reduced as expected. Butane-2,3-diol must be converted to the sulfonate analog in order to deoxygenate. Such deoxygenations could possibly be applied to lignocellulosics

Organic feedstocks to provide organic feedstocks [5]. We might note that electroreductive cleavages of substituted 9,10-anthraquinone esters have been of interest in biochemical research studies [6,7].

3.9.1 Cleavage of Ethers

Diphenyl ethers, phenoxynaphthalenes and dinaphthyl ethers can be cleaved at very negative potentials in a one-electron process [8].

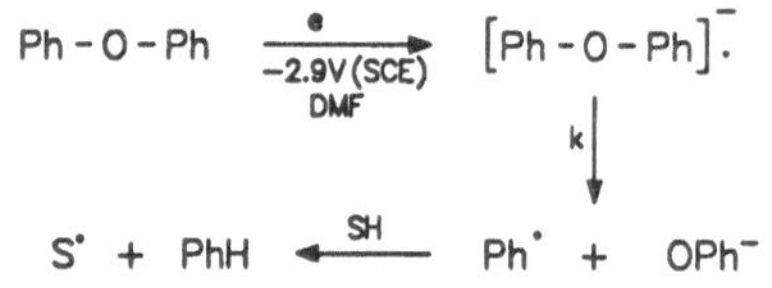

$$Ph-O-Ph \xrightarrow[\substack{-2.9V(SCE) \\ DMF}]{e} [Ph-O-Ph]^{\overline{\cdot}}$$

$$S^{\cdot} + PhH \xleftarrow{SH} Ph^{\cdot} + OPh^{-}$$

Picolyl group Radical anions of 1- and 2-phenoxynaphthalines survive longer than one minute in very dry dimethylformamide. A rare type of cathodically cleaved ether bond is one with the picolyl group [9]. The picolyl group is a protective group for alcohols:

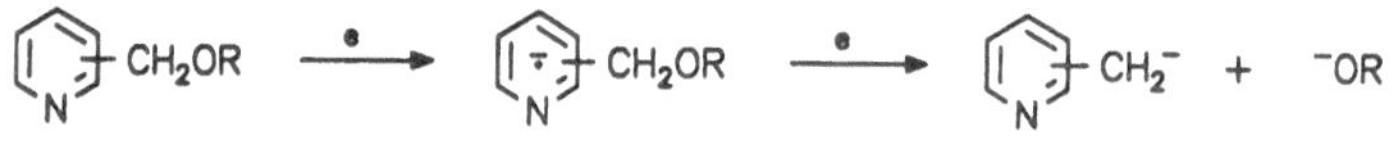

3.10 Electrocarboxylations

3.10.1 Electroreduction of Carbon Dioxide

Formic acid, formaldehyde and methanol

Carbon dioxide can be electrochemically reduced to give hydrogen containing products. The electrochemical conversion of CO_2 to formic acid, formaldehyde, and methanol is a very active research field [1]. As a biomass derived material it is receiving very special attention [1c]. Electro-

Formate ion reduction of CO_2 to formate ion takes place via adsorbed species [2]

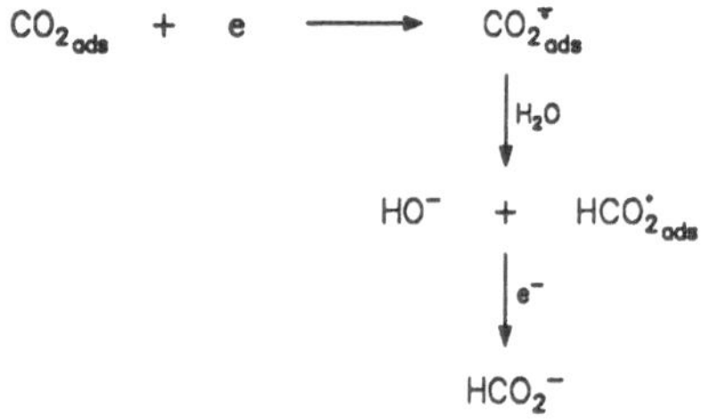

$$CO_{2\,ads} + e \longrightarrow CO_{2\,ads}^{\overline{\cdot}}$$

$$\downarrow H_2O$$

$$HO^- + HCO_{2\,ads}^{\cdot}$$

$$\downarrow e^-$$

$$HCO_2^-$$

Reduction to formic acid can be almost complete, but formic acid cannot be practically reduced to methanol. The reduction of formaldehyde to methanol proceeds via the dehydrated molecule [3]:

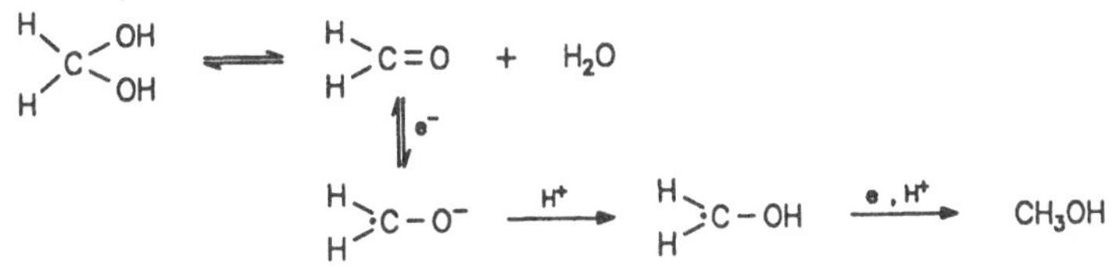

$$H\underset{H}{\overset{\diagdown}{C}}\overset{OH}{\underset{OH}{\diagup}} \rightleftharpoons \underset{H}{\overset{H}{\diagdown}}C=O \ + \ H_2O$$

$$\downarrow e^-$$

$$\underset{H}{\overset{H}{\diagdown}}C-O^- \xrightarrow{\ H^+\ } \underset{H}{\overset{H}{\diagdown}}C-OH \xrightarrow{\ e,H^+\ } CH_3OH$$

It has been reported that at ruthenium electrodes the reduction of CO_2 may go all the way to methane [4]

Methane

$$CO_2 \ + \ 8e \ + \ 8H^+ \xrightarrow[0.1M\ H_2SO_4]{} CH_4 \ + \ 2H_2O$$

Bard [5] and Sammels [6] studied the reduction of carbon dioxide at copper electrodes and obtained high faradaic efficiencies. Gas diffusion electrodes were recommended for high reduction rates. Metal impregnated polytetrafluoreothylene-bonded gas diffusion electrodes were used in aqueous solutions [7] (pH 1–5). The catalytic activity of these electrodes for the reduction of carbon dioxide to formic acid was in the order Pb > In > Sn. A complete process to produce oxalic acid on a large scale was developed using a consumable zinc anode [8]. Zinc oxalate was formed and this was treated with sulfuric acid to yield oxalic acid and zinc sulfate.

Oxalic acid

At low temperatures CO_2 has been converted to methane and ethane at Al, Co, Mn and In electrodes [9]. The gas phase reduction of CO_2 to hydrocarbons was studied at metal/SPE interfaces using Nafion 117 and 417 membranes as the SPE, the polymer cells possessing the configuration,

Methane and ethane

$$CO_2/Cu/H^+/ \underset{membrane}{Nafion} / \underset{electrolyte}{aqueous} / Pt,H_2$$

The metal, Cu, was deposited onto the SPE by electroless deposition using as reductants $NaBH_4/NaOH$ or $N_2H_4/NaOH$. Many metals were thus tested as electrocatalysts for the conversion of CO_2 to hydrocarbons [10]. It appears that large scale applications need more research and development studies. SPE methodology may enable gas-phase electrolysis to overcome the inherent mass-transfer limitations imposed by low solubilities of gaseous reactants.

3.10.2 CO_2 as Electrophile

As an electrophile CO_2 combines with organic anion radicals and anions. Numerous such carboxylations have been reported [11]. Acetylenes, olefins, ketones, alkyl halides and *N*-heterocyclics are good substrates for electrocarboxylations. These substrates are usually reduced at the cathode before carbon dioxide. The generated organic anion radical is attacked by the electrophilic CO_2 as is indicated in the following example [11a]:

Electrocarboxylations

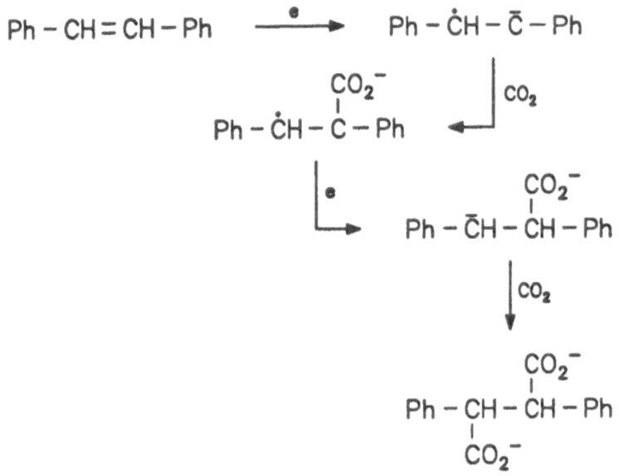

$$Ph-CH=CH-Ph \xrightarrow{\;e\;} Ph-\overset{\cdot}{C}H-\overset{-}{C}-Ph$$

$$Ph-\overset{\cdot}{C}H-\underset{|}{\overset{\overset{\displaystyle CO_2^-}{|}}{C}}-Ph \xleftarrow{\;CO_2\;}$$

$$\downarrow e$$

$$Ph-\overset{-}{C}H-\underset{|}{\overset{\overset{\displaystyle CO_2^-}{|}}{C}}H-Ph$$

$$\downarrow CO_2$$

$$Ph-\underset{\underset{\displaystyle CO_2^-}{|}}{\overset{\overset{\displaystyle CO_2^-}{|}}{C}}H-CH-Ph$$

Carboxyla-
tions benzylic
allylic
structures

Organic halides such as those of benzylic and allylic structures are electrocarboxylated more efficiently than other organic halides [12]. A number of organohalides were carboxylated in a V-shaped electrolyzer equipped with a consumable magnesium rod anode surrounded by a stainless grid cylinder as cathode [13]. Bromobenzene is carboxylated to benzoic acid at mercury cathodes apparently via the intermediate phenylmercury radical [14].

Azomethines, azo compounds and N-heterocyclics are readily electrocarboxylated [15]. Perichon and coworkers electrocarboxylated benzyl chloride, 3-bromofuran and 1-bromodecane in aprotic solvents in undivided cells and obtained the corresponding carboxylic acids [16]. Advantage was taken of sacrificial magnesium anodes. The cell reactions were described as follows:

$$Mg \longrightarrow Mg^{2+} + 2e \qquad\qquad anode$$
$$RX + 2e + CO_2 \longrightarrow RCO_2^- + X^- \qquad\qquad cathode$$
$$Mg + RX + CO_2 \longrightarrow RCO_2^- + Mg^{2+} + X^- \qquad overall$$

Electrocarboxylations of cyclic ketene-S-S-acetals, substituted vinyl ketones [17] and benzophenones [18] have been carried out efficiently:

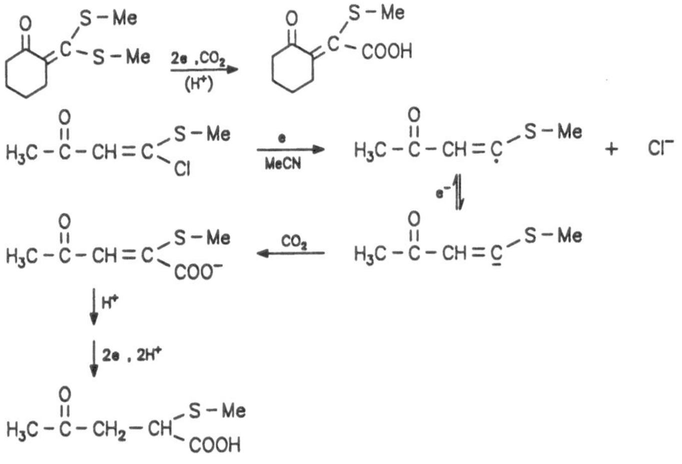

It is possible sometimes to effect carboxylations by electrogenerated $CO_2^{\bar{\cdot}}$ radical anions if the organic substrate is reduced at more negative potential than the CO_2. Butadiene and ethylene are thus carboxylated [19]:

$$H_2C=CHCH=CH_2 \ + \ 2CO_2^{\bar{\cdot}} \ \longrightarrow \ {}^-OOC-CH_2CH=CHCH_2COO^-$$

$$H_2C=CH_2 \ + \ 2CO_2^{\bar{\cdot}} \ \longrightarrow \ {}^-OOC-CH_2CH_2-COO^-$$

We also refer here to the possibilities of electrosynthesis utilizing CO, inasmuch as CO is a reduced product of CO_2. Electroreduction of CO in DMF solution affords the interesting structurally squaric acid and in MeOH solution the formate [20]:

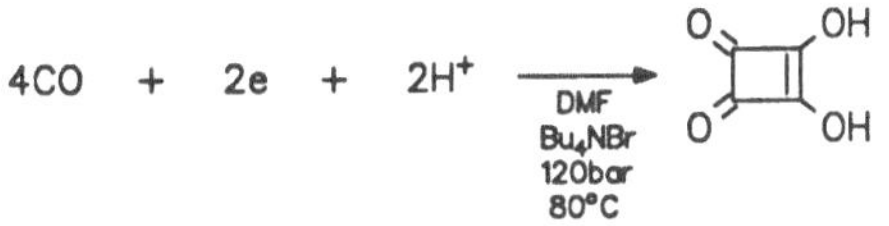

$$4CO \ + \ 2e \ + \ 2H^+ \ \xrightarrow[\substack{DMF \\ Bu_4NBr \\ 120bar \\ 80°C}]{}$$

$$CH_3OH \ + \ e \ \longrightarrow \ CH_3O^- \ + \ 1/2\,H_2$$

$$\xrightarrow[\,CO\,]{} \xrightarrow[\,MeOH\,]{} \ CH_3O^- \ + \ HCOOMe$$

Electroreduction using CO and alcohols afford dialkyl carbonates if NH_4Br or $LiBr$ are the electrolytes, while with Bu_4NBr as electrolyte formates are produced as major products [21].

3.10.3 Photoelectrochemical Reduction of Carbon Dioxide

The process of photosynthesis, on which all life one earth depends, is formally expressed in textbooks most simply as:

Photosynthesis

$$nCO_2 \ + \ nH_2O \ \xrightarrow[\substack{chlorophyl \\ enzymes}]{sunlight} \ n(CH_2O) \ + \ nO_2\uparrow$$

This expression denotes that water reduces carbon dioxide under the assistance of light-energy; and via the mediation of chlorophyl and enzymes the carbohydrates are produced in the cells of the leaves of plants [22]. In this very complex natural process the first fundamental event is now believed to be in effect a light-assisted indirect, chlorophyll-mediated electron transfer from water to carbon dioxide. A very complex sequence of enzyme-catalyzed chemical reactions leads to the formation of the carbohydrates (despite the fact that the photosynthesis as a reaction is thermodynamically not possible (not spontaneous).

Electrochemically viewed, the reduction of CO_2 by H_2O cannot be spontaneous, as is apparent by considering the relevant potentials for the half-cell reactions and the overall cell reaction:

$$CO_2 + 2H^+ + 2e \longrightarrow HCO_2H \qquad E^\circ = -0.61V$$

$$H_2O \longrightarrow 1/2O_2 + 2H^+ + 2e \qquad E^\circ = -0.82V$$

$$E^\circ_{cell} = -1.43V$$
$$\Delta G^\circ = 5.72eV, \quad \Delta G^\circ > 0$$

The electrocatalytic or variously mediated reduction of CO_2 provides a stimulus for interesting electrochemical research. One area of such research is concerned with solar cells. In a recent solar cell, CO_2 is reduced to methanol by oxidizing leucoindigo (LIn) to indigo (In) in a six-electron overall cell reaction [23]:

Solar cells

$$CO_2 + 3LIn \xrightleftharpoons{\text{sunlight}} CH_3OH + 3In + H_2O \quad (6e)$$

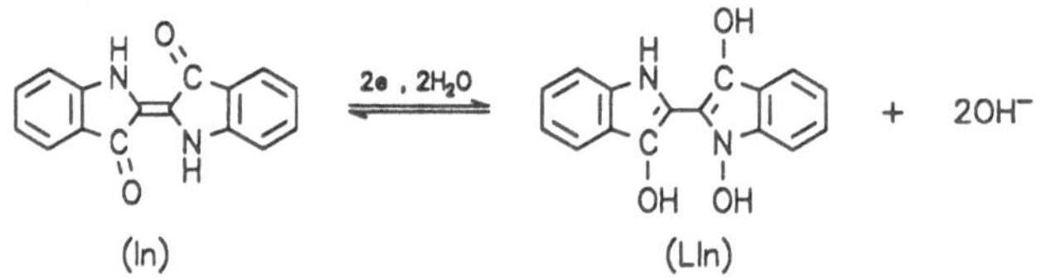

Artifical photosynthesis

Artificial photosynthetic processes are actively studied for the purpose of converting solar energy to chemical energy. The light-assisted electron transfer in such processes would occur across a membrane in a manner similar to natural photosynthesis. Recently Khaled and coworkers studied the electrooxidation of carotenoids in CH_2Cl_2 medium. Carotenoids may play an active role in photosynthesis via the carotenoid radical cation. The oxidation was found to involve two electrons with β-carotene as the substrate [24].

3.11 Nitriles and Oximes

In acidic solutions nitriles and oximes are reduced to their corresponding amines. In basic solutions cleavage of the C—CN bond occurs [1]:

$$RCN \xrightarrow[\text{(acid)}]{4e,\ 4H^+} RCH_2NH_2$$

$$RCN \xrightarrow[H_2O]{2e\ \text{(base)}} RH + CN^- + OH^-$$

$$R'R''C{=}NOH \xrightarrow{H^+} [R'R''C{=}NOH]H^+$$

$$[R'R''C{=}NOH]H^+ \xrightarrow{2e,\ 2H^+} R'R''C{=}NH_2^+$$

$$R'R''C{=}NH_2^+ \xrightarrow{2e,\ 2H^+} R'R''CHNH_3^+$$

Udupa studied the reduction of lawronitrile and stearonitrile on a catalytic nickel black cathode [2]. The reduction to amines was most efficient, probably as a result of the high surface area provided by the deposited nickel on the graphite substrate plate. Tallec and coworkers [3] studied the electroreductive cyclization of methyl benzhydroxyiminoformy(S) prolinate and obtained the two epimers RS and SS of pyrrolo [1,2a] 3-phenyl 1,4-diketopierazine with the SS epimer predominating (optical yield 34%)

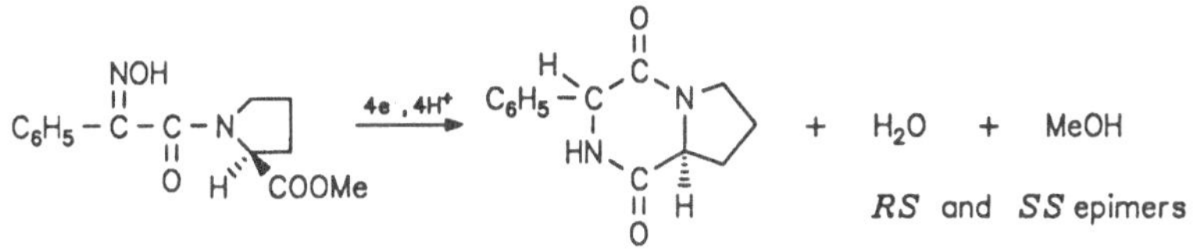

3.12 Various Cathodic Couplings

3.12.1 Acylations

Electrochemical acylations take place in aprotic media containing organic substrate and acetic anhydride as acylating agent. First, the organic substrate is reduced at the cathode to produce an anion radical which is attacked by acetic anhydride as exemplified below [1]:

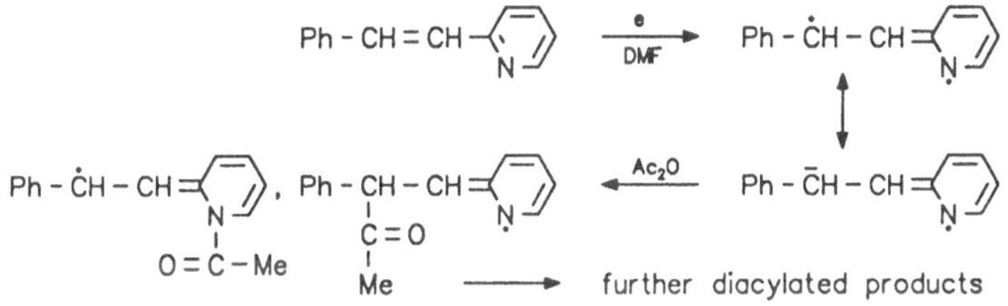

Benzophenone is thus acylated by reduction at a mercury cathode in MeCN/Et$_4$NBr/Ac$_2$O solution to give the ketoacetate analog in 66% yield [2]. Aryl olefins have been acylated in aprotic solutions via their anion radicals or the dianions in the presence of an acid anhydride or an acylating electrophilic solvent [3]. Methyl methacrylate [4] and the carotenoid canthaxantin [5] were thus acylated. The acylation of 1,2-acenaphthenedione gave two products, depending on the acylating agent [6], Scheme 3.25

Aryl olefins acylation

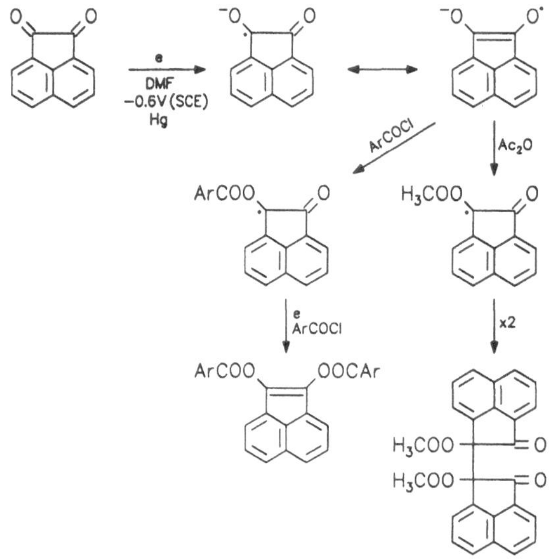

Scheme 3.25

3.12.2 Organophosphorus Compounds

Phosplines Cathodic couplings with trivalent organophosphorus compounds are efficient reactions [7]. Halodiphenylphosphines may be cleaved by reduction at platinum or monel electrodes in dry acetonitrile (treated with $P_2O_5/$ in a divided cell to give exclusively the tetraphenylphosphine compound, as shown in Scheme 3.26:

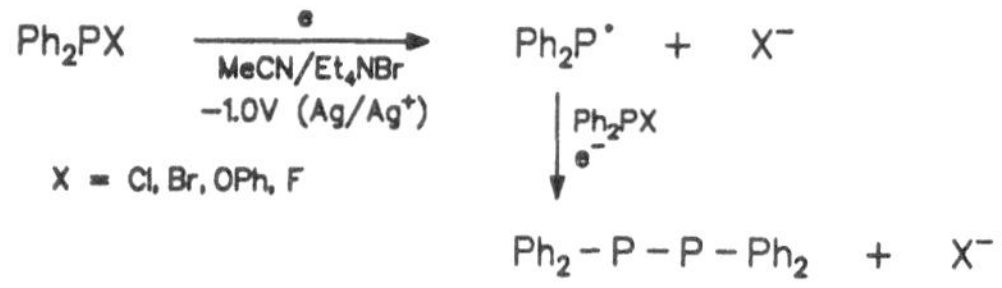

Scheme 3.26

Bicycloalkanes Electroreduction of certain olefinic enol phosphates in DMF at a mercury
Trichloro- cathode gives, by double cyclization, bicycloalkanes [8] in 50–60% yields.
phosphonates Cathodic reduction of trichlorophosphonates leads to the synthesis of their dichloro alkylated analogs [9], Scheme 3.27:

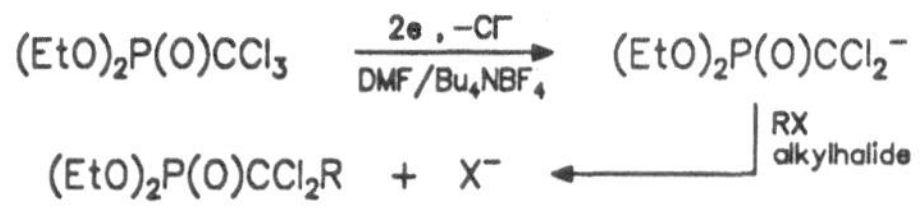

Scheme 3.27

For this synthesis the alkyl halide, RX, should have a cathodic potential greater than $-1.2\,V$ vs Ag/AgI so that only the phosponate compound is reduced.

Alkylations of halodiphenylphosphines are effected in undivided cells fitted with magnesium (sacrificial) anodes and stainless steel cathodes, in aprotic solvents. Perichon and coworkers developed such methods for the synthesis of many phosphines from organic halides and chlorophosphines:

$$Ph_2PCl \xrightarrow[DMF/Bu_4NBF_4]{2e} Ph_2P^- + Cl^-$$

$$Ph_2P^- + RX \longrightarrow Ph_2PR + X^-$$
$$RX = PhCH_2Cl, CH_3(CH_2)_3Br \text{ etc.} \qquad 30-90\%$$

Electroreductive couplings via formation of P—C and S—C bonds have been studied in CH_2Cl_2 media. They have been synoptically shown as follows:[11]

$$R-W-H \xrightarrow{e} R-W^- + H^{\cdot}$$
$$\downarrow R'X$$
$$RX = alkylhalide$$
$$W = P, S \text{ or } O \qquad R-W-R' + X^-$$

* Anmerkung: Die Gleichung von Kapitel 3.12.3, Seite 95, befindet sich in der Datei Kyriac53 *

3.12.3 Arene-Metal Complexes

Arenes are activated towards hydrodimerization by forming chromium tricarbonyl complexes [12]. Stilbene forms such a complex and hydrodimerizes at less negative potential than it does in the uncomplexed form [12]:

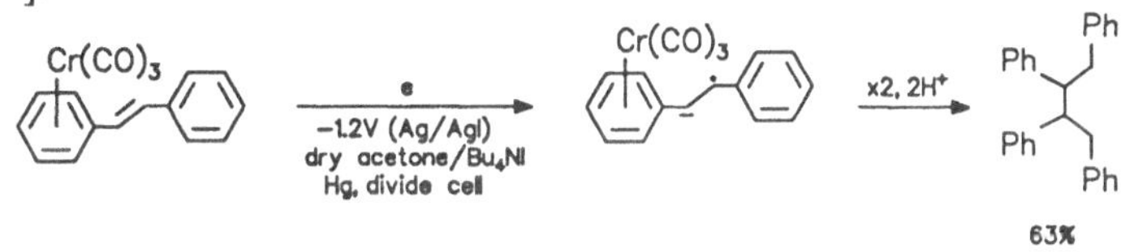

In contrast to stilbene, phenanthrene prefers to hydrogenate. Probably, complexation with the ring electrons stabilizes the anion radical and this tends to favor hydrodimer formation. In this connection Saveant's observation that traces of water may speed the dimerization of certain olefins comes to mind. The effect of water was ascribed to a possible complex intermediate, as is formally shown below, which by stabilizing the radical anion speeds up the dimerization reaction [13]:

$$A + e \longrightarrow A^{\overline{\cdot}} \xrightarrow{H_2O} A(H_2O)^{\overline{\cdot}} \text{ complexe}$$
$$\downarrow$$
$$dimer$$

3.12.4 Cathodic Reactions (Couplings) of Organometalics

Electrochemical couplings of organometallic compounds are known reactions [14]. For example, triphenyltin chloride couples with alkyl halides at a mercury cathode. Adsorbed species are involved in the overall coupling reaction [14a]:

$$Ph_3SnX \underset{Hg}{\overset{e}{\rightleftharpoons}} Ph_3SnHg^{\bullet}_{ads} + X^-$$
$$\longrightarrow (Ph_3Sn)_2Hg$$

$$RX \xrightarrow{e} R^{\bullet}_{ads} + X^-$$
$$R^{\bullet} + Ph_3SnHg^{\bullet} \longrightarrow Ph_3SnHgR$$

Carbon-silicon bonds can be formed by electroreductive coupling of suitable organic compounds with silylating agents such as $HSiCl_3$ [15].

Tomilov studied the synthesis of 3-(tricholorosilyl)propionitrile and proposed an ionic mechanism [16]:

$$Cl_3SiH \xrightarrow[\substack{MeCN \\ graphite}]{e} Cl_3Si^- + H^{\bullet}$$

$$Cl_3Si^- + H_2C = CHCN \xrightarrow{H^+} Cl_3SiCH_2CH_2CN$$

Organosilicon compounds   Yoshida [17] and coworkers reported the synthesis of organosilicon compounds (aryl, allyl, vinyl halides). The corresponding silicon compounds were obtained in 50–70% yields by electrolysis in DMF solutions containing silylating agents as shown below:

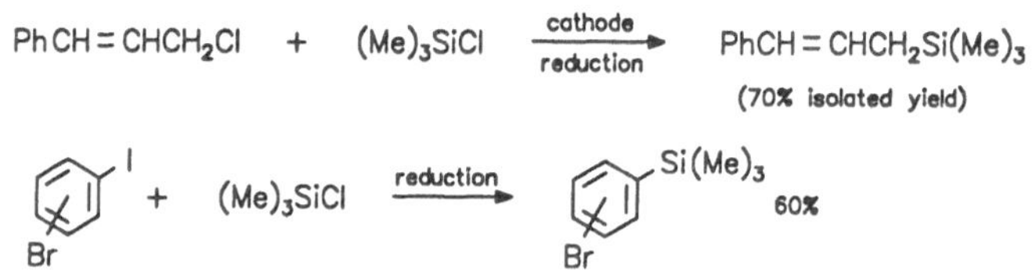

$$PhCH = CHCH_2Cl + (Me)_3SiCl \xrightarrow[reduction]{cathode} PhCH = CHCH_2Si(Me)_3$$
$$(70\% \text{ isolated yield})$$

These electrolyses were performed in $DMF/0.2MEt_4NOTS$ solutions containing the organic halide and the silylating agent, in an H-type cell fitted with platinum electrodes, and at currents of $10\,mA/cm^2$. The electrolyzed solution was mixed with ice and the products were recovered by extraction with hexane or ether, and were isolated by flash chromatography or by distillation.

C—H bond cleavages   Bukhtiarov and coworkers demonstrated that in aprotic electrolysis media and at transition metal electrodes, certain unsaturated hydrocarbons are reduced by cleavage of C—H bonds [18]. Hydrogen and organic radical anions are thus generated at the cathode. Cyclopentadiene, indene and fluorene were reduced at Pt, Ni, Co, Fe, Cr, V and Cu electrodes in $MeCN/Et_4NCl$ solutions. After electrolysis the catholyte was treated with chlorotrimethylsilane, and the trimethylsilyl derivatives were produced in good yields:

$$C_5H_6 \xrightarrow{\ e\ } C_5H_5^- + H^\cdot$$
cyclopentadiene

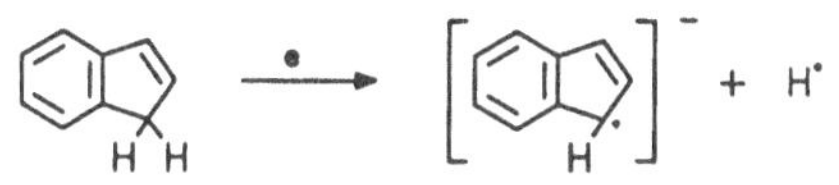

$$C_5H_5^- + (Me)_3SiCl \xrightarrow{-Cl^-} C_5H_5Si(Me)_3$$

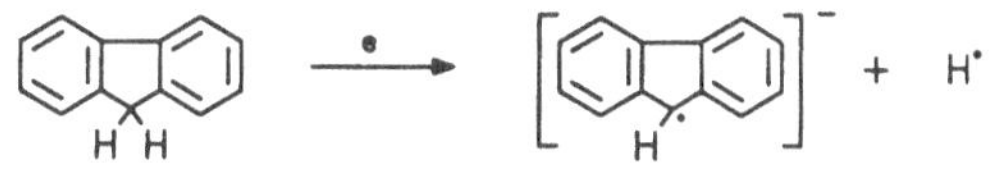

If the electrolyzer is equipped with a magnesium anode organometallic compounds are formed which upon treatment with alkyl halides or chlorosilanes are converted to the corresponding derivatives:

**Organo-
magnesium
compounds**

$$Mg \xrightarrow{-2e} Mg^{2+}$$
$$Mg^{2+} + 2C_5H_5^- \longrightarrow Mg(C_5H_5)_2$$
$$Mg(C_5H_5)_2 + 2RX \longrightarrow 2RC_5H_5 + MgX_2$$

RX = organohalide
R = Et, Bu, PhCH$_2$
X = Cl, Br, I

Such cathodic reactions are also possible with compounds containing N—H bonds, such as ureas, thioureas and *N*-acetylurea. If a magnesium anode is used in the electrolysis the produced organomagnesium compound upon treatment with butyl iodide affords the butyl derivative [18b]:

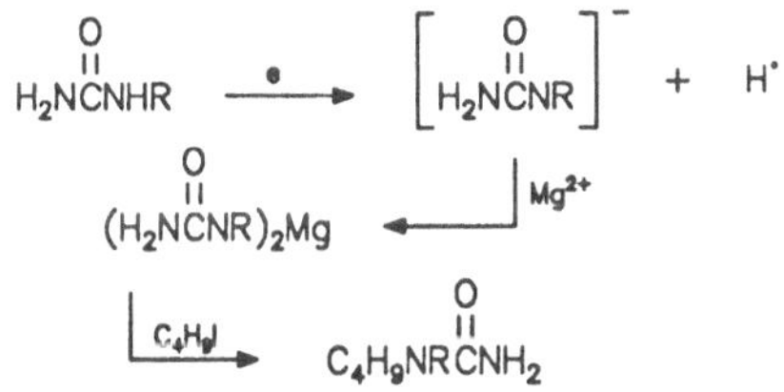

In this connection we consider the possibilities of *paired* organometallic synthesis. Bukhtiarov and coworkers [19] studied this type of electrosynthesis using transition metal electrodes for the formation of triphenylsiloxy derivatives. Current efficiencies of 50–100% were realized. This electrosynthesis is formally indicated below:

**Organo-
metallic
compounds
(transition
metals)**

cathode, $nPh_3SiOH \xrightarrow[\substack{MeCN \\ Q_4N^+X^-}]{ne^-} nPh_3SiO^- + H^\cdot$

anode, $M - ne \longrightarrow M^{n+}$

solution, $nPh_3SiO^- + M^{n+} \longrightarrow M(OSiPh_3)_n$

M = Ti, Fe, Co, Zn, Cr

References

Section 3.2 Carbon-Hydrogen Bond Formations

1. Coleman JA, Wagenknecht JH (1981) J Electrochem Soc 128:322
2. Baizer MM (1969) US Patent 3193408
3. Jansson R (1984) Chem Eng News, Special Report. November 19, p 43
4. Kyriacou D, Jannakoudakis D (1986) Electrocatalysis for organic synthesis. Wiley-Interscience, New York
5. Petrovich J, Anderson JR, Baizer MM (1966) J Org Chem 31:3897
6. Farnia G, Sandova C (1988) J Chem Soc Perkin Trans 2:247; Tezovek JR, Mark HB (1970) J Phys Chem 74:1627
7. Pletcher D, Razak M (1981) Electrochim Acta 26:819
8. Popp FD, Schultz HP (1962) Chem Rev 62:22
9. Horanyi G (1986) Electrochim Acta 31:1095
10. Cassadei MA, Pletcher D (1988) Electrochim Acta 33:117; Lain MJ, Pletcher D (1987) Electrochim Acta 32:99

Section 3.3 Carbonyl Compounds

1. Zuman P, Barus D, Ravolova KA (1968) Discuss Faraday Soc 45:202
2. Leary KJ, Babiuski AD (1976) Abstracts of Papers, Meeting of the Electrochemical Society. Washington, August, p 694
3. Filardo G, LaRosa F, Alfeo C, Cambino S, Silvestri G (1983) J Appl Electrochem 13:403
4. van Der Plase JF, Barendrech E, Zeilmaker H (1980) Electrochim Acta 25:1471
5. Little RD et al. (1988) J Org Chem 53:2287
6. Sibillo S, d'Inkan E, DePort L (1987) Tetrahedron Lett 28:55
7. Sugino K, Nonaka T (1968) Electrochim Acta 13:613
8. Ferles M (1975) Coll Czech Chem Commun 40:2183
9. Tissot P, Surbek JP, Gülacar FO (1981) Helv Chim Acta 64:1570
10. Lain MJ, Pletcher D (1987) Electrochim Acta 32:109
11. Stradius J, Benders J, Kadysh V (1986) Electrochim Acta 31:637
12. Smirnov VA et al. (1986) Doklady Akad Nauk, Eng Ed 288:203
13. Oryoruk H, Pekmez K, Yildiz A (1987) Electrochim Acta 32:574
14. Troup L, Tokec L (1967) J Am Chem Soc 89:4789
15. Barber JJ (1985) US Patent 4517062
16. King JH (1979) J Chem Techn, Biotechn 29:741
17. Horanyi G, Torkos K (1982) J Electroanal Chem 136:301
18. Pienemann T, Schäfer HJ (1989) Dechema Monog Electroorg 112:367
19. Shono T and Kise N (1990) Tetrahedron Lett 31:1303
20. Khnenitskaya E, Razgonyava ID, Solodar SL (1985) Zh Obs Khim 54:1495
21. Bukhtiarov AV et al. (1990) J Gen Chem USSR 59:1505
22. Kunugi Y, Nonaka T, Tien-H (1990) Electrochim Acta 35:1167

Section 3.4 Carboxylic Acids and Esters

1. Vilambi NRK, Chin D-T (1987) J Electrochem Soc 134:3074
2. Oi R, Shimakawa C, Shimakawa Y, Takenaka S (1987) Bull Chem Soc (Japan) 60:4193
3. Alkire RC, Tsai JE (1982) J Electrochem Soc 129:1158; Chin E-T, Cheng CY (1985) J Electrochem Soc 132:2605

4. Iversen PE, Lund H (1967) Acta Chem Scand 21:389
5. Dillon CS (1948) British Patent 611674; (1949) Chem Abst 43:4589; Wagenknecht J (1972) J Org Chem 37:1513
6. Lund H (1963) Acta Chem Scand 17:972; Horner L, Holn H (1977) Liebigs Ann 2036; Deprez D, Margraff R, Bizot J, Pulicani JP (1987) Tetrahedron Lett 28:4679
7. Tan Z et al. (1989) Chem Abst 111:30216z
8. Oi T, Takenaka S (1989) Chem Abst 110:221484c
9. Kumar N, Tuck DG, Watson KD (1987) Can J Chem 65:740; Kyriacou D (unpublished work)
10. Kucheico SI, Turova N Ya, Shreider VA (1986) J Gen Chem USSR 55:3130
11. Said FF, Tuck DG (1982) Ing Chim Acta 59:1
12. Kumar R, Tuck DG (1985) J Organomet Chem 281:47
13. Baizer MM, Anderson JM (1964) J Electrochem Soc 111:233
14. Nishiguchi I, Hirashima T (1983) Angew Chem, Int Ed Eng 22:52
15. Aray K, Shaw C, Nozawa K, Kawai K, Nakagima S (1987) Tetrahedron Lett 28:441
16. Fox DP, Little RD, Baizer MM (1985) J Org Chem 50:2202

Section 3.5 Organohalides

1. Stackelberg M, Stracke W (1941) Z Electrochem 53:118
2. Elving PJ, Pullman B (1961) Adv Chem Phys 3:1; Casanova BS, Eberson L (1974) In: Patai S (ed) Chemistry of the carbon-halogen bond. Interscience, New York
3. Ulery HE (1969) J Electrochem Soc 116:1201; Bard AJ, Merz A (1979) J Am Chem Soc 101:2939; Andrieux CP, Gallardo I, Saveant JM, Su KB (1986) Ibid. 108:638; Koch A, Henne BJ, Bartak RE (1987) J Electrochem Soc 134:3062; Benefice-Malouef S, Blanoce H, Calas P, Comeyaras A (1988) J Fluorine Chem 39:125
3a. Kyriacou D (unpublished work)
4. Fry AJ, Mitnick MA (1969) J Am Chem Soc 91:6207
5. Miller LL, Riekena E (1969) J Org Chem 34:3359
6. Wawzonek S, Duty RD (1961) J Electrochem Soc 108:1135
7. Wawzonek S, Wagenknecht JH (1963) J Electrochem Soc 110:420
8. Covitz FH (1967) J Am Chem Soc 89:540
9. Gilch H (1966) J Polymer Sci 4:1351
10. Fritz HP, Kornrumf W (1978) Liebigs Ann Chem 1416
11. Ref. 3, section 3.5
12. de Lua C, Inesi A, Rampazzo L (1983) J Chem Soc, Perkin Trans 2:1821
13. Rifi MR (1971) J Org Chem 36:1018
14. Covitz FH (1967) J Am Chem Soc 89:4442
15. Fry AJ, Chang LL (1976) Tetrahedron Lett 17:645
16. Fry AJ, Britton WE (1973) J Org Chem 38:2620
17. va Tilborg WJM, Plomb R, de Ruiter R, Smit CJ (1980) Recl Trav Chim 99:206
18. Durandetti S, Sibille S, Perichon J (1991) J Org Chem 56:3255
19. Chaussard J, Folest J-C, Nedelek JY, Perichon J, Sibille S, Troupel M (1990) Synthesis 369; Nedelec JY, Mouloud HAH, Folest J-C, Perichon J (1988) J Org Chem 53:4721
20. Sibille S, d'Inkan E, Leport L, Perichon J (1986) Tetrahedron Lett 27:3132
21. Conan A, Sibille S, d'Inkan E, Perichon J (1990) J Chem Soc Chem Commun 48
22. Silvestri G, Gambino S, Filardo G (1991) Acta Chem Scand 45:987–992; Lu YW, Nedelec JT, Folest JC, Perichon J (1990) J Org Chem 55:2503; Sibille S, Meharek S, Perichon J (1989) Tetrahedron 45:1448
23. Folest JC, Nedelec JY, Perichon J (1988) Synth Common 18:1491
24. Gishbrech JP, Lund H (1985) Acta Chem Scand B39:823

25. Bèrubè D, Renaud RN (1983) Electrochim Acta 28:1367
26. Lund H, Haboth E (1976) Acta Chem Scand B30:895
27. Cipris D (1987) J Appl Electrochem 8:537
28. Casadei MA, Inesi A, Morocci FM, Occhialini D (1989) Tetrahedron 45:6885
29. Voigtlander R, Matschriner H, Krzreminski C, Biering H (1985) J Prakt Chem 327:649
30. Bukhtiarov AV, Colyshin VH, Tomilov AP, Kuz'min OV (1988) J Gen Chem (USSR) 58:1296
31. Hansen PE, Berg A, Lund H (1976) Acta Chem Scand B30:267
32. Kristensen LH, Lund H (1979) Acta Chem Scand B33:735
33. Wawzonek S, Wagenknecht JH (1963) J Electrochem Soc 110:420
34. Barba F, Guidaro A, Zapata A (1982) Electrochim Acta 27:1335
35. Kyriacou D (unpublished work); Jannson R (1984) Chem Eng News, Nov. 19, p 43
36. Kyriacou D, Edamura F, Love J (1980) US Patent 4217185
37. Saveant JM (1980) Acc Chem Rev 18:223 and 13:323; Saveant JM (1988) Bull Soc Chim, No 2, 225; Bunnet JF (1985) Acc Chem Res 18:212
38. Alder RW (1980) J Chem Soc, Chem Commun 1184
39. Oturan MA, Pinson J, Saveant JM, Thiebault A (1989) Tetrahedron Lett 30:1373
40. Thobie-Coutier, Degrand C (1991) J Org Chem 56:5703; Degrand C, Prest R (1990) J Electroanal Chem 282:218
41. Brisoe MW, Chamber RD, Silvester MJ (1988) Tetrahedron Lett 29:12995

Section 3.6 Nitrocompounds

1. Fry AJ (1982) In: Patai S (ed) The chemistry of the nitro and azo compounds. Wiley, London, p 139, etc.
2. Lund H (1983) In: Baizer MM, Lund H (eds) Organic electrochemistry. Dekker, New York, p 285
3. Cyr A, Huot P, Belot G, Lessard J (1990) Electrochim Acta 35:147
4. Gunawardena NE, Pletcher RD (1983) Acta Chem Scand B37:549
5. Sayo H, Tsukitani Y, Masui M (1968) Tetrahedron 24:1717
6. Ogumi Z, Inaba M, Quashi S, Uchida M, Takehara Z (1988) Electrochim Acta 33:365
7. Haber F (1898) Z Electrochem 4:506
8. Beck F, Gabriel W (1985) Angew Chem 24:77
9. Ravichandran C et al. (1989) J Appl Electrochem 19:49
10. Stutts KJ, Scortichini CL, Repucci CM (1989) J Org Chem 54:3744
11. Marquez J, Pletcher D (1981) Electrochim Acta 26:1751
12. Hazard R, Jubault M, Tallec A (1983) Can J Chem 61:2359
13. Kuee S, Lund H (1969) Acta Chem Scand 23:2711
14. Chibani A, Hazard R, Jubault M, Tallec A (1987) Bull Soc Chim 795
15. Wagenknecht JH, Johnson GV (1987) J Electrochem Soc 134:2754
16. Christensen L, Iverson PE (1979) Acta Chem Scand B33:352
17. Cault C, Moinet C (1989) Tetrahedron 45:3429
18. Iversen PE, Christensen TB (1977) Acta Chem Scand B31:733
19. Shono T et al. (1983) J Org Chem 48:2103
20. Petsom A, Lund H (1980) Acta Chem Scand B34:614
21. Bawer WJ, Evans H (1988) J Org Chem 53:5234
22. Bukhtiarov AV et al. (1989) Dokladi Chemistry Proceedings of the Academy of Sciences 304:35
23. Ref. 4, section 3.6

Section 3.7 N-Heterocyclics and other Heteroatom Structures

1. Lund H (1968) Discus Faraday Soc 45:193; Lund H (1970) Adv Heterocycl Chem 12:213; Lund H, Tabacovic I (1984) Adv Heterocycl Chem 36:173

2. O'Reiley JE, Elving PJ (1971) J Am Chem Soc 93:1871
3. Armand J, Chekir K, Pinson J (1978) Can J Chem 56:1894
4. (1983) Chem Abst 98:34503e
5. Navda J, Givadinovitch H, Devaud M (1983) J Chem Res Synop 1992
6. Janic B, Elving PJ (1968) Chem Rev 68:295
7. Boullares L, Bellec C, Pinson J (1987) Can J Chem 65:1619
8. Martigny P, Lund H (1979) Acta Chem Scand B33:575
9. Falsig M, Iversen PE (1977) Acta Chem Scand B31:15
10. Armand J, Bellec C, Boullares L, Pinson J (1983) J Org Chem 48:2847
11. Barrado E, Pardo R, Batanero RS, Alberola A, Laguna MA, Pulido FJ (1988) Electrochim Acta 33:171
12. Pragst F, Henrion A, Abraham W, Michael G (1987) J Prakt Chem 329:1071
13. Webber A, Osteryoung J (1982) J Electrochem Soc 129:273
14. Tsveniashivili VS et al. (1984) J Gen Chem (USSR) 54:546
15. Mangi A, Oloman CW (1987) J Appl Electrochem 17:532
16. Kyriacou D (1981) Basics of electroorganic synthesis. Wiley-Interscience, New York, p 107
17. Kyriacou D (unpublished work); also ref. 16, section 3.7
18. Holst HH, Lund H (1979) Acta Chem Scand B33:233; Hazard R, Tallec A (1976) Bull Soc Chim 433

Section 3.8 Organosulfur Compounds

1. Saveant JM (1969) CR Acad Sci 258:585; Mairanovsky VG (1976) Angew Chem, Int Ed Eng 15:218; Degrand C, Lund H (1979) Acta Chem Scand 33B:512; Iversen PE, Lund H (1974) Acta Chem Scand 28B:827; Montenegro MI (1986) Electrochim Acta 31:607
2. LeGuillanton G, Do QT, Simonet J (1986) Tetrahedron Lett 27:2261; (1989) Bull Soc Chim, No 3, 433
3. Donahue J, Oliver JW (1969) Anal Chem 41:753
4. Falsig M, Lund H (1980) Acta Chem Scand B34:591; Chambers JQ (1974) J Org Chem 39:51
5. Barnard D, Evans M, Higgins G, Smith J (1961) Chem Ind 20
6. Fleizer B, Sanecki P (1982) Electrochim Acta 27:429
7. Pragst F, Kaltofen B (1982) Electrochim Acta 27:1181
8. Langer LK, Schmidtke ST, Volz J (1987) Chem Ber 120:67
9. Falsig M, Lund H, Nadjo L, Saveant JM (1980) Acta Chem Scand B34:685
10. Degrand C, Lund H (1979) Acta Chem Scand B33:512
11. Lund H, Kristense LH (1979) Acta Chem Scand B33:495
12. Johansson BL, Persson B (1978) Acta Chem Scand B32:431
13. Darchen A et al. (1986) J Heterocyc Chem 23:1603
14. Voss J, Wiegand G, Huelmeyer K (1985) Chem Ber 118:4806
15. Bukhtiarov AV, Mikheev VV, Lebeder AV, Tomilov AP, Kuz'min OV (1988) J Gen Chem, (USSR) 58:605
16. Simonet J et al. (1989) Electrochimica Acta 34:1615
17. Buktiarov AV, Lebedev AV, Tonilov AP (1989) J Gen Chem, (USSR) 58:2420
18. Le Guillanton G, Do QT, Simonet J (1989) Bull Soc Chim, No 3, 433

Section 3.9 Some Specific Cleavages of C—O Bonds in Hydroxy Compounds

1. a. Lund H, Doupeux H, Mickel MA, Mousset G, Simonet J (1974) Electrochim Acta 19:629; b. Shono T, Matsumura Y, Tsubata K, Sugihara Y (1979) J Org Chem 44:4508; c. Sopher, WW, Utley JHP (1981) J Chem Soc Chem Commun 134; d. Islam N (1987) Tetrahedron 43:2741

2. Nonaka T, Baizer MM (1983) Electroch Acta 28:661
3. Eggert G, Heilbaum J (1986) Electrochim Acta 31:1448
4. Chum HL, Baizer MM (1985) The electrochemistry of biomass and derived materials, ACS Monograph 183, p 161
5. Ellis KG, Islam N, Sopher DW, Utley JHP, Chum HC (1987) J Electrochem Soc 134:3058
6. Blankespor RL, Hsung R, Schutt P (1988) J Org Chem 53:3232
7. Kempt DS, Reizek J (1977) Tetrahedron Lett 1031; Pan SS, Peterson L, Bochur NR (1981) Mol Pharmacology 19:184
8. Thonton TA et al. (1989) J Am Chem Soc 111:2434
9. Widitz S, Schäfer HJ (1983) Acta Chem Scand B37:475

Section 3.10 Electrocarboxylations

1. a. Lapidus AL, Ping YY (1981) Russian Chem Rev 50:63; b. Russel PG, Srinivasan N, Steinberg M (1977) J Electrochem Soc 124:1329; c. Ref. 4, section 3.9; d. Kyriacou G, Anagnostopoulos A (1992) J Electroanal Chem 322:233; (1992) Ibid. 328:233
2. Paik W, Anderson T, Eyring H (1969) Electroch Acta 14:1217
3. Schiffrin DI (1974) Discus Faraday Soc 56:75
4. Frese KW, Leach S (1985) J Electrochem Soc 132:259
5. DeWulf DW, Jim T, Bard JA (1989) J Electrochem Soc 136:1688
6. Cook RL, MacDuff RC, Sammels AF (1989) J Electrochem Soc 136:1982
7. Mahmood MN, Mashender D, Hartly CJ (1987) J Appl Electrochem 17:1159
8. Fisher J, Lehmann T, Heitz E (1981) J Appl Electrochem 11:743
9. Azuna M, Watanaba M (1989) J Electroanal Chem 260:441
10. Cook RL, MacDuff RD, Sammels AF (1990) J Electrochem Soc 137:187
11. a. Wawzonek S, Wearing D (1959) J Am Chem Soc 81:2067; b. Berkey R, Runner ME (1955) J Electrochem Soc 102:235; c. Tyssee DA, Baizer MM (1974) J Org Chem 39:2823; d. Lamy E, Nadjo L, Saveant JM (1979) Nouv J Chim 3:21
12. Fuchs P, Hess V, Holst HH, Lund H (1981) Acta Chem Scand B35:185
13. Chaussard J, Troupel M, Robin Y, Jacob G, Juhasz JP (1989) J Appl Electrochem 19:345
14. Matue T et al. (1982) Denk Kagku Oyobi Kogyo 50:732
15. Weinberg NL, Hollfamann AK, Reddy TR (1971) Tetrahedron Lett 277
16. Sock O, Troupel M, Perichon J (1985) Tetrahedron Lett 26:1509; McHarek S, Heintz M, Troupel M, Perichon J (1989) Bull Soc Chim, No 1, 95
17. Janietz S, Rüttinger HH, Matschiner H (1988) J Prakt Chem 330:147
18. (1985) Japanese Patent 6024386; (1985) CA. 103:13560c
19. van Tilborg WJM, Smit OJ (1981) Recl Trav Chim 100:437; Neikam WC (1967) US Patent 3344045
20. Baizer MM (1984) Tetrahedron 40:935
21. Cipris D (1980) J Electrochem Soc 127:1045
22. Meyers TJ (1989) Acc Chem Res 22:163
23. Ogura K, Yoshida I (1987) Electrochim Acta 32:1191
24. Khaled M et al. (1990) J Phys Chem 94:5164

Section 3.11 Nitriles and Oximes

1. Lund H (1959) Acta Chem Scand 13:249; Nonaka H (1967) J Electrochem Soc 114:1255; Lund H (1968) Tetrahedron Lett 3651
2. Krishnan V, Mahalingem H, Udupa HVK (1979) J Chem Tech Biotechn 29:703
3. Boulmedais A, Jubault M, Tallec A (1989) Tetrahedron 45:5497

Section 3.12 Various Cathodic Couplings

1. Lund H, Degrand C (1979) Acta Chem Scand B33:57
2. Layoff T (1979) J Org Chem 26:383
3. Engels R, Schäfer HJ (1978) Angew Chem, Int Ed Eng 17:460
4. Lund H, Degrand C (1977) Tetrahedron Lett 3593
5. Hall EAH, Moss CP, Utley JHP, Weedon BCL (1976) J Chem Soc Chem Commun 586
6. Guirado A, Barba F, Hurshouse M, Arcas A (1989) J Org Chem 54:3205
7. Hall TJ, Hargis JH (1986) J Org Chem 51:4185
8. Gassman PG, Lee C (1989) J Am Chem Soc 111:739
9. Cristophe LeMenn J, Sarrazin J (1989) J Chem Research 26
10. Folest JC, Nedelec JY, Perichon J (1987) Tetrahedron Lett 28:1885
11. Petrosyan VA et al. (1989) Doklady Chemistry 302:283
12. Lin Chiu Y, San'ana AEG, Utley JHP (1987) Tetrahedron Lett 28:1349
13. Saveant JM (1983) Acta Chem Scand B37:365
14. Tuck DG (1979) Pure Appl Chem 51:2005
14a. Feasson C, Devaud M (1982) J Chem Research 152
15. Coriu R, Dabosi G, Martineau M (1980) J Organometal Chem 188:63; Benkeser R, Tinker C (1988) J Organometal Chem 196:139; Yoshida J, Murata T, Isoe S (1987) Tetrahedron Lett 27:373
16. Tomilov AP et al. (1985) J Gen Chem (USSR) 54:2145
17. Yoshida J, Muraki K, Funahash H, Kawabata N (1986) J Org Chem 51:39966
18. a. Bukhtiarov AV, Golyshin VN, Tomilov AV, Lebeder AV (1989) J Gen Chem (USSR) 59:366; b. (1989) Ibid. 59:372
19. Bukhtiarov AV, Golyshin VN, Rodnikov IA, Melekhin PG, Makarov YV (1989) J Gen Chem USSR 59:953

4 Indirect Electroorganic Reactions (Indirect Electrolysis)

4.1 Introduction

A formal definition of direct and indirect electroorganic reactions was given in Chapter 1. The purpose of the present chapter is to provide a fundamental understanding of indirect electrolysis for organic synthesis. In Steckhan's view [1] indirect electroorganic synthesis is a "modern chapter in organic electrochemistry." Conceptually, almost any electrosynthesis may be considered feasible by either direct or indirect electrolysis. However, in practice there are some distinct differences between these two general types of electrochemical processes. The practical significance of the difference would depend on the specific goal of the project undertaken and the scale of the process in the laboratory or in the plant. Both direct and indirect electrochemical processes have come into existence since the birth of electrochemical technology. Indirect electrolysis processes have been recently considered from both theoretical and practical perspectives by Steckhan [1] (1986), Alkire [2] (1986) and Pitchaichanarong [3] (1990). We may note that many apparently chemical processes in which redox reagents are used are actually electrochemical processes if we consider that the redox reagents were prepared in some electrochemical cells, although in places remote from the actual place of use. (The scholarly M. Baizer used to call these processes "cryptoelectrochemical".)

Redox electrocatalysis per se and electron transfer reactions in general have been theoretically dealt with by Saveant [4] and by Eberson [5]. (Springer-Verlag plans to begin a series of volumes in electrochemistry including indirect electrolysis under the editorship of Professor E. Steckhan.)

4.1.1 Basic Mechanisms

Homogeneous and heterogeneous reactions

Indirect electroorganic reactions are viewed as electrocatalytic homogeneous reactions if the mediator reacts catalytically with the organic substrate in solution [1,6,7]. If the mediator is immobilized in some way at the electrode surface, the reaction occurs in effect as a heterogeneous indirect process. In either case the basic mechanisms are simply depicted as follows:

$$M \xrightarrow[\text{cathode}]{e^-} M^{\overline{\cdot}}$$
$$M^{\overline{\cdot}} + R \longrightarrow R^{\overline{\cdot}} + M$$

$$M \xrightarrow[\text{anode}]{-e^-} \overset{+\cdot}{M}$$
$$\overset{+\cdot}{M} + R \longrightarrow \overset{+\cdot}{R} + M$$

M = mediator
R = substrate

The overall reaction rate and the catalytic efficiency would depend on the chemical reaction rate of the mediator with the organic substrate and the number of cycles of mediator regeneration (turnovers). Another type of indirect reactions is one in which very reactive inorganic or organic species are generated in situ such as $Cl^\cdot$, $H^\cdot$, $O_2 \cdot^-$, OH, $:CCl_2$, $\cdot CH_2NO$ and $\cdot CH_2OH$. These species react with organic substrates which may or may not be electroactive under the electrolysis conditions. A very special **Special feature** feature of indirect electrolysis is that *exotic* reagents may be generated in situ as and when needed for a particular organic synthesis. Table 4.1 lists a number of redox couples that are potential mediators for indirect electro-organic synthesis.

Table 4.1 Common Redox Mediators

Ce(IV)/Ce(III)	I^+/I^-, I_2/I^-
Ti(III)/Ti(II)	Br^+/Br^-, Br_2/Br^-
Ti(IV)/Ti(III)	Ag(II)/Ag(I)
Sn(IV)/Sn(II)	Mn(III)/Mn(II)
Cr(III)/Cr(II)	Hg(II)/Hg(I)
Os(VIII)/Os(VI)	Pd(II)/Pd(0)
Cr(VI)/Cr(III)	V(V)/V(IV)
Cl^+/Cl^-,Cl_2/Cl^-	Cu(II)/Cu(I)

Some of the mediators in Table 4.1 have been used on an industrial scale (with in situ or ex situ regeneration of the mediator). Alkali and alkaline earth amalgams and active metallic oxides on oxide covered electrodes may also be included in this list of mediators (heterogeneous electrocatalysis). Several organic mediators are also known [8,9]. The most im- **Mediator** portant properties of a mediator redox catalyst are: **properties**

1. chemical stability and
2. rapid chemical reversibility.

In general, inorganic redox systems of metallic ions and oxyanions in simple or complexed forms are more stable and more reversible than organic redox mediators. It must be noted that the thermodynamic potential by itself may not qualify a redox system as a mediator. For example, both redox couples Ag^{2+}/Ag^+ and Co^{3+}/Co^{2+} are powerful oxidizing agents. However, Co^{3+} ions are unstable in aqueous media and therefore the Co^{3+}/Co^2 couple can not be useful in uncomplexed form as a mediator [9a].

Fundamental electron transfer mechanisms

Two fundamental electron transfer mechanisms are generally considered. One mechanism (1) involves a "pure" electron transfer between the redox reagent and the organic substrate without any other chemical event taking place during the electron exchange. The other mechanism (2) requires both electron transfer and some simultaneous chemical (or concerted) reaction such as extraction or expulsion of a hydrogen atom. Mechanism (1) and mechanism (2) are, respectively, referred to as "outer sphere" and "inner sphere" electron transfer processes [9b]. Formal examples are shown in Scheme 4.1a,b

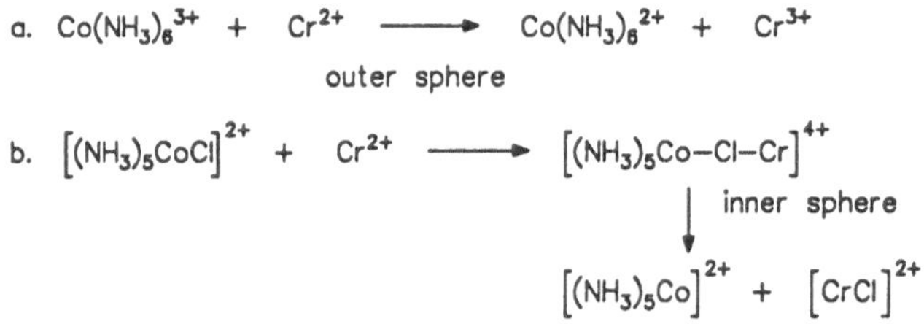

a. $Co(NH_3)_6^{3+}$ + Cr^{2+} $\longrightarrow$ $Co(NH_3)_6^{2+}$ + Cr^{3+}

outer sphere

b. $\left[(NH_3)_5CoCl\right]^{2+}$ + Cr^{2+} $\longrightarrow$ $\left[(NH_3)_5Co-Cl-Cr\right]^{4+}$

$\downarrow$ inner sphere

$\left[(NH_3)_5Co\right]^{2+}$ + $\left[CrCl\right]^{2+}$

Scheme 4.1a,b

In an outer sphere process, the electron moves from the reductant to the oxidant without any alteration of the primary coordination spheres. In an inner sphere process, a spontaneous chemical reaction must occur for the electron transfer to take place, as is shown in Scheme 4.1b by the formation of the *chloride bridge* containing precursor complex. In an electrolytic reactor these mechanisms may be visualized, schematically, as follows:

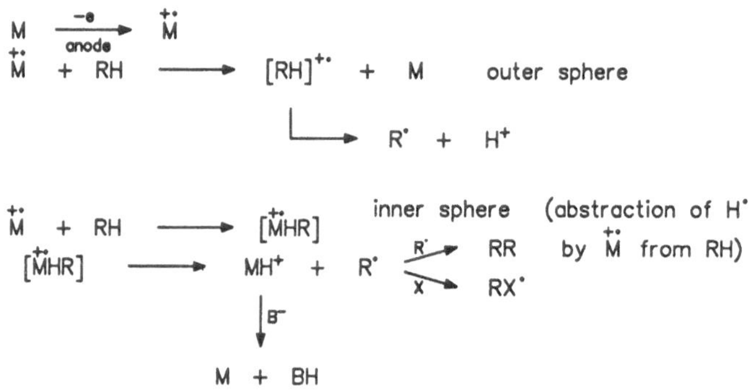

M $\xrightarrow[\text{anode}]{-e}$ $\overset{+\cdot}{M}$

$\overset{+\cdot}{M}$ + RH $\longrightarrow$ $[RH]^{+\cdot}$ + M outer sphere

$\qquad\qquad\qquad \longrightarrow$ $R^\cdot$ + H^+

$\overset{+\cdot}{M}$ + RH $\longrightarrow$ $[\overset{+\cdot}{M}HR]$ inner sphere (abstraction of $H^\cdot$

$[\overset{+\cdot}{M}HR]$ $\longrightarrow$ MH^+ + $R^\cdot$ by $\overset{+\cdot}{M}$ from RH)

$R^\cdot \overset{R^\cdot}{\longrightarrow} RR$

$R^\cdot \underset{X^\cdot}{\longrightarrow} RX^\cdot$

$\downarrow B^-$

M + BH

In organic electrochemistry the terms *bonded* and *nonbonded* ET processes are also used for outer- and inner-sphere mechanisms. Electron transfer mechanisms were recently considered on the basis of the *Marcus Theory* [10]. An inner sphere electron transfer mechanism with the very effective organic mediator triarylamine was studied by Steckhan and coworkers [11]:

Triarylamine mediator

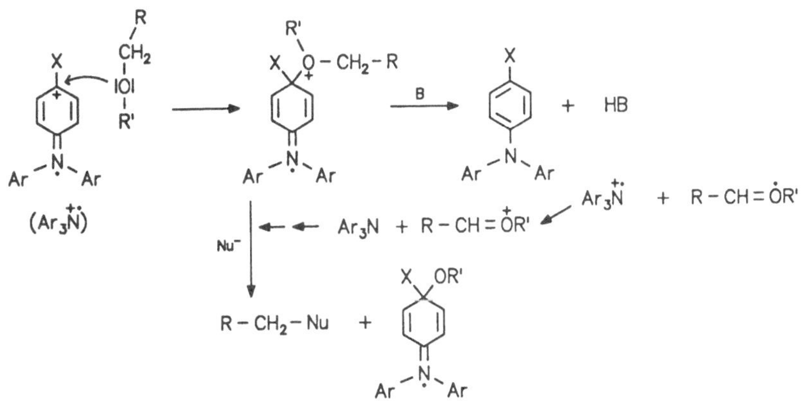

$$R-CH_2-Nu \quad + \quad \text{(structure)}$$

It is sometimes observed in practice that a given redox couple is an effective mediator for a given indirect oxidation or reduction; while an electrode set at the same potential which is characteristic of the redox couple cannot achieve the same oxidation or reduction. This *apparent paradox* is *not* a thermodynamic violation. It is only a kinetic consequence, common to many electrochemical reactions, organic and inorganic. For example, the electrolytic production of chlorine by electrolysis of brines is possible because kinetic factors suppress the thermodynamically favored oxygen evolution from the oxidation of water at the anode as they also do in the Kolbe reaction (Chapter 2).

Kinetic factors

4.2 Oxidative Indirect Reactions

High-valent transition metal ions, simple or complexed are common mediators for indirect oxidations. Also organic radical cations of aromatic and heteroaromatic substances and some trialkyl amines may be very effective oxidative mediators in certain cases. Positive halide ions, (haloniums), I^+, Br^+, have been shown to be good mediators for the oxidation of alcohols. With halonium ion mediators a bonded-type mechanism apparently operates, since a transitory chemical bond is formed between the halogen mediator and the organic substrate during the oxidation course.

Oxidative mediators

4.2.1 Metallic Ions and Oxyanions Mediators

The oxidation of side chains in aromatics has been demonstrated with transition metal mediators, on a laboratory and industrial scale of synthesis. Thus toluene has been oxidized to benzaldehyde on a plant scale

Oxidation of toluene

(Zurich) using the redox couple mediator Ce^{4+}/Ce^{3+}. Regeneration of the redox catalyst is done in situ or ex situ (internally or externally) for industrial processes [1–5]. Acetoxylation of toluene on the methyl group is effected in the presence of cobaltic acetate as a mediator [1]. The cobaltic acetate complex is generated in situ, as is formally depicted below:

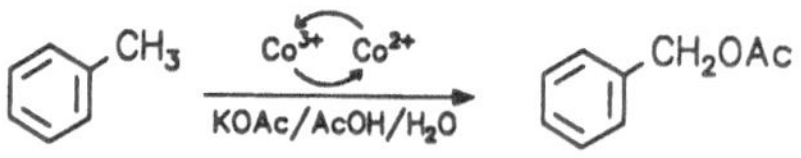

Oxidation of alhenes

Ce(IV)/Ce(III) mediator

Indirect oxidations of alkylbenzenes by recycled (ex situ) $NH_4Ce(NO_3)_6$ in aqueous or methanolic acetic acid media was studied by Torii [2]. Anisaldehyde was thus obtained from p-methoxytoluene in 94% yield. The electrogeneration of the Ce(IV) salt was easily effected by electrolysis in methanol solution containing the Ce(III) complex, yellow crystals, which were recovered from the reaction mixture after solvent evaporation and extraction of anisaldehyde with benzene.

Oxidation of alkylbenzenes

Oxidation of alcohols, cyclic ketomes and C—H bonds

Ruthenates are studied as potential oxidation mediators. The oxidant species Ru(VI) is fixed at the electrode surface (heterogeneous electrocatalyst) and is continuously regenerated thereon by deelectronation of the reduced species Ru(V). Thus alkenes have been oxidized using lead and bismuth ruthenates [3]. It is assumed that the nonstoichiometric lead ruthenates are the active catalysts. Organic ruthenium(IV) complexes were found to be very effective mediators for the oxidation of alcohols, cyclic ketones and C—H bonds adjacent to aromatic or olefinic group, such as p-xylene or toluene which are oxidized to terephthalic acid and benzoic acid respectively [4,5]. The Ru(IV) oxidant was used in the complex form $[Ru(trpy)(bpy)(0)]^{2+}$, (tryp is 2,2′,2″-terpyridine, and bpy is 2,2′-bipyridine). For example, the electrocatalytic oxidation of the isopropyl alcohol to acetone is expressed as follows:

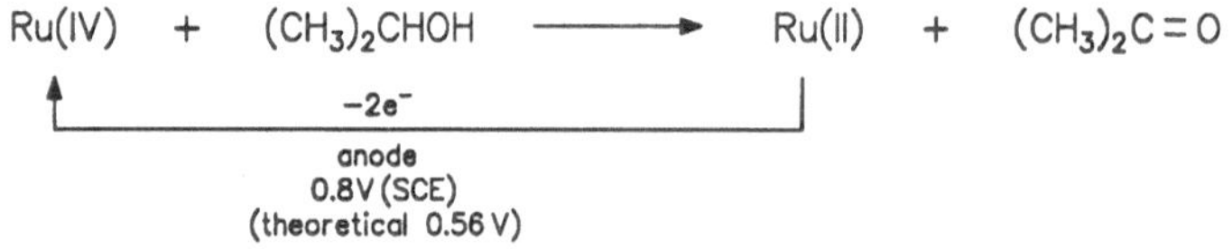

Electrocatalytic oxidation of isopropyl alcohol to acetone

Electrocatalytic Oxidation of Isoropyl Alcohol to Acetone

This electrolysis was conducted in a divided cell equipped with platinum electrodes and a SCE as reference for the anode. The oxidation of isopropanol was carried out in aqueous solution (pH 6.8, HPO_4^-/HPO_4^-) at 25 °C under nitrogen. The anolyte contained the

> alcohol and the Ru(IV) complex in the millimolar radio of 1000/1 and
> the anode potential was set to 0.8 V vs SCE which was sufficient
> for the regeneration of the catalyst. Hydrogen was evolved at the
> cathode. Product yields were determined by gas chromatography and
> were in the range of 90–100%.

Torii [6] reported an interesting double mediator system for the oxidation **Double**
of alcohols and aldehydes. For example, the oxidation of secondary **mediator**
alcohols to the corresponding ketonic products occurs as depicted formally **systems**
in Scheme 4.2

$$Cl^- \xrightarrow{anode} Cl^+ + 2e$$

$$2Cl^+ + RuO_2 \xrightarrow{2H_2O} RuO_4 + 2Cl^- + 4H^+$$

$$RuO_4 + 2RCH(OH)R' \longrightarrow 2R\overset{\overset{\displaystyle O}{\|}}{C}R' + RuO_2 + 2H_2O$$

Scheme 4.2

This electrolysis medium is an aqueous-organic two-phase system. Another **Hydroxylation**
double mediator system is the OsO_4/OsO_3-$K_4Fe(CN)_6/K_3Fe(CN)_6$ com- **of olefinic**
bination. This system was used for the hydroxylation of olefinic com- **compounds**
pounds [7], as depicted in Scheme 4.3

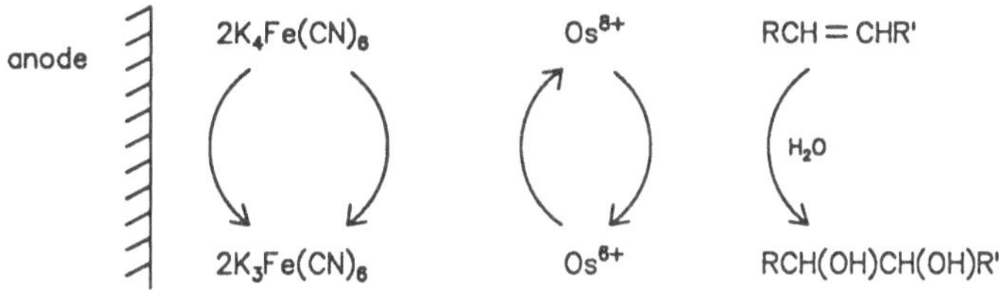

Scheme 4.3

Interesting examples of external regeneration of the redox catalyst are
the systems Cr^{6+}/Cr^{3+}, Ce^4/Ce^3 and Co^{3+}/Co^{2+} in acidic solutions. Saccharin was produced by oxidative conversion of *o*-toluenesulfonamide by **Saccharin**
electrolysis using the Cr^6/Cr^{3+} couple in aqueous sulfuric acid [8], and a
lead dioxide anode.

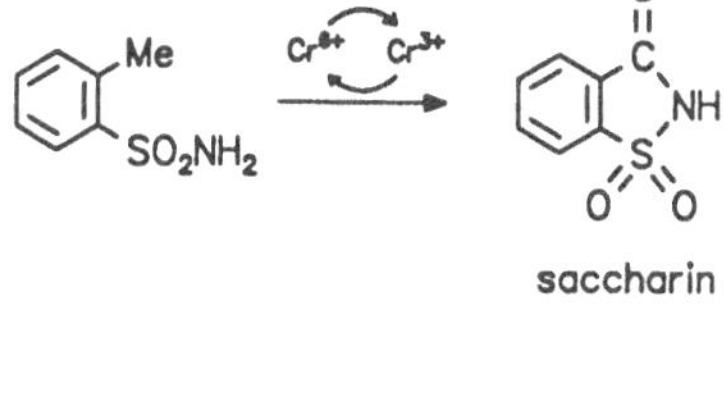

saccharin

Oxidation of toluene and anthracene

The oxidation of toluene to benzaldehyde by ex situ regenerated ceric ion is very efficient (99% yield) [9]. The process is used commercially (Zurich). Anthracene is industrially oxidized to anthraquinone by electrogenerated chromic acid (Holiday, U.K.). A paste consisting of small anthracene particles is oxidized in sulfuric acid by the dichromate and the resulting chromic sulfate is reoxidized in an electrolysis cell and is used again [10].

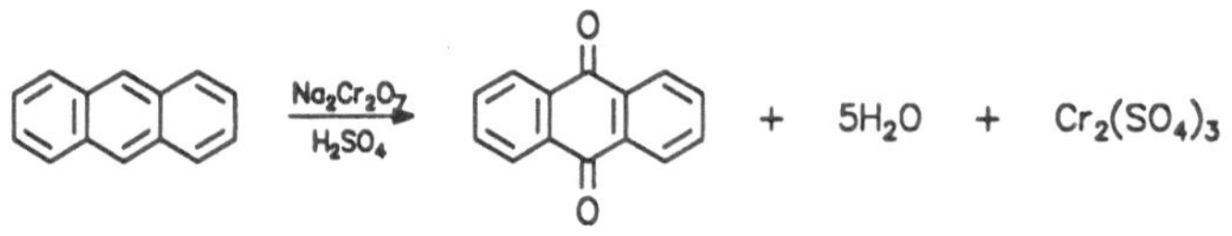

Carboxymethyl and nitromethyl radicals

Electrochemically generated Mn(III) ions in acetic acid are oxidative mediators producing reactive carboxymethyl and nitromethyl radicals capable of reacting with olefins and aromatics [11–13]. The Mn(III) species is directly generated at the anode and reacts in solution with acetic acid or nitromethane to yield the corresponding radicals $[CH_2COOH]^{\cdot}$ and $[CH_2NO_2]^{\cdot}$ which enter into reaction as shown in Scheme 4.4.

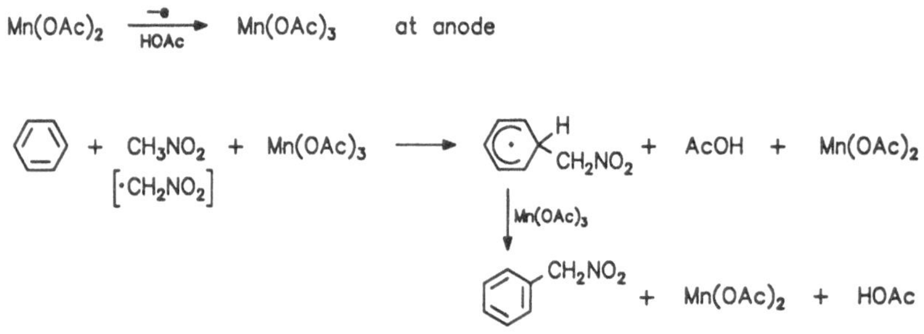

Scheme 4.4

4.2.2 Paired Oxidative Processes

Oxidation of toluene

It may be sometimes possible to develop efficient *indirect paired* electrosynthetic processes. An example is the oxidation of toluene to benzaldehyde [14], as shown in Scheme 4.5. The anodic mediator for the oxidation of toluene is the Mn^{3+}/Mn^{2+} couple while at the cathode the couple V^{5+}/V^{4+} is generating $OH^{\cdot}$ radicals which also oxidize toluene so that both half-cell reactions lead to the mediated production of the desired aldehyde, Scheme 4.5

$$Mn^{2+} \xrightarrow[HOAc/H_2O]{-e} Mn^{3+} \quad \text{at anode}$$

$$C_6H_5CH_3 + 4Mn^{3+} \xrightarrow[H_2O]{} C_6H_5CHO + 4Mn^{2+} + 4H^+ \quad \text{in solution}$$

$$V^{5+} + e \longrightarrow V^{4+}$$
$$O_2 + 2H_2O + 2e^- \longrightarrow H_2O_2 + 2OH^- \qquad \text{at cathode}$$

$$V^{4+} + H_2O_2 \longrightarrow V^{5+} + OH^- + OH^\cdot$$
$$C_6H_5CH_3 + OH^\cdot \xrightarrow{O_2} C_6H_5CHO \qquad \text{in solution}$$

Scheme 4.5

The reaction via OH$^\cdot$ radicals most probably involves chemical steps as indicated below:

$$C_6H_5CH_3 + OH^\cdot \longrightarrow C_6H_5CH_2^\cdot + H_2O$$
$$C_6H_5CH_2^\cdot + O_2 \longrightarrow C_6H_5CH_2OO^\cdot$$
$$\downarrow$$
$$C_6H_5CHO + OH^\cdot$$

Thus in theory the efficiency of this cell reaction should be 200% in terms of product yield. Another paired electrosynthesis example was recently reported [15]. Benzene is oxidized to hydroquinone at the cathode compartment indirectly, while at the anode, (PbO$_2$/H$_2$SO$_4$), benzene is oxidized to benzoquinone. The pertinent reactions are shown in Scheme 4.6. The electrolysis is performed in a divided cell. The anolyte is aqueous sulfuric acid containing 10 ml of benzene per 100 ml of solution and the catholyte aqueous 0.1 M acetate solutions with some acetonitrile and 10 ml of benzene. Oxygen is bubbled through the catholyte during electrolysis in order to generate the desired OH$^\cdot$ radicals for the oxidation of benzene. The current efficiency is above 90%.

Paired oxidations of benzene

OH$^\cdot$ radicals from O$_2$

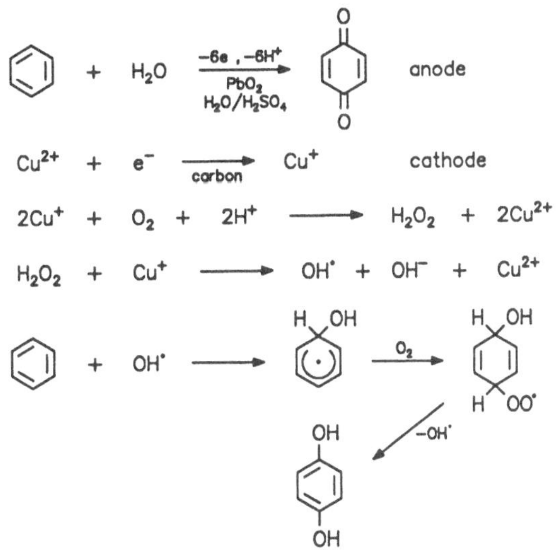

Scheme 4.6

Hydroxyla-tions of sertain organic comfounds Fenton's reagent

Electroreduction of molecular oxygen in the presence of Fe^{3+} ions in aqueous solutions generates Fenton's reagent for in situ hydroxylations of certain organic compounds [16–18]. Recently, a study was made concerning the oxidation of aromatics by Fenton's reagent. It was concluded that the reaction is actually very complex for both batch and continuous processes [19]. The oxidizing agent is the $OH^{\cdot}$ radical resulting from the reduction of dioxygen to H_2O_2 and reduction of H_2O_2 by Fe^{2+},

$$O_2 + 2e \xrightarrow{2H^+} H_2O_2$$
$$Fe^{3+} + e \longrightarrow Fe^{2+}$$
$$Fe^{2+} + H_2O_2 \longrightarrow Fe^{3+} + OH^- + OH^{\cdot}$$

Graft copolymeri-zation

The couples Cu^{2+}/Cu^+ and V^{4+}/V^{3+} may also be used for the generation of $OH^{\cdot}$ radicals [20,21]. The Fe^{3+}/Fe^{2+} couple may be most conveniently generated by the in situ dissolution of a consumable iron anode. This technique has been successfully used in a recent electroinitiated graft copolymerization of cellulose with acrylonitrile [22]. Cellulose radicals were generated and reacted with acrylonitrile to form the copolymer product.

4.2.3 Halide Ions as Mediators

Halonium ions

Molecular halogens and hypohalites or positive halonium ions, I^+, Br^+ may be very easily generated to act as mediators in situ for a variety of oxidative reactions. Small amounts of $NaOCl$ generated in aqueous solutions are sufficient to oxidize hydroxy compounds such as menthol, isoborneol 2-octanol and other secondary alcohols to their corresponding ketones in high yields [1]. Iodonium ion, I^+, is especially able to oxidize primary alcohols and aldehydes to esters [2], and secondary alcohols to ketones [3], as illustrated in Scheme 4.7.

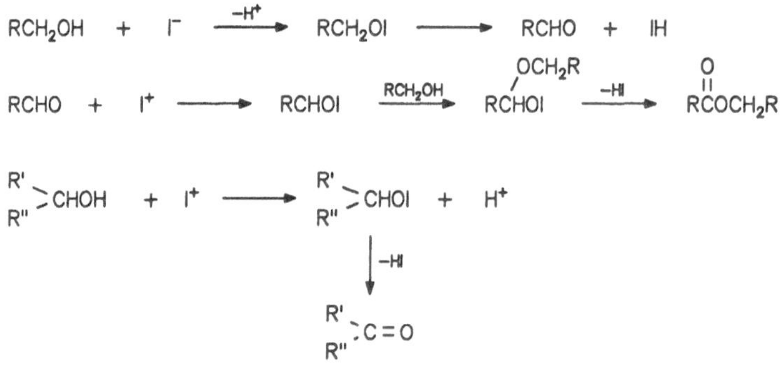

Ketones and esters

Scheme 4.7

Anodic electrolysis of methonolic solutions of aldehydes in the presence of KI or KBr in undivided cells transforms the aldehyde hemiacetals to esters by oxidation of the hemiacetal by the electrogenerated iodonium ion, **Esters**

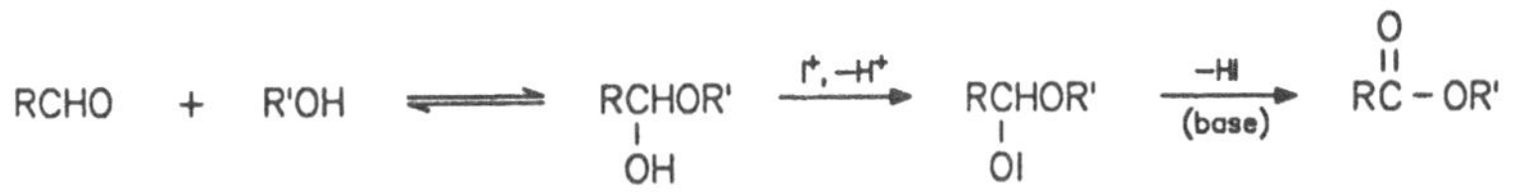

$$RCHO \; + \; R'OH \; \rightleftharpoons \; \underset{OH}{\overset{}{RCHOR'}} \; \xrightarrow{I^+, -H^+} \; \underset{OI}{\overset{}{RCHOR'}} \; \xrightarrow[(base)]{-HI} \; \overset{O}{\overset{\|}{RC}} - OR'$$

Benzyl benzoate (a pharmaceutical) has thus been prepared by electrolysis in aqueous solution containing benzyl alcohol and potassium iodide [4]. Aromatic aldehydes are oxidized by catalytic amounts of KI in methanol-NaOMe medium to give the corresponding methyl carboxylates using only theoretical amounts of electricity [5]. Aliphatic aldehydes with α-hydrogens are not suitable substrates because they undergo aldol condensations. **Methyl carboxylates**

In the presence of NH_3 nitriles can be obtained via the aldimine intermediates which are in equilibrium with the aldehyde, Scheme 4.8 **Nitriles**

$$RCHO \; \underset{-H_2O}{\overset{+NH_3}{\rightleftharpoons}} \; RCH = NH \quad \text{aldimine}$$

$$RCH = NH \; + \; I^+ \; \longrightarrow \; RCH = NI \; + \; H^+$$

$$RCH = NI \; \xrightarrow[\substack{base \\ (NaOMe)}]{-HI} \; RCN$$

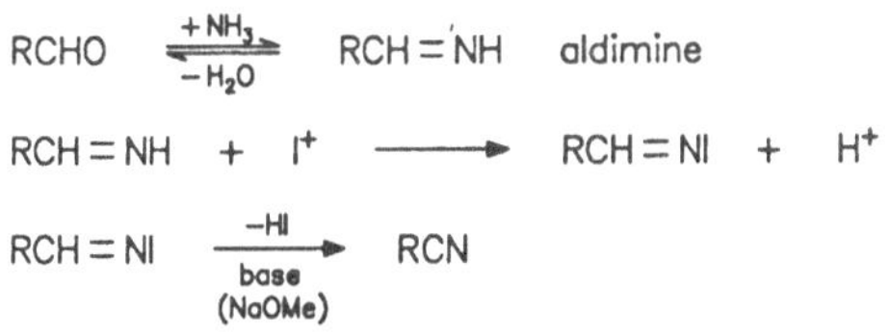

Scheme 4.8

The NaOMe facilitates removal of HI from the iodo intermediates. (Caution: formation of the explosive $NI_3(I_2+NH_3)$ is possible.) **Warning, NI_3**

Indirect epoxidations using the couples Cl_2/Cl^- and Br_2/Br^- were recently studied from an electrochemical engineering aspect by Alkire and Köhler [6]. Aqueous dispersions of 1-hexene were electrolyzed in undivided cells with parallel-plate electrodes in the presence of the halide catalysts. Epoxide current efficiencies of 65% were attained. Steckhan and coworkers [7] domonstrated that it is possible to methoxylate N-protected aminoacid esters in the α-position relative to nitrogen by indirect electrolysis in methanol solution using chloride ion as primary mediator. For example, the methoxylation of *N*-benzoyl glycine methyl ester occurs by an overall process involving the in situ generated MeOCl reagent as indicated in Scheme 4.9. The process is analogous to the chemical process in which *tert*-butyl hypochlorite would be employed in at least stoichiometric amounts. **Aminoacid esters**

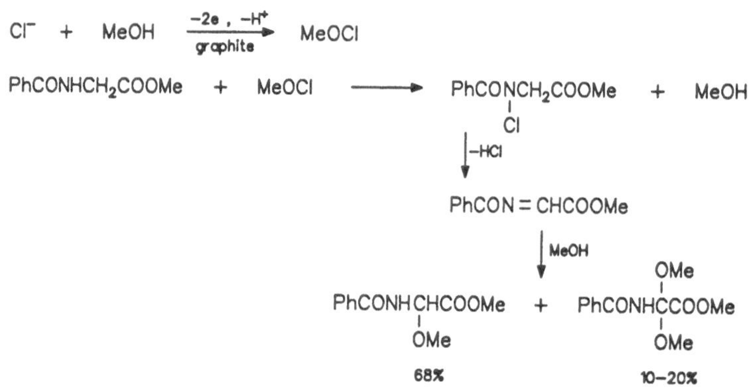

$$Cl^- + MeOH \xrightarrow[graphite]{-2e,\ -H^+} MeOCl$$

$$PhCONHCH_2COOMe + MeOCl \longrightarrow \underset{\underset{Cl}{|}}{PhCONCH_2COOMe} + MeOH$$

$$\downarrow -HCl$$

$$PhCON=CHCOOMe$$

$$\downarrow MeOH$$

$$\underset{\underset{OMe}{|}}{PhCONHCHCOOMe} + \underset{\underset{OMe}{|}}{PhCONHCCOOMe}$$
$$\overset{OMe}{|}$$

$$68\% \qquad\qquad 10-20\%$$

Scheme 4.9

Direct anodic oxidation of these compounds is difficult because of the presence of the electron-withdrawing ester group in the molecule. The indirect oxidation is much easier than the direct and the electrolysis procedure can be simply done in undivided cells under constant current by applying $7-10\,V$ across the cell. The substrate is dissolved in methanol containing some sodium chloride and sodium perchlorate. The methanol is easily evaporated under reduced pressure and the products are recovered by extraction with chloroform from an aqueous solution.

**Halooxyanions
Oxidation
of glucose**

Indirect oxidations using halide salts or halooxyanions have found some industrial applications. The oxidation of glucose to gluconic acid [8] and of furans to 2,5-dihydrofurans [9] in the presence of Br^- ion as oxidative mediator are industrial processes. In the production of calcium gluconate the bromine is generated at the anode from the bromide ion while in solution glucose is oxidized by bromine to gluconic acid [10]:

$$2C_6H_{12}O_6 \xrightarrow[CaCO_3]{2Br^- \overset{\frown}{} Br_2} Ca(C_6H_{11}O_7)_2$$

**Coupled
products**

Activated methylenes (carbon acids) are indirectly oxidized via their carbanion ions to give coupled products [11]. The electrolysis is carried out in undivided cells in acetonitrile-potassium iodide solution with platinum or graphite electrodes. The potassium iodide provides the needed reagents, metallic potassium and iodine, which react as shown in Scheme 4.10:

$$2K^+ + 2e \longrightarrow 2K \qquad cathode$$

$$2K + 2CH_2X_2 \longrightarrow 2K^+ + 2\bar{C}HX_2 + H_2$$
$$(X = ester\ group)$$

$$2I^- - 2e \longrightarrow I_2$$

$$2\bar{C}HX_2 + I_2 \longrightarrow X_2CHCHX_2 + 2I^-$$

overall $2CH_2X_2 \longrightarrow X_2CHCHX_2 + H_2$

Scheme 4.10

This technique takes advantage of two opposite indirect reactions, an anodic oxidation and a cathodic reduction, and may therefore be viewed as a paired *doubly indirect* electrosynthesis process.

Polysaccharides have been indirectly oxidized using the IO_4^-/IO_3^- or the Cr(IV)/Cr(III) redox couples as mediators. The mediated oxidation of starch to dialdehyde starch was practiced for some time on an industrial scale [12]. The IO_4^- was regenerated either in situ or ex situ. The process was discontinued apparently because the product demand was limited.

Oxidation of poly-saccharides

The periodate-iodate redox mediatory system has been considered for the oxidation of a number of biomass-derived materials. An optimization study was conducted for the "in-cell" indirect conversion of 2.3-butanediol to acetaldehyde in a divided cell and with lead dioxide as anode [13]. The needed periodate is formed at the anode while the diol is oxidized in solution to give 80% yields of acetaldehyde. Direct anodic oxidation gave low aldehyde yields. In order to minimize catalyst losses and improve product recovery Yoshida and coworkers developed a technique using hydrobromides of polymeric amines such as poly(4-vinylpyridine) in the form of beads which are stirred between anode and cathode. The immobilized oxidant is most probably the Bro$^-$ groups on the surface of the beads [14].

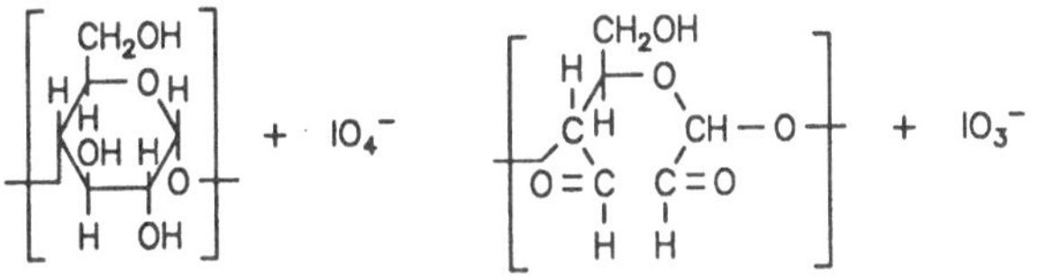

Oxidation of biomass-derived materials

Immobilized oxidant

Indirect oxyselenylation and deselenylation for the one-step transformation of olefins into allylic alcohols and ethers has been accomplished with the Br$^+$/Br$^-$ redox couple as mediator. By such a process a reaction with the olefin in the presence of an alcohol takes place [15]:

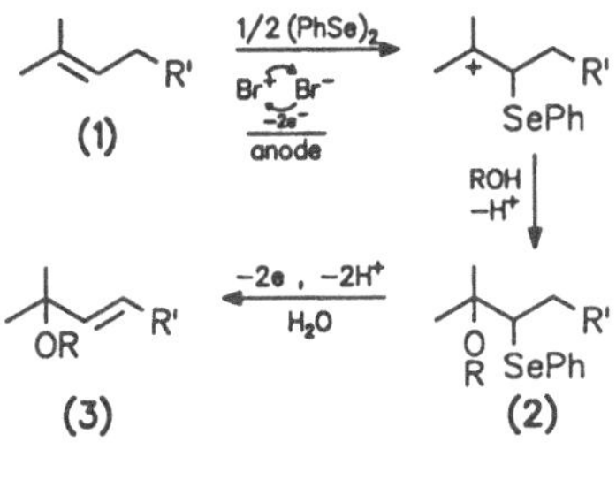

Oxyseleny-lation and deselenylation

Scheme 4.11

The essential step in this reaction, Scheme 4.11, is the formation of the intermediate (2) which by further oxidation leads to compound (3). Torii et al. reported the conversion of citronellol to the cyclic ether, *dl*-rose [15] by electrolysis in methanolic solution containing $(PhSe)_2$ and Et_4Br, and hydrolysis of the intermediate methoxylated product,

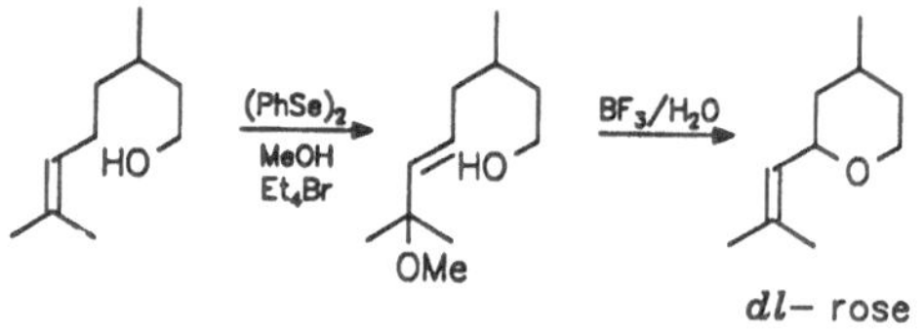

dl− rose

Conjugated aryl olefins are transformed to α-bromoketals by indirect oxidation in methanol/bromide solutions in undivided cells. The electrode reactions are shown in Scheme 4.12 [16],

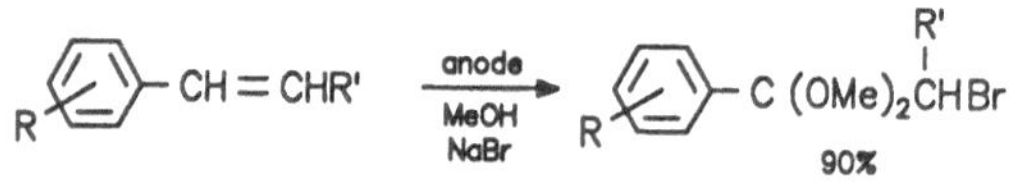

$$2Na^+ + 2e^- \xrightarrow{cathode} 2Na$$

$$2Na + 2MeOH \longrightarrow 2NaOMe + H_2$$

Scheme 4.12

4.2.4 Nitrate Ion as Mediator

Oxidation of alcohols   The nitrate ion may act as an effective mediator in the oxidation of some alcohols that are difficult to oxidize directly at the anode. For example, 2-butanol requires a potential about 2.4 V versus the Ag/Ag$^+$ to deelectronate at a platinum anode. In the presence of nitrate ion high yields of the ketonic product are obtained. A potential of 1.6 V at which the $NO_3^.$ radical is generated is sufficient for the oxidation of the alcohol [1]. Hexanol is oxidized by $NO_3^.$ radicals to the ketonic product in 80% current yield whereas the yield in the absence of NO_3^- is only 5%.

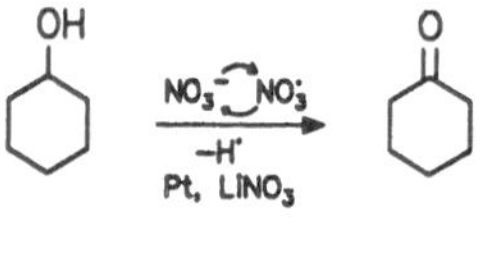

4.2.5 Organic Mediators for Oxidations

Electrogenerated organic cation radicals may act as electron transfer oxidants in a catalytic manner provided the redox couple is stable and reversible in the particular electrolysis system. The couple tri-*p*-bromo-phenylamine and its cation radical is an effective electrocatalytic couple for indirect oxidations of organic compounds [1]. An example is the homogeneous electrocatalytic decarboxylation of n-alkylcarboxylates (n = 6, 7, 12) in the presence of tris-(4-bromophenyl)amine in acetonitrile solution. Esters and nitriles are thus produced:

Production of esters and nitriles

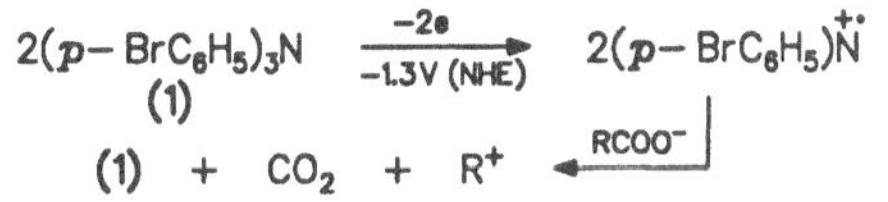

$$2(p-BrC_6H_5)_3N \xrightarrow[-1.3V \ (NHE)]{-2e} 2(p-BrC_6H_5)\overset{+\cdot}{N}$$
$$(1)$$

$$(1) \ + \ CO_2 \ + \ R^+ \ \xleftarrow{RCOO^-}$$

$$R^+ \ + \ RCOO^- \longrightarrow \ RCOR$$

$$\text{also} \quad R^+ \ + \ CH_3CN \longrightarrow \ RCH_2CN \ + \ H^+$$

Steckhan and Schmidt prepared several triarylamines and tested them as redox catalysts [2]. Typical of such amines are the following:

Triarylamines

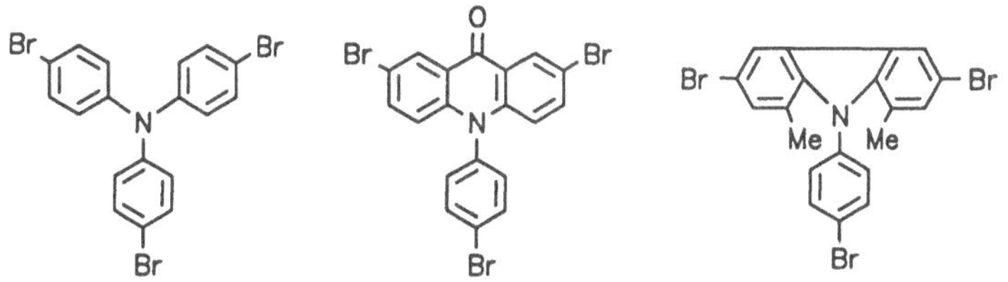

$$(1) \qquad\qquad (2) \qquad\qquad (3)$$

Standard potentials of these materials are in the range 0.75 to 1.96 V, depending on the structure and ring substituents.

Anodic cyclization by indirect electrolysis of 2-hydroxychalcones to obtain flavonoids has been accomplished by electrolysis in methanol-dichloromethane solvent in divided cells in the presence of tris-(4-bromo-phenyl)amine as catalyst [3].

Flavonoids

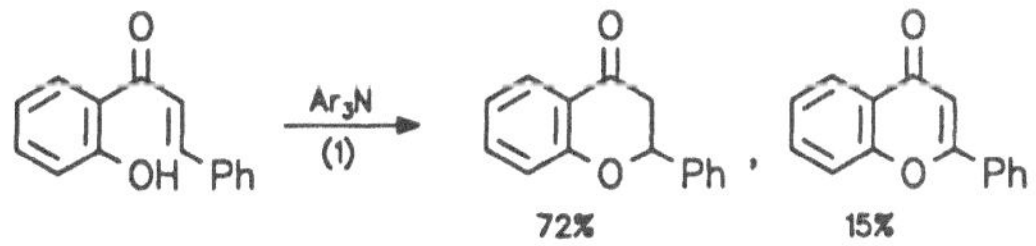

Another effective organic mediator for the oxidation of alcohols to alde-hydes and ketones is the compound *N*-hydroxyphthalimide. Electrolysis is conducted in acetonitrile containing pyridine and sodium perchlorate. Thus 2-butanol gave 88% yield of ethylmethyl ketone [4].

Aldehydes and ketones

It is to be noted that oxidations catalyzed by the aforementioned organic mediators can be carried out under very mild conditions which allow successful selective oxidations of sensitive substrates and facile cleavages of protective groups as for example in the deprotection of the aldehyde shown below [5,6]:

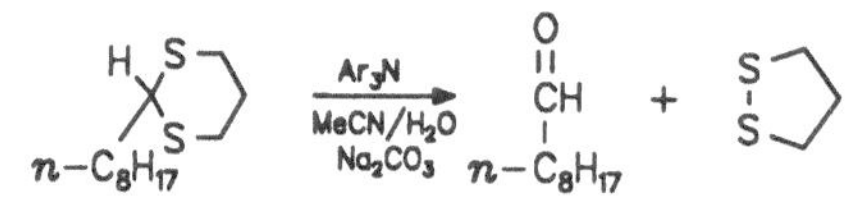

Benzaldehydes Benzylic alcohols are oxidized to their corresponding benzaldehydes in good yields in the presence of catalytic amounts of a triarylamine mediator and a weak base, such as Na_2CO_3 or 2,6-lutidine, to facilitate proton abstraction from the intermediate products [6], as is indicated in Scheme 4.13:

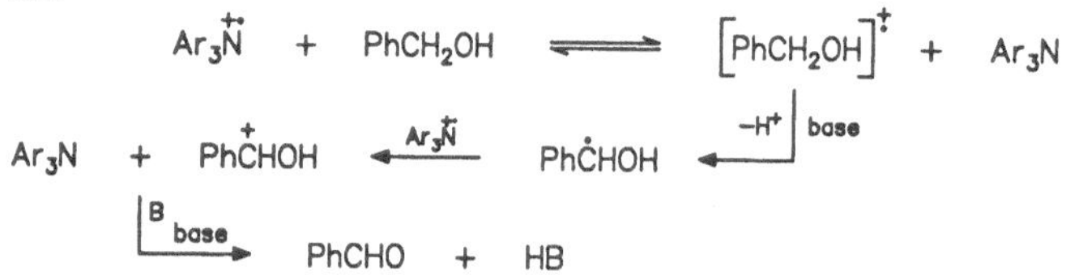

Scheme 4.13

It is of interest to note that esters and acetals, which are very difficult to oxidize directly at the anode, are easily oxidized by the action of triaryl-

Bonded-type lamine cation radicals. These reactions occur most likely by a *bonded-*
mechanism type (inner sphere mechanism), as proposed by Steckhan [7] in the α-methoxylation of aliphatic ethers, Scheme 4.14:

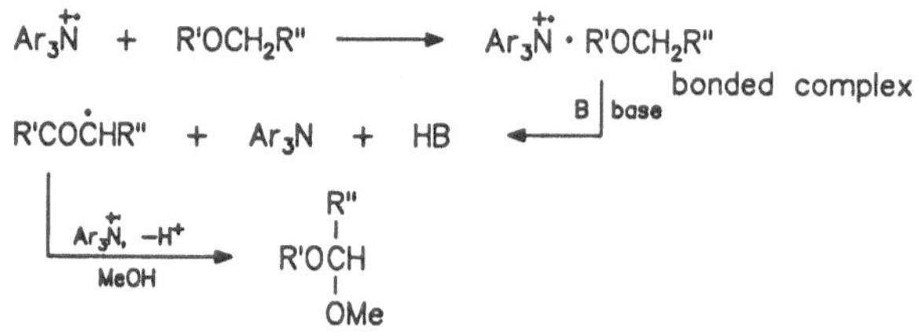

Scheme 4.14

According to Eberson, such *bonded* or *inner-sphere* mechanisms would operate most effectively when there is a large potential difference between the substrate and the mediator [8].

Electroactive Some attention has been given to potential applications of electroactive
organic films organic films as oxidative mediators [9,10]. Such films are functionalized
as oxidative polypyrroles, polyanilines and polymeric nitroxides. The last are good
mediators one-electron transfer catalysts [10]. The polymer (**2**) made from the nitroxide monomer (**1**) is an efficient mediator for the oxidation of amines [10].

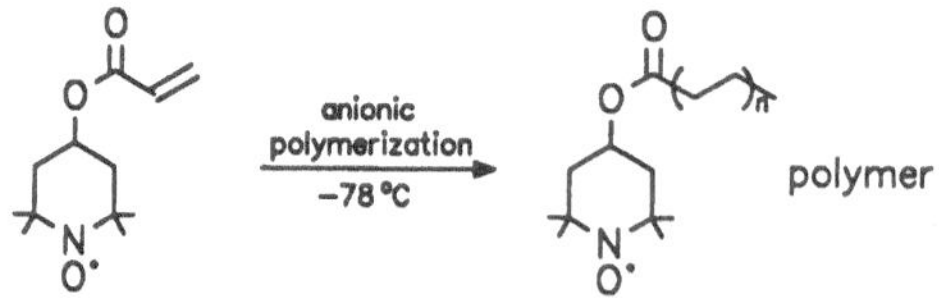

The catalytic activity of the polymer(2) is depicted below for the oxidation of benzylamine. The electrolysis is carried out in acetonitrile-lithium perchlorate medium with a graphite fiber cloth electrode coated with a monolayer of polymer(2). This heterogeneous electrocatalytic oxidation gave benzaldehyde and benzonitrile with 78% and 22% current efficiencies, respectively.

Polymer coated electrode

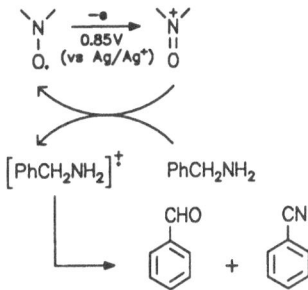

A polyaniline modified stainless steel electrode was found to be an effective electrode for the oxidation of hydroquinone to benzoquinone [9]. Electroactive organic films with electrocatalytic properties are easily tested for their effectiveness by cyclic voltammetry. The electrode is coated with a thin layer of the polymeric organic film and cyclic voltammograms are recorced with the film-modified electrode using a suitable electrolysis medium. If the film is reversibly electroactive two peaks of equal height appear as shown in Figure 4.1. The peaks will change in some way when an organic substrate that is oxidized or reduced by the meidator film added to the solution. The effective film-life may also be tested by periodic voltammetric scans during prolonged electrolysis conditions.

Benzoquinone

Polyaniline modified electrode

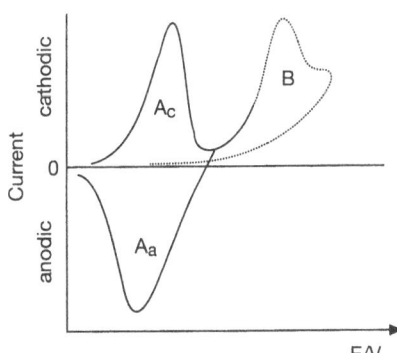

Fig. 4.1. Cyclic voltammogram of a hypothefical electroactive film *A*: in the absence of organic subtrate *B*: in the presence of organic substrate (oxidizable in this case)

4.3 Indirect Reductive Reactions

Homogeneous electrocatalysts

In indirect reductive methodology low-valent metallic ions and oxyanions are the usual mediators or homogeneous electrocatalysts. In aqueous or aqueous-organic media the electroreduction of high-valent ions to the lower valency states occurs preferably before hydrogen evolution at the cathode.

4.3.1 Metallic Ions as Catalysts

Reductive Mediators Ti^{4+}/Ti^{3-} couple

Ordinary metallic ions in their lower oxidation states, such as the ions Fe^{2+}, Cr^{2+}, Sn^{2+} and Ti^{3+} are effective reductive mediators. Even in cases of direct heterogeneous electrolysis the overall electrochemical process may be improved by the addition of appropriate amounts of the mediators. Mediated electrochemical reductions may sometimes eliminate the need of the reducing agents lithiumaluminum hydride or tributyltin that are necessary for certain difficult chemical reductions of organic compounds. Thus an N—OH function can be selectively reduced to the N—H function by electrolysis in aqueous media in the presence of the Ti^{4+}/Ti^{3+} couple [1].

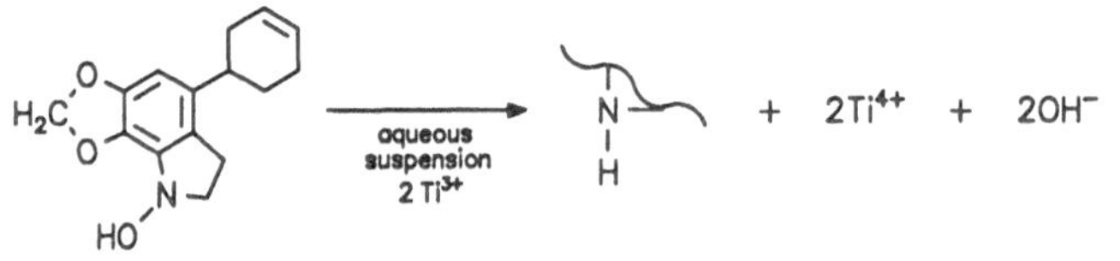

The compounds *p*-nitrosophenol and *m*-nitrotoluene are indirectly reduced to p-aminophenol and m-toluidine, respectively in the presence of the Ti^{3+}/Ti^{4+} redox couple [2] The titanium couple enhances the yield of aniline from the direct reduction of nitrobenzene in acidic solution [3]:

$$PhNO_2 \ + \ 4e \ + \ 4H^+ \ \xrightarrow[\text{cathode}]{Cu} \ PhNHOH \ + \ H_2O$$

$$also \quad Ti(IV) \ + \ e \ \rightleftharpoons \ Ti(III)$$

$$PhNO_2 \ + \ 6Ti(III) \ + \ 6H^+ \ \longrightarrow \ PhNH_2 \ + \ 6Ti(IV) \ + \ 2H_2O$$

$$PhNHOH \ + \ 2Ti(III) \ + \ 2H^+ \ \longrightarrow \ PhNH_2 \ + \ 2Ti(IV) \ + \ H_2O$$

Aniline Amino acid synthesis

The Ti^{3+}/Ti^{4+} couple was employed successfully in electroaminations for the synthesis of amino acids. Aspartic acid was obtained by electrolysis of hydroxylamine in the presence of Ti^{3+} and maleic acid [4]. This couple was used to prepare dl-phenylalanine by electrolyzing cinnamic or 3,4-dihydrocinnamic acid in $2MH_2SO_4$ solution via generation of aminyl radicals according to the method used by Sammels [5], as shown below:

$$Ti^{4+} + e \xrightarrow[\text{Hg cathode}]{} Ti^{3+} \xrightarrow{NH_2OH} Ti^{4+} + OH^- + NH_2^{\cdot} \quad \text{aminylion}$$

$$PhCH = CHCOOH + NH_2^{\cdot} \xrightarrow{Ti^{3+}, H^+} PhCH_2\underset{NH_2}{CH}COOH + Ti^{4+}$$

Aminyl ions thus generated add to certain unsaturated systems to give diamino compounds [6]:

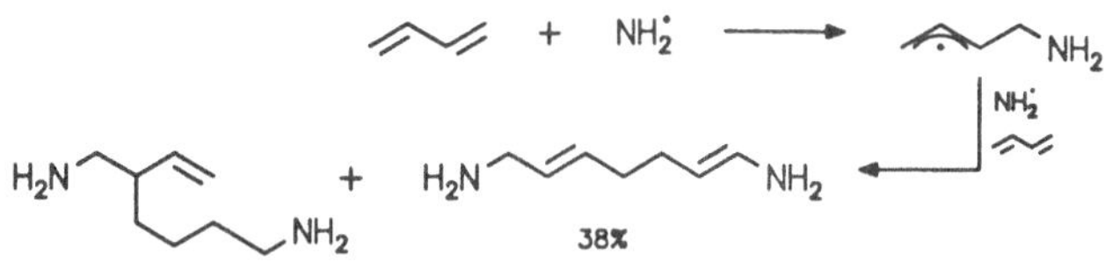

Aminyl ions

Yeager and coworkers [7] recently reported on the catalytic effects of heat-treated mixtures of polyacrylonitrile with Co(II) or Fe(II) salts for the electrolytic reduction of oxygen in both alkaline and acidic solutions. An extension of this catalysis system to organic electrosynthesis might be contemplated. Titanous salts can be used as mediators in micellar electrolysis systems [9,10]. In such media the electrolysis is facilitated by the enhanced miscibility of the organic substrate with the aqueous phase, although sometimes the micelles formed are of a type that inhibit rather enhance the desired effect, depending on whether they are anionic, cationic or monionic. It was reported that the cyanation of methoxybenzenes was favored by cationic micelles but inhibited by the other types [11].

Wolf and Steckhan studied the mediated reduction trichloromethyl carbinols and their corresponding ethers in H_2O/DMF media containing chromous chloride as a mediator. It was thus possible to obtain Z-monochlorovinyl compounds in good yields in one step.

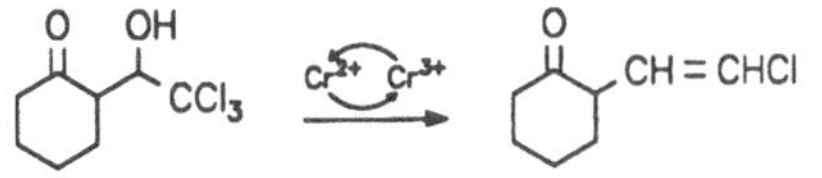

Z-mono-chlorovinyl compounds

The chromous ion was generated in situ and ex situ. These reductions are greatly facilitated by the trichloromethyl group in the α-position with respect to OH or the ether oxygen position in the starting molecules. Pinacolizations (reductions of ketones) are assisted by chromous ion – which, although is not able to reduce the ketonic carbonyl, forms a CrIII-carbonyl complex which is easily reduced at the cathode [13]

A natural reductive mediator is the biologically important cobalt-containing vitamin B12. This cobalt complex and some much simpler models thereof have been found to be reductive catalysts for dehalogenations and couplings of organic compounds [15]. Under appropriate irradiation conditions they also catalyze additions and alkylations [15].

Vitamin B$_{12}$

Reductive reactions using cobalt complexes as mediators may be useful for the synthesis of some natural, biologically interesting compounds. Thus cleavage of C—Br bonds in certain olefins or acetylenes by elec-

Cobalt complexes mediators

Cyclopentanoid products

trolysis with the Cobalt(I)/Cobalt(III) complex redox couple as mediator leads to the formation of cyclopentanoid products [16], as indicated in Scheme 4.15. The cobaloxime(I) is generated at the cathode from cobaloxime(III) at $-1.45\,V$ (vs Ag/Ag$^+$).

cobaloxime (III)
(Ref. 16)

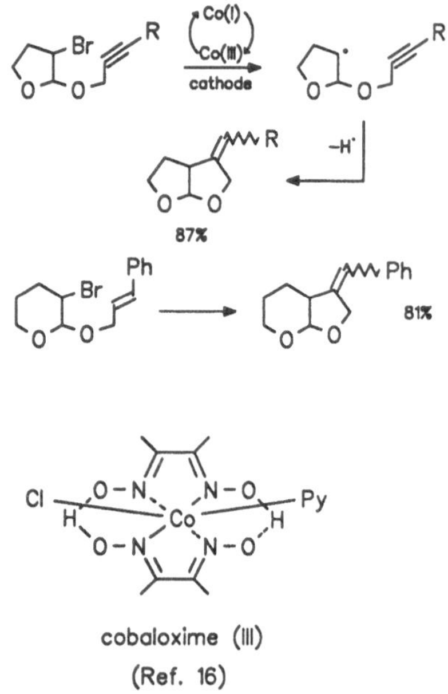

Scheme 4.15

These electrolyses are performed in divided cells equipped with platinum electrodes in methanol solution containing tetraethylammonium tosylate as electrolyte. The catholyte contains the bromo compound and cobaloxime(III) in about 3:1 ratio. A constant current of $15\,mA/cm^2$ is passed through the solution under an inert gas atmosphere. The products are extracted with hexane-ether solvent from the diluted with cold water catholyte solution. They are purified chromatographically and characterized spectroscopically by the usual techniques and elemental analysis. It is noted that these cyclizations can be effected chemically in aprotic solvents with trialkyltin hydride as reductant.

Symmetrical ketones

Symmetrical ketones were obtained by electrocarboxylation catalyzed by nickel bipyridyl complexes, Ni^{0}bpy CO, by electrolysis in DMF or N-methylpyrrolidone containing an organic halide [17,18] in one-compartment cells with a consummable magnesium anode:

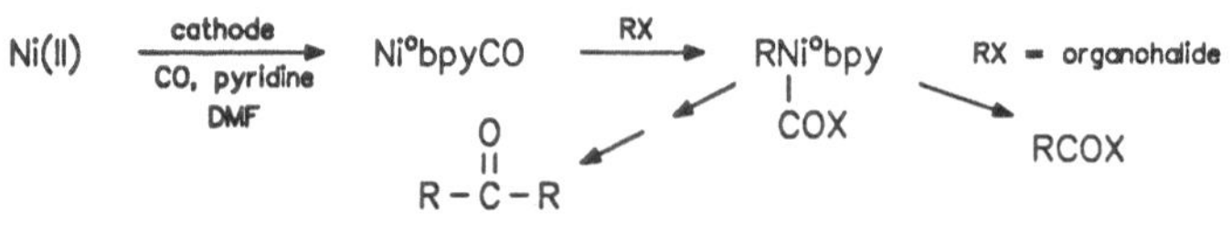

Aliphatic, allyl and benzyl halides afford mainly ketonic products while **Acyl analogs**
aryl and vinyl halides are transformed to their acyl analogs.

Substituted olefins have been prepared via catalytic couplings of aromatic **Substituted**
halides and alkenes with in situ generated organonickel-phosphine com- **olefins**
plexes [19] as catalysts:

$$ArX \; + \; RCH=CH_2 \xrightarrow{\;Ni(II)P(Ph)_3\;} RCH=CHAr \; + \; HX$$

Dinitrosyliron complexes have been used as catalysts for cyclodimeriza- **Iron complexes**
tions of conjugated dienes [20]. The iron complexes are generated in situ
by reduction of $FeCl_3$ in the presence of NO in a propylene carbonate
solution:

$$FeCl_3 \; + \; NO \xrightarrow{\;cathode\;} Fe(NO)_2$$

$$Fe(NO)_2 \; + \; 2 \;\text{(diene)} \longrightarrow complex$$

4.3.2 Reductions with Amalgams

Reductions with amalgams of the alkali and alkaline earth metals and the
more powerful reductants, the tetraalkylammonium ($Q_4\dot{N}(Hg_5)$) amalgams
are a special type of indirect electrochemical reductions [21–23]. These
amalgams are produced by electrolysis at very negative potentials, using
mercury pool cathodes. The produced amalgams are believed to be formed
stoichiometrically as indicated with the organic Q_4N^+ cation:

$$Q_4N^+ \; + \; e \; + \; 5Hg \longrightarrow Q_4\dot{N}(Hg)_5$$

Also tin, bismuth and lead form similar amalgams. Graphite forms *intercalation* compounds between its layers which may act like amalgams in
certain reductions [24]. Amalgams are able to transfer electrons to organic
compounds that are difficult to electronate, such as aryl fluorides, aliphatic
ketones, benzene, benzofurans, aryl esters and aryl amides. The amalgam
can be continuously regenerated by applying the required potential to the
mercury cathode. Therefore, the reduction can be viewed as an electrocatalytic mediated heterogeneous reductive process. These reductions fall
in the general category of *dissolving metal* reductions or *Birch-type* reductions in which direct electron transfer takes place from the metal or the
$Q_4N^\cdot$ radical species to the organic substrate in contact with the amalgam
or with the metal in liquid ammonia:

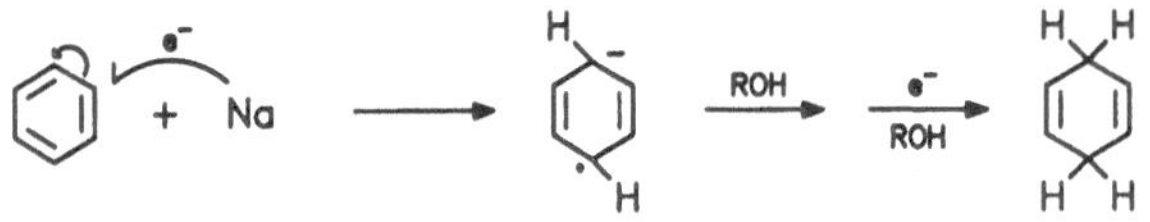

4.3.3 Reductions with Diborane Formed in Situ

Alcohol production

Recently Shundo and coworkers reported an indirect electrochemical method for the transformation of carboxylic acids and olefins to their corresponding alcohols [25]. This method uses sodium borohydride which by anodic oxidation is converted to diborane which reduced the carboxyl group. Also, reaction of diborane with an olefinic bond, followed by oxidation in alkaline H_2O_2 solution, transforms the olefin to the saturated hydroxy compound. These reactions proceed as follows:

$$2BH_4^- \xrightarrow[\text{Pt anode}]{-2e} B_2H_6 + H_2$$

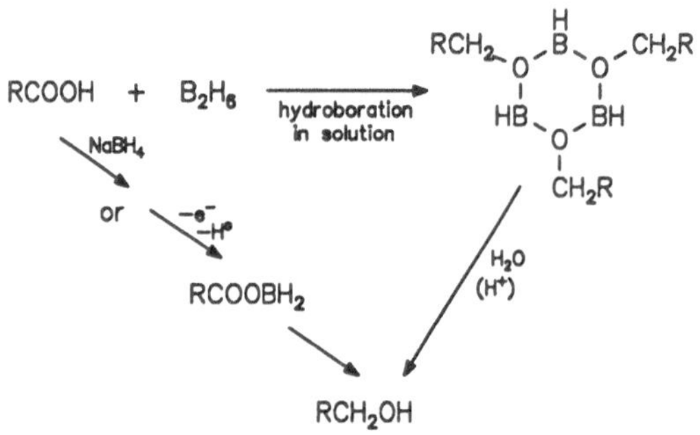

$$3RCH{=}CH_2 + B_2H_6 \xrightarrow{\text{in solution}} (RCH_2CH_2)_3B \xrightarrow[(OH^-)]{H_2O_2} 3RCH_2CH_2OH$$

(BH_3 adds across olefinic bond)

Electrolysis is performed in an undivided cell under constant current of about $50\,mA/cm^2$ using a platinum foil ($2\,cm \times 2\,cm$) as anode and a stainless steel plate as cathode. The borohydride anion is very easily oxidized at the anode (-0.045 V vs SCE) to give diborane. The carboxylic acid or olefin is dissolved in anhydrous diglyme, containing a relative molar excess of sodium borohydride. The electrolysis is carried out at 25 °C under nitrogen. The products are extracted from an acidified solution with ether. Evaporation of the washed and dried ether leaves the product which is purified by distillation under reduced pressure or by silica-gel chromatography. Typical yields are as follows:

$$CH_3(CH_2)_6COOH \longrightarrow CH_3(CH_2)_7OH \quad 70\%$$

$$CH_3OOC(CH_2)_4COOH \longrightarrow CH_3OOC(CH_2)_5OH \quad 49\%$$

$$PhCOOH \longrightarrow PhCH_2OH \quad 65\%$$

$$CH_3(CH_2)_8COOH \longrightarrow CH_3(CH_2)_9OH \quad 72\%$$

Electrolysis of olefins is performed similarly in diglyme-NaI-$NaBH_4$ medium, giving the following typical yields of the corresponding alcohols:

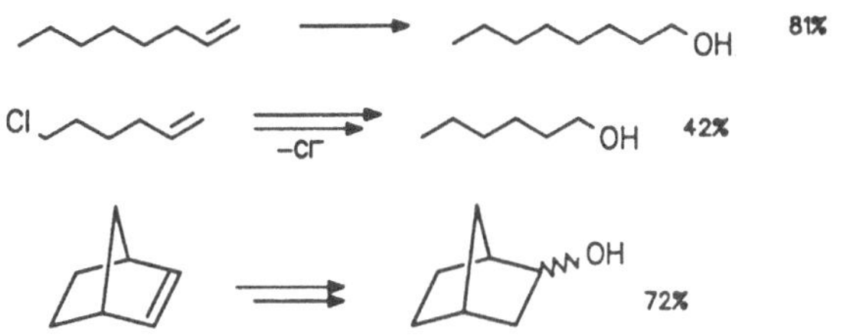

(Conventionally such reductions are effected by reactions with metallic hydrides and catalytic hydrogenation.)

4.3.4 Organic Mediators

Anthracene

Organic mediators for electrochemical reductions are substances that can be reversibly reduced at the cathode to yield anion radicals that are stable in the electrolysis medium and also able to transfer an electron to the organic substrate. Typical such mediators are: azobenzenes, anthracenes, benzonitriles, phenanthridines, some organometallic complexes and metalloporphyrins. For example, anthracene is a mediator for the hydrogenolysis of halobenzenes and halopyridines [1]:

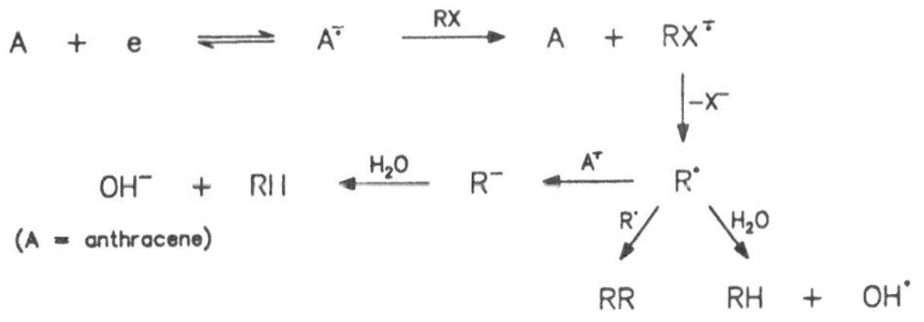

Anthracene effectively mediates the reduction of benzyl and aryl halides by transferring electron in steps to produce radicals $B^{\cdot}$ and anions B^-, from the benzyl or aryl halides [3]. The anions act both as bases and as nucleophiles. Couplings of $B^{\cdot}$ with B^- are insignificant:

$$A + e \longrightarrow A^{\overline{\cdot}}$$

$$A^{\overline{\cdot}} + BX \xrightarrow{k_1} A + B^{\cdot} + X^- \qquad BX = \text{benzyl halide}$$

$$A^{\overline{\cdot}} + B^{\cdot} \xrightarrow{k_2} A + B^-$$

Azobenzene

Azobenzene is a good redox catalyst in acetonitrile solution for organic halides such as bromobenzophenone.

Diphenyl-diselenides and ditellurides

Diphenyldiselenides and ditellurides are good reductive mediators for the transformation of α,β-epoxy compounds to β-hydroxy carbonyl compounds [4]. Also, α,β-epoxyketones and α,β-epoxy esters and nitriles are easily converted to aldols and β-hydroxy analogs in the presence of these mediators. A formal typical example is depicted below:

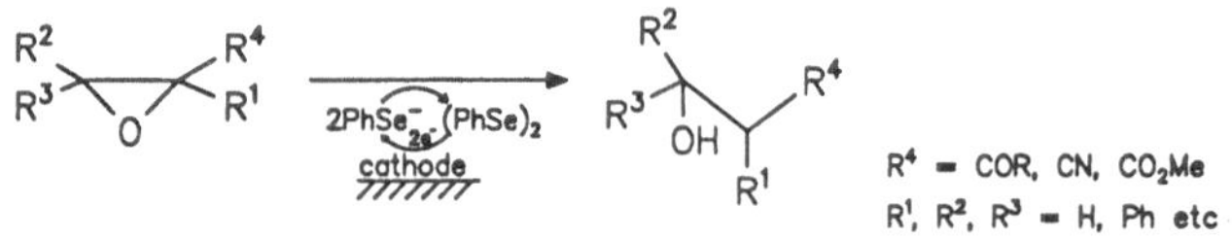

R^4 = COR, CN, CO_2Me
R^1, R^2, R^3 = H, Ph etc

DMP⁺

Dimethylpyrrolidinium ion (DMP⁺) is a mediator for reductions at lead electrodes [5]. This electrocatalysis is heterogeneous in the sense that the reduction of DMP⁺ forms an amalgam-type (DMPPb₅) monolayer at the surface of the lead electrode. This monolayer amalgam acts catalytically for the reduction of alkyl halides – but not for other substrates. Organorhodium complexes are effective reductive mediators [6]. Bis(bipyridine) rhodium is an efficient two-electron transfer agent for the electrochemical generation of NADH from NAD⁺. Direct reduction of NAD⁺ gave mainly dimers of NAD.

Organor-hodium complexes

Biological redox catalysts

In concluding this section we refer to the biological redox catalysts such as the various metalloporphyrins, chlorophyls, hemoglobin and vitamin B_{12} which are indispensable catalysts for life processes in plants and animals [7,8]. The catalytic activity of these organometallic substances is attributed to the reversible electron transfer process of the transition metal (Mg, Fe, Co) in the N-macrocyclic complex.

References

Section 4.1   Introduction

1. Steckhan E (1986) Angew Chem, Int Ed Eng 25:683; Gross-Brinkhaus KH, Steckhan E, Schmidt W (1983) Acta Chem Scand B37:499
2. Alkire R, LaRoche R, Cera G, Stadtherr M (1986) J Electrochem Soc 133:290
3. Pichaichanarong P, Spotnitz RM, Kreh RP, Goldfard SM, Lundquist JT (1990) Chem Eng Commun 94:119
4. Andrieux CP, Celis L, Saveant JM (1990) J Am Chem Soc 112:786
5. Eberson L (1980) Acta Chem Scand B34:481; (1984) Acta Chem Scand B38:439
6. Kyriacou D, Jannakoudakis D (1986) Electrocatalysis for organic synthesis. Wiley-Interscience, New York
7. Wendt H (1984) Electrochim Acta 29:1513
8. Schmidt W, Steckhan E (1978) J Electroanal Chem 89:221
9. Shono T, Matsumura Y, Hayaski J, Mizoguchi M (1979) Tetrahedron Lett 165

9a. Ghaladi EA, Belccadi S (1988) J Appl Electrochem 18:27

9b. Taube H, Myers H, Rich RT (1983) J Am Chem Soc 75:4118

10. Eberson L, Saik SS (1990) J Am Chem Soc 112:4484; Andrieux CP, Celis L, Medielle M, Pinon J, Saveant JM (1990) J Am Chem Soc 112:3509; Kochi JK (1990) Acta Chem Scand 44:409

11. Steckhan E, Dapperhed S, Grose Brinkhaus H, Sech T (1991) Chem Ber 124:255

Section 4.2   Oxidative Indirect Reactions (4.2.1 and 4.2.2)

1. Koel WJ (1969) US Patent 3448021

2. Torii S et al. (1982) J Org Chem 47:1647

3. Van Veen JAR, Van Der Eijk, DeRuter R, Huizinga S (1988) Electrochim Acta 33:1

4. Moyer BA, Thompson MS, Meyer TJ (1980) J Am Chem Soc 102:2310

5. Thompson MS, De Giovani WF, Moyer BA (1984) J Org Chem 49:4972

6. Torii S, Inokutchi T, Sugiera T (1986) J Org Chem 51:55

7. Mayell JS (1968) Ind Eng Chem, Proc Res Dev 7:129

8. Kuhn AT, Birkett M (1979) J Appl Electrochem 9:777

9. Ibl N, Kramer K, Pinto L, Robertsen R, (1974) AIChE Symposium Series 18575:45

10. Chung RH (1978) In: Kirk-Othmer Encyclopedia of Chemistry and Technology, vol 2, 3rd edn. Wiley, New York, p 702

11. Heiba EI, Desau RM, Rodewald PG (1974) J Am Chem Soc 96:7977

12. Kurz ME, Chen TYR (1978) J Org Chem 43:239

13. Bellamy AJ (1971) Acta Chem Scand B33:208

14. Jow JJ, Lee TC (1987) J Appl Electrochem 17:753

15. Ito S, Katayama R, Kunai A, Sasaki K (1989) Tetrahedron Lett 30:205

16. Steckhan E, Wellman J (1976) Angew Chem Int Ed Eng 15:294

17. Tomat F, Rigo A (1976) J Appl Electrochem 6:257

18. Wellman J, Steckhan E (1977) Chem Ber 110:3561

19. Tzedakis T, Savall A, Clifton MJ (1989) J Appl Electrochem 19:911

20. Feszar B, Sobkoviak A (1983) Electrochim Acta 28:1351

21. Kinoshita T, Harada J, Ito S, Sasaki K (1983) Angew Chem Int Ed Eng 22:502

22. Bhadani SN, Ansari Q (1988) Makromol Chem Rapid Commun 9:171

Section 4.2.3   Halide Ions as Mediators

1. Stevens RV, Chapman KT, Weller HN (1980) J Org Chem 45:2030

2. Shono T, Matsumura Y, Hayski J, Mitsoguchi M (1979) Tetrahedron Lett 165

3. Shono T et al. (1985) J Org Chem 50:4917

4. Dixit G, Rastogi R, Zutshi K (1982) Electrochim Acta 27:561

5. Okimoto M, Chiba T (1988) J Org Chem 53:218

6. Alkire R, Köhler J (1988) J Appl Electrochem 18:405

7. Ginzel KD, Brungs P, Steckhan E (1989) Tetrahedron 45:1691

8. Barba F, Garsia PA, Guirado A, Zapata A (1982) Carbohyd Res 105:158

9. Iwasaki T, Nishitani T, Horikawa H, Inoue I (1982) J Org Chem 47:3799

10. Jackson C, Kuhn AT (1973) In: Kuhn AT (ed) Industrial electrochemical processes. Elsevier, Amsterdam, p 153

11. Thomas HG, Lux E (1972) Tetrahedron Lett 10:965

12. Mehltretter CL, Rankin JC, Watson PR (1957) Ind Eng Chem 49:350

13. Yoshiama TC, Nonaka T, Baizer MM, Chou TC (1985) Bull Chem Soc (Japan) 58:201

14. Yoshida J, Hashimoto J, Kawaba N (1982) J Org Chem 47:3575

15. Torii S, Uneama K, Ono M, Barnou T (1981) J Am Chem Soc 103:4606

16. Nikishin GI, Elinson MH, Makhova IV (1988) Tetrahedron Lett 29:1603

Section 4.2.4   Nitrate Ion as Mediator

1. Leonard JE, Scholl PC, Steckel TP (1980) Tetrahedron Lett 21:4695

Section 4.2.5   Organic Mediators for Oxidations

1. Schmidt W, Steckhan E (1978) J Electroanal Chem 89:215; Gunic J, Tabakovic I (1990) Electrochim Acta 35:225; Lund H, Tabakovik I (1984) Adv Heterocycl Chem 36:235
2. Schmidt W, Steckhan E (1980) Chem Ber 113:577
3. Sanicarin Z, Tabakovic I (1986) Tetrahedron Lett 27:407
4. Masui M, Ushima T, Ozaki S (1983) J Chem Soc Chem Commun 479
5. Platen M, Steckhan E (1978) Angew Chem, Int Ed Eng 17:673
6. Platen M, Steckhan E (1980) Tetrahedron Lett 21:511; Platen M, Steckhan E (1984) Liebigs Ann Chem 1563
7. Cinzel KD, Steckhan E (1987) Tetrahedron 43:5597
8. Eberson L (1982) Adv Phys Org Chem 18:79
9. Derongier A, Moutet JC (1989) Acc Chem Res 22:249
10. MacCorquodale F, Crayston JA, Walton JC, Worsfold DJ (1990) Tetrahedron Lett 31:771

Section 4.3   Indirect Reductive Reactions
Section 4.3.1–4.3.3 Metallic Ions as Catalysts

1. Fegori G, Lund H (1976) Acta Chem Scand B30:651
2. Unathraraman PN, Udupa HVK (1983) J Electrochem Soc (India) 32:51
3. Kargin YM, Kamalova GA (1989) J Cen Chem (USSR) 58:2063
4. Farnia G, Sandova G, Viamello E (1978) J Electroanal Chem 88:147
5. Cook RI, Sammels AF (1989) J Electrochem Soc 136:1845
6. Chiba T, Takata Y (1978) Bull Chem Soc (Japan) 51:1481
7. Gupta S, Tryk D, Bae I, Aldred W, Yeager E (1989) J Appl Electrochem 19:19
8. Ref. 4, section 4.3.1
9. Udupa HVK (1980) Electrochim Acta 25:1083
10. Eberson L, Helgee B (1975) Acta Chem Scand B29:451
11. Laurent E, Rauniar G, Tomala M (1982) Nouv J Chim 6:515
12. Wolf R, Steckhan E (1986) J Chem Soc Perkin Trans 1:733
13. Sopher DW, Utley JHP (1979) J Chem Soc Chem Commun 1087
14. Fournier F, Bertbelot J, Paskal YL (1984) Tetrahedron 40:339
15. Lexa RSD, Saveant JM, Sauplet JP (1979) J Electroanal Chem 100:159; Margel S, Anson FC (1978) J Electrochem Soc 125:1232
16. Torii S, Inokuchi T, Yukawa T (1985) J Org Chem 50:5875
17. Troupel M, Rollin V, Perichon J, Fauvarque JF (1981) Nouv J Chim 5:621
18. Fauvarque JF, Jutand D, Francois M (1988) J Appl Electrochem 18:109
19. Rollin Y, Meyer G, Troupel M, Fauvarque JF, Perichon J (1983) J Chem Soc Chem Commun 793
20. Le Roy E, Huchette D, Mortreaux A, Petit F (1980) Nouv J Chim 4:173
21. Settineri WJ, McKeever D (1975) In: Weinberg NL (ed) Technique of electroorganic synthesis Wiley-Interscience, New York, p 397
22. Kariv-Miller E, Lawin PB, Vajtner Z (1985) J Electroanal Chem 195:435
23. Simonet J, Lund H (1977) J Electroanal Chem 75:719
24. Birch AJ, Nasipari D (1959) Tetrahedron 6:148; Horner L, Newman H (1965) Chem Ber 98:3462
25. Shumdo R, Matsubara Y, Nishiguchi I (1992) Bull Chem Soc (Japan) 65:530

Section 4.3.4   Organic Mediators

1. Simonet J, Michel NA, Lund H (1974) Acta Chem Scand B29:900
2. Avaca LA, Gonzales ER, Ticianelli EA (1983) Electrochim Acta 28:1473
3. Degrand C, Prest R (1987) J Org Chem 52:5224
4. Inokucuchi T, Kusomoto M, Torii S (1990) J Org Chem 55:1548
5. Lawin PB, Hutson AC, Kariv-Miller E (1989) J Org Chem 54:526
6. Wienkamp R, Steckhan E (1982) Angew Chem, Int Ed Eng 21:782
7. Dolphin D, Felton RH (1974) Acc Chem Res 7:26
8. Lexa D, Saveant JM (1983) Acc Chem Res 16:235

In this chapter we overview some special topics of importance to organic electrosynthesis. Only a very brief general discussion of each topic is possible within the space of this book.

5.1 Electrolysis in Two-Phase Solvent Systems

Immiscible solvents, emulsions, dispersions, and micelles

Phase-transfer catalyst (PTC)

Benzaldehyde

Some electrochemical syntheses may be advantageously conducted in two-phase electrolysis media that are composed of two immiscible solvents, usually water and an organic solvent, such as dichloromethane or nitrobenzene. Emulsions, dispersions and micellar electrolysis media have been used as two-phase media for various synthetic applications. The recent reviews by Feess and Wendt [1] and reports by Pletcher [2] and Eberson [3] and others [4–11] contain valuable information regarding this methodology. Two-phase electrolyses usually require the presence of a phase-transfer catalyst (PTC) such as a quaternary ammonium salt. A typical example is the indirect oxidation of benzyl alcohol to benzaldehyde, in CH_2Cl_2/H_2O emulsion. The process is depicted in Scheme 5.1

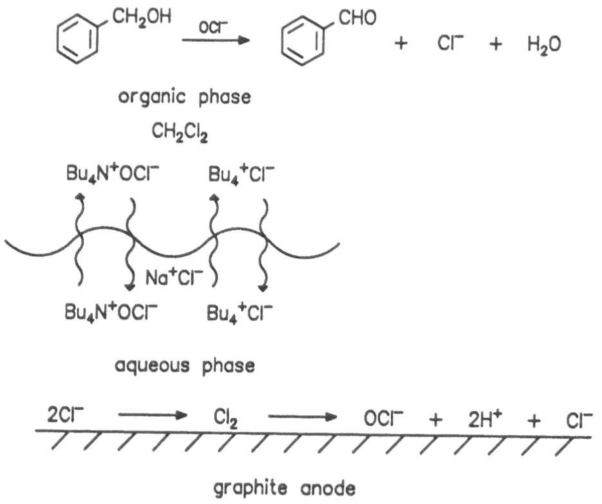

Scheme 5.1

The benzyl alcohol is oxidized by the ClO^- ion which *shuttles* between the organic and the aqueous phase by associating with PTC. In a typical experiment [5] the two-phase system consisted of dichloromethane and 0.1 M NaCl aqueous solution, 70 ml of each. The concentration of benzyl alcohol in the CH_2Cl_2 solvent was 0.5 M, and the PTC was Bu_4NSO_4. The anode potential was set at 1.3 V versus SCE, which is sufficient to oxidize Cl^- to Cl_2. Current was passed under vigorous stirring of the emulsion (mechanical stirring 500 rpm). Benzaldehyde is produced with higher than 80% current efficiency. Naphthalenes are chlorinated in $CH_2Cl_2/H_2O/$ NaCl emulsions and afford their monochlorinated analogs in 40–90% yields [10]. It is possible in certain cases to achieve higher yields by two-phase electrolysis [12]. Thus optimum yields of N,N,N',N'-tetramethyl-$1,1'$-naphthadine (TMN) were obtained by electrolysis of N,N-dimethyl-1-naphthalamine (DMN) in water/nitrobenzene/NaClO4 emulsion. The optimization was ascribed to preferential removal of the generated protons into the aqueous phase thus minimizing formation of the inactive $DMNH^+$, Scheme 5.2

Chlorination of naphthalenes

$$DMN \longrightarrow DMN^{+\cdot} + e$$

$$2\,DMN^{+\cdot} \longrightarrow TMN + 2H^+ \quad \text{aqueous phase}$$

$$TMN \rightleftharpoons TMN^{2+} + 2e$$

$$DMN + H^+ \longrightarrow DMNH^+ \quad \text{inactive}$$

$$TMNH^{2+} + H_2O \xrightarrow[\text{(overnight)}]{\text{slow}} TMN + 2H^+ + 1/2O_2$$

Scheme 5.2

Styrene was polymerized in a two-phase medium consisting of styrene and formamide with zinc salts as electrolytes [13].

Polymerization of styrene

The electrochemical reduction of 2-ethylanthraquinone (EAQ) to 2-ethylanthrahydroquinone (EAHQ) in a two-phase system consisting of a mixture of 2-ethylbenzene and tributyl phosphate, and aqueous 2 M sodium hydroxide was recently studied under well-defined conditions by Tobias and coworkers [14]. It was found that the direct electrochemical reduction of EAQ was a feasible process as an alternate step in the industrial production of hydrogen peroxide. The manufacture of hydrogen peroxide involves two chemical steps:

EAQ to EAHQ

Hydrogen peroxide manufacture

1. EAQ + H_2 $\xrightarrow{\text{reduction}}$ EAHQ

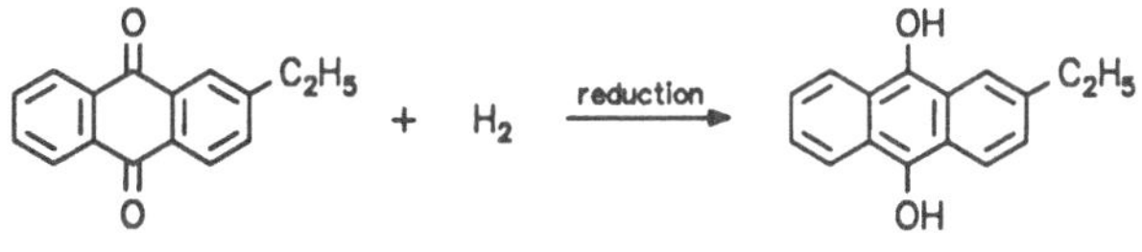

2. EAHQ + O_2 $\longrightarrow$ H_2O_2 + EAQ
    (org.)   (aqu.)   (org.)

    spontaneous oxidation

Direct electroreduction of EAQ took place at a lead cathode at the boundary of the organic/aqueous/electrode system. The electrolysis was carried out in a cell divided with a Nafion cation exchange membrane under controlled potential and an inert atmosphere. The catholyte was magnetically stirred so as to disperse the organic phase in the aqueous 2 M NaOH phase. The anolyte was aqueous NaOH solution and the anode a nickel foil. Current efficiencies of about 100% and current densities of $30 \, mA/cm^2$ were attained with cylindrical electrodes under vigorous turbulance electrolysis conditions. A cell voltage of 1.8 V was sufficient to start the electrolysis as shown by immediate red coloration of the catholyte solution. The overall reaction at the cathode was a two-electron process, expressed as follows:

$$2Na^+ + EAQ + 2e \xrightarrow[\text{cathode}]{Pb} Na_2EAHQ$$
$$\text{aqueous} \quad \text{organic}$$

5.2 Solid Polymer Electrolyte – Electrolysis

Organic membranes

A solid polymer electrolyte, SPE, is an organic membrane, a polyelectrolyte, which while acting as a cell divided allows the passage of either anions or cations through it so that the electric current can be carried across the cell without the presence of a soluble electrolyte (supporting electrolyte).

Metal particles

It is possible to prepare SPE membranes with selective electrocatalytic properties by incorporating metallic particles on one or both sides of the membrane [5,6]. Thus far the Nafions 415, 417 and 315 have been given most consideration. A combination of cathode and anode is fabricated with the SPE membrane *sandwiched* between anode and cathode so that the cell is in effect divided by the SPE membrane. The metal particles can be deposited on the membrane either by electroplating or by chemical reduction of chloroplatinic acid with hydrazine [7]. Nafion-SPE are polyfluorinated ion exchange membranes based on perfluoroethylene-perfluoropropane oxide copolymers [4,8] with chemical structures represented as

$$(-OCF_2 - CF_2)_x OCF_2 CF_2 SO_3 H$$

$$\text{and} \quad (-CF_2 CF_2)_x CF_2 CF_2 \underset{\underset{(-OCF_2 CF)O(CF_2)_n \overset{O}{\overset{\|}{C}} - OR}{\big|}}{CF_3}$$

SPE composite electrode

Poly(ethyleneoxide) and other materials are also investigated for uses as SPE [9]. A Nafion-415 cation exchange membrane modified with Cu-Pt was successfully used as a SPE *composite* electrode for the reduction of

nitrobenzene to aniline [7]. Current efficiencies of 95% were attained. **Reduction of**
Two sheets of expanded titanium mesh covered with platinum gauze were **nitrobenzene**
firmly pressed onto both sides of the SPE composite electrode to function
as *current feeders*. The catholyte was a 30% nitrobenzene in methanol and **Current**
the anolyte pure water. Thus no supporting electrolyte was needed in this **feeders**
system. The water was the proton source for the formation of aniline as
the nitrobenzene was electronated at the cathode side of the SPE.

Metal-SPE-type electrodes have much larger active areas than ordinary **Oxidations of**
metal electrodes of the same apparent geometric area. It was observed **isopropanol,**
that the oxidation of CO on a Au-SPE composite electrode was about 100 ***n*-octanol, and**
times faster than at a gold electrode of similar geometric area [10]. The **propargyl**
oxidations of isopropanol, *n*-octanol and propargyl alcohol were studied **alcohol**
using SPE-cells. High yields of oxidized products were obtained [11].
Electrocatalytic oxidations of alkyl alcohol vapors at Au-Nafion 117 com- **Oxidation of**
posite electrodes were recently reported [12]. Poly(aniline)-Nafion and **alkyl alcohol**
poly(3-methylthiophene)-Nafion composite films were prepared by a one- **vapor**
step electrolysis in propylene carbonate using Li$^+$-Nafion and H$^+$-Nafion
and the monomer precursors of the electronically conducting polymers
poly(aniline) or poly(3-methylthiophene) [13]. Such materials may find
uses as SPE for some special synthetic applications. Figure 5.1 represents
typical SPE cells.

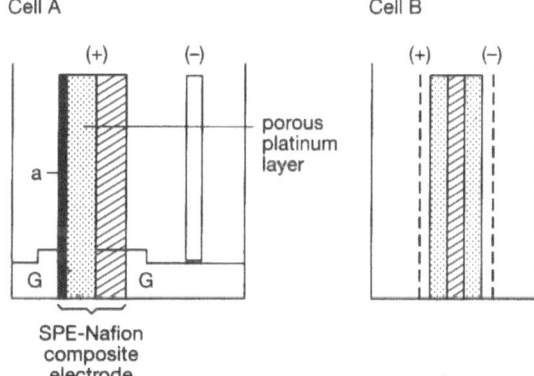

Fig. 5.1. Schematics of SPE-Composite electrode Cells
Cell A: With one side Pt-SPE composite electrode as anode. a: current collector, a
platinum screen firmly pressed against the porous platinum layer deposited on the SPE
membrane (nafion). *G*: teflon gaskets
Cell B: With Pt-SPE composite electrodes on both sides of the SPE membrane. One side
acting as anode and the other as cathode

5.3 Reactions via Electrode Dissolution

**Organome-
tallics**

Under certain electrolysis conditions the electrode itself may become one of the reactants or a special catalyst in the overall synthetic reaction [1]. Electrogenerated radicals may attack the metal electrode thus causing formation of organometallics containing carbon-metal bonds. Such possibilities are more common at cathodes than at anodes. Tetraalkyl lead compounds are produced by attack of alkyl radicals on the lead electrode (Nalco process [2]):

$$RCl \; + \; Mg \; \longrightarrow \; RMgCl \; \longrightarrow \; R^- \; + \; MgCl^+$$

$$4R^- \; \xrightarrow{-4e} \; 4R^{\bullet} \; \xrightarrow{Pb} \; R_4Pb$$

**Tetraethyl
lead**

Ethyl iodide produces radicals at a lead cathode which attack the electrode to form tetraethyl lead:

$$C_2H_5I \; \xrightarrow[Pb]{e} \; C_2H_5^{\bullet} \; + \; I^-$$

$$Pb \; + \; 4C_2H_5^{\bullet} \; \longrightarrow \; Pb(C_2H_5)_4$$

**Alkylation of
aromatic rings**

**Aluminum
oxalate**

Aluminum anodes in the presence of alkyl halides afford upon anodic dissolution Friedel-Crafts catalysts for in situ chemical alkylation of aromatic rings [3]. Electroreduction of CO_2 in DMF/0.1 M Bu_4NBr and dissolution of an aluminum anode result in the formation of aluminum oxalate [4] with 38% current yield.

$$2CO_2 \; + \; 2e \; \xrightarrow{Hg} \; C_2O_4^{2-}$$

$$\underset{anode}{Al} \; \longrightarrow \; Al^{3+} \; + \; 3e$$

$$2Al^{3+} \; + \; 3C_2O_4^{2-} \; \longrightarrow \; Al_2(C_2O_4)_3$$

**Carboxylations
of alkenes,
ketones,
organohalides,
aldehydes and
imines**

**Monoalky-
lation**

The dissolvable anode methodology makes possible the use of undivided cells and the application of high pressures needed for certain electrosynthetic processes [1]. Cathodic carboxylations of a wide variety of organic substrates have been reported in recent years using metallic anodes which thermodynamic and kinetic factors allow to dissolve upon anodic polarization in organic solvents. Carboxylations of alkenes, ketones, organohalides, aldehydes and imines are conveniently performed in cells with dissolvable metallic anodes. The dissolved metals may be recovered by electrolytic reduction or by precipitation with OH^- and subsequent isolation as metallic oxides. Activated methylene compounds are very easily monoalkylated in undivided cells fitted with consumable magnesium anodes, in DMF/Bu_4NBF_4 media [5]. Constant current electrolysis in an undivided cell with a magnesium rod anode and a cylindrical stainless steel cathode gave high yields (80–85%) of alkylated products according to the following reaction scheme:

$$\text{anode,} \quad Mg \xrightarrow{\;-2e\;} Mg^{2+}$$

$$\text{cathode,} \quad RX \xrightarrow{\;2e^-\;} R^- + X^-$$

$$R^- + R' - CH_2 - A \longrightarrow RH + R' - \overset{-}{C}H - A$$

$$R' - \overset{-}{C}H - A + RX \longrightarrow R' - \overset{\overset{\displaystyle R}{|}}{C}H - A + X^-$$

$$R' - CH_2 - A + 2RX + 2e \longrightarrow R' - \overset{\overset{\displaystyle R}{|}}{C}H - A + RH + 2X^-$$

overall cathodic reaction

— —

RX = EtBr, PhCH$_2$Cl

R'CH$_2$A = active methylene substrate eg. PhCH$_2$CO$_2$Et, PhCH$_2$CN

Alkoxides of metals and metalloids have been prepared electrolytically using metals or metalloids as anodes [6]. Tetraethyl silicate is produced by electrolysis in ethanol solution and a ferrosilicon anode. Tetramethylsilane and dimethyldichlorosilane were obtained by electrolysis in LiCl-CH$_3$Cl-cellosolve with Si-Cu or Si-Ca anodes by dissolution of the silicon-containing electrode [7]. **Alkoxides, silicates, and silanes**

Cationic polymerizations are initiated by anodic dissolution of certain sacrificial anodes [8]. The polymerization of styrene lactones and other monomers was effected by electrolysis in CH$_2$Cl$_2$ or CH$_3$NO$_2$ solvents with Bu$_4$NClO$_4$ as electrolyte using mercury or aluminum sacrificial anodes. (Cu, Ag, Sn, Pb were not effective for polymer initiation.) **Cationic polymerication**

Benzoaniline was electrocarboxylated to give the corresponding α-amino acid in a filter-press type cell with semicontinuous renewal of sacrificial electrodes [9]. **α-amino acids**

We have already referred to the electrochemical preparation of metallic salts of aliphatic carboxylic acids and the preparation of metallic alkoxides by anodic dissolution of metal electrodes. The preparation of metal thiolates of Zn, Cd, Hg, Pb and Sn by anodic dissolution of these metals has been recently reported [10]. **Metal thiolates**

Dissolution of elemental sulfur and silicon may result in the formation of organoelemental bonds as shown below [10]:

$$S + 2e + 2H^+ \longrightarrow H_2S$$

$$2\,H_2C = CHCN + H_2S \longrightarrow (NCCH_2CH_2)_2S$$

$$Si + 4Cl^- \longrightarrow SiCl_4 + 4e$$

$$SiCl_4 + 4C_2H_5OH \longrightarrow Si(OC_2H_5)_4 + 4HCl$$

Electrochemical preparation of tetradentate Schiff-base complexes of Zn(II), Cd(II) and Ni(II) have been effected using the appropriate metal as the consumable anode in acetonitrile containing the ligand (H$_2$L) and tetrabutylammonium perchlorate as electrolyte [11], as indicated below: **Tetradentate Schiff-base complexes**

$$H_2L + 2e \xrightarrow{Pt} L^{2-} + H_2 \quad \text{cathode}$$

$$M_{metal} + L^{2-} \longrightarrow ML + 2e \quad \text{anode}$$

$$H_2L =$$

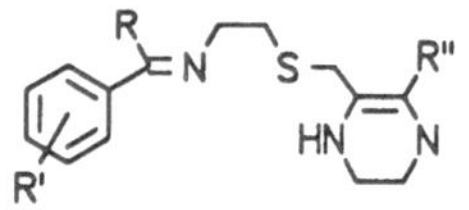

(R, R', R" = H, Me, OMe, OEt etc)

The complexes were isolated as powders, monomeric with zinc and cadmium and at least dimeric with nickel, as implied in the abbreviated structures below:

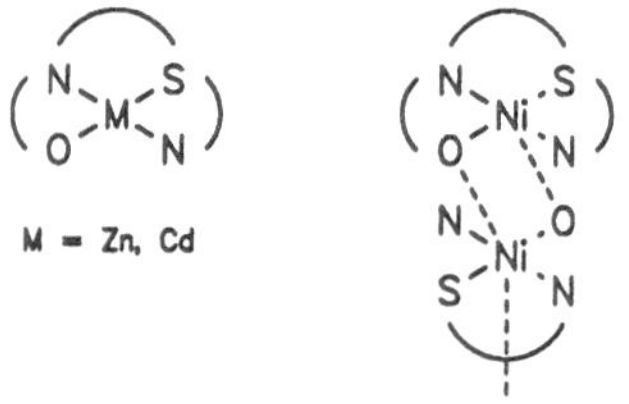

M = Zn, Cd

A review concerning consumable anode methodology has been published recently [12].

5.4 Electropolymerizations. Conductive Polymers

5.4.1 General View

Although it has been known for quite a long time that organic monomers can polymerize by electrolytic methods there is no reference in the open literature to any industrial electropolymerization process. Electropolymerizations differ from the usual polymerization in that the former processes occur at the electrode surface and the electrode may become coated with the polymer. The process may thus stop unless the coating polymer is electrically conductive. Therefore conditions must be used that allow rapid diffusion of the initially generated low molecular weight polymer species to diffuse away from the electrode surface so that further polymerization may continue in solution.

Acrylonitrile monomer   Electropolymerizations have been carried out by both anodic and cathodic initiation [1,2,3]. An anionic-initiated polymerization such as one with the acrylonitrile monomer, A, may be described as follows:

$$A + e \longrightarrow A^{\cdot -}$$

$$2 A^{\cdot -} \longrightarrow (A_2)^{2-}$$

$$(A_2)^{2-} + A \longrightarrow (A_3)^{2-} \longrightarrow \longrightarrow \longrightarrow (A_n)^{2-}$$

$$(A_n)^{2-} + A \longrightarrow (A_n)^- + A^{\cdot}$$

$$(A_n)^- + A \longrightarrow (A_{n+1})^-$$

$$(A_n)^- + A \longrightarrow (A_n) + A^{\cdot}$$

A free radical polymerization of styrene, S, can be initiated by a free radical species as one derived from the anodic oxidation of a carboxylate ion:

Styrene monomer

$$RCOO^- \xrightarrow{-e} R^{\cdot} + CO_2$$

$$R^{\cdot} + S \longrightarrow RS^{\cdot} \xrightarrow{S} RSS^{\cdot} \text{ etc.}$$

In electrolytic polymerization it is relatively easy to control the molecular weight distribution and possibly the polymer structure by manipulating the current-time parameter as in the "living polymer" process in which the reaction can be terminated by a single polarity reversal [1].

Anodic polymerizations of aromatics are well-known reactions. Benzene and *p*-methoxy derivatives are polymerized by an one-electron initiated process [3]:

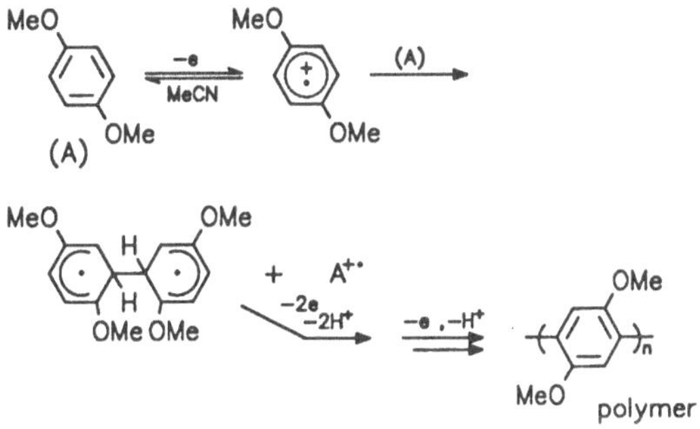

Control of molecular weight distribution and polymer structure

Anodic polymerization of benzene in liquid sulfur dioxide yields poly-*para*-phenylene films which are electroactive if the electrolyte is Bu_4NBF_4 but non-electroactive if the electrolyte is Bu_4NClO_4. Electropolymerization of *ortho*- and *meta*-toluidine in aqueous sulfuric acid yields polymer films which have electrochromic properties [5].

Electroactive and non-electroactive films

Anodic polymerization of dithienylbenzenes in CH_2Cl_2/Bu_4NBF_4 solution gave polymeric films that could be cycled between yellow (reduced) and blue (oxidized) forms [5a]:

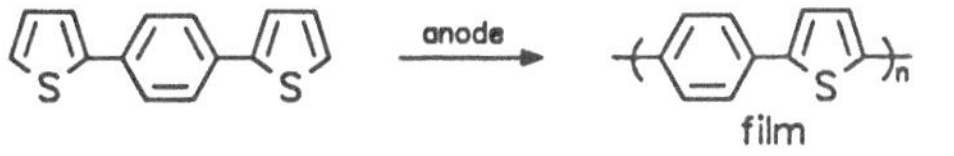

Films with electrochromic properties

Electrooxidation of naphthalene in dry acetonitrile or in 1,2-dichloroethane with Q_4BF_4 salts as electrolytes gives the 1,4-polynaphthalene [6]. Hydroquinone is polymerized by oxidation in $CH_3/NO_2/Q_4NClO_4$ solution [7]. The polymerization proceeds cationically:

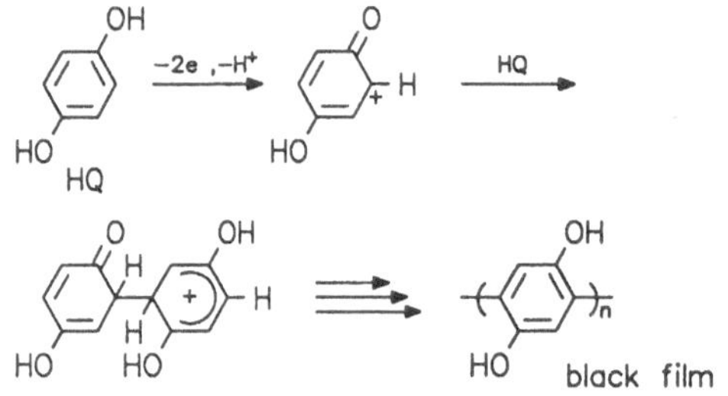

In acidic media the electroactivity involves the protonated polymer:

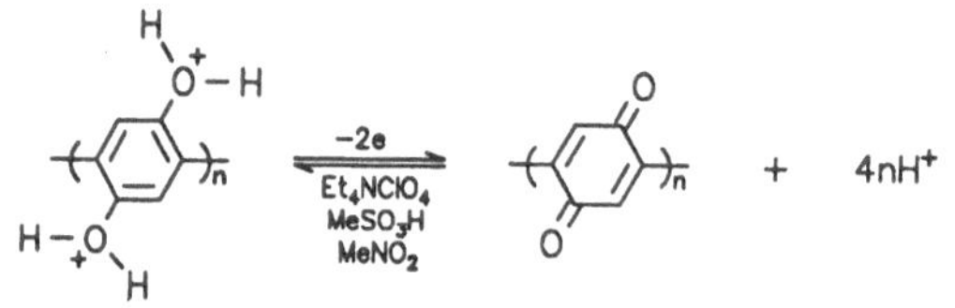

Vinylpyridine polymer   At a silver electrode in aqueous solution electrochemically initiated polymerization of 4-vinylpyridine was proposed to take place as depicted below [8]:

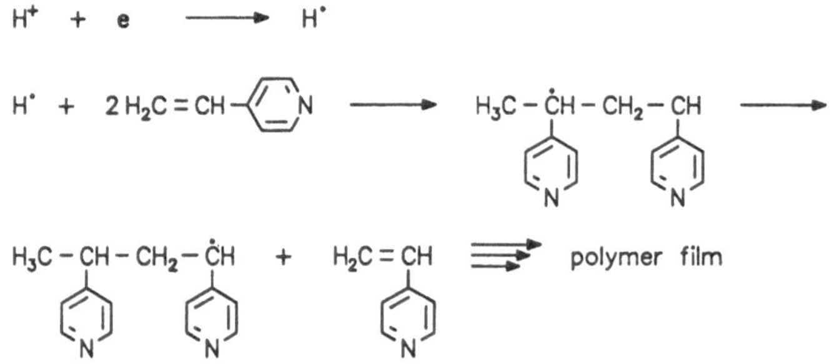

Vinylpyridine dimer   This electropolymerization occurs at the potential region of -0.1 to $-1.0\,V$ (vs Ag/AgCl). If the applied potential is more negative than $-1.5\,V$ the vinylpyridine molecules are directly electronated to give the anion radicals which dimerize to give dianions and are then protonated to afford the neutral dimer product:

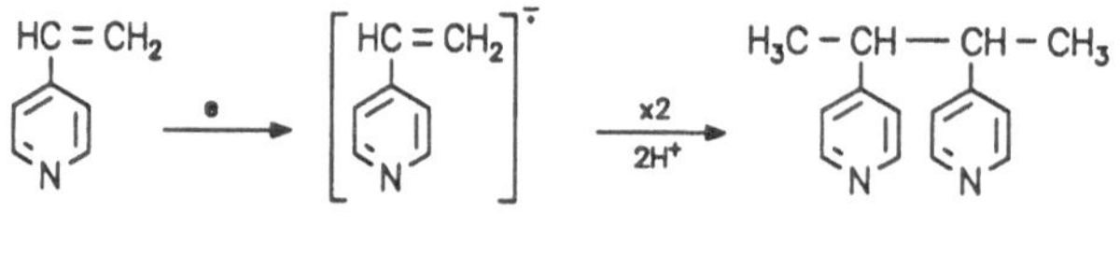

5.4.2 Conductive Polymers

In recent years – and particularly as a result of the developing semiconductor technology, the search for organic materials exhibiting electronic conductivity has become an active subject of study by both chemists and physicists. Following Little's article [9] (1964) the interest in *organic metals* grew exponentially. Polyacetylene was the first polymeric synthetic metal [10,12]. Following this work verious polymers were synthesized exhibiting electronic conductivities in their oxidized or reduced states [11].

Organic metals

Polymeric organic conductors are classified in five systems [12]:

Polymeric organic conductors

a.

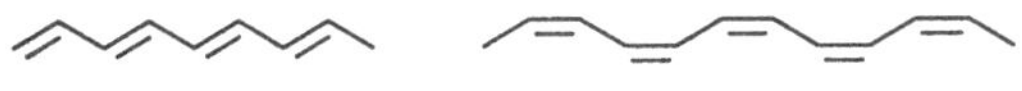

simple polyacetylene

made at −10 °C   made at −78 °C

b. 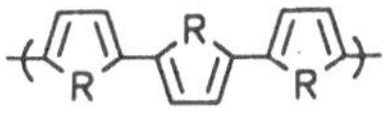 R = NH, S, O

c. 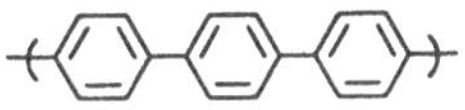 polyparaphenylene

d. 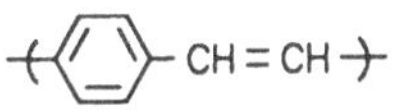 mixed

e. 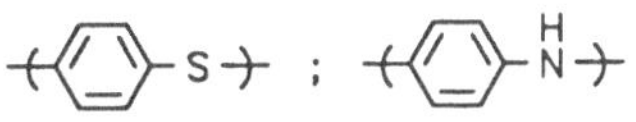 heteroatom in main chain

It is apparent that all the polymers in these systems posses conjugated π-electron systems – otherwise they could not be conductive. In their pure state they are usually not conductive, but some of them may have semiconductive properties. They become conductive upon "doping" by suitable chemical or electrochemical oxidations. By the process of doping they undergo injection of electrical charges into their main chains. These charges may be either positive or negative or only charges of one kind. Polypyrrole, polythiophene, polyaniline and derivatives are easily prepared by anodic oxidation methods [11,14,15]. Doping may be accomplished by intercalation and by controlled potential electrochemical procedures. Conductive organic polymers may be useful for a number of practical applications, e.g. light weight energy storage devices (solar batteries) and in electronic circuitry applications [16].

"Doping"

Electrochemical synthesis of conductive polymers allows preparation of layers of polymers of the desired thicknesses and shapes and dimensions generally not possible by chemical methods [17].

Thus far most promising conductive polymers have been those made from the monomers of pyrrole, thiophene, aniline and benzene. These

Solar batteries and electronic circuitry applications

Polypyrrole the best known conductive polymer

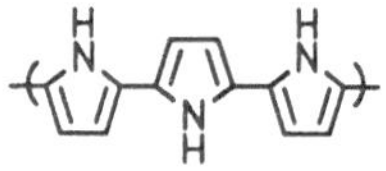

polymers can be easily prepared electrolytically in organic or aqueous solutions. Conductivities are usually in the semiconductor range [10^{-9} to $(100\ \Omega\ \mathrm{cm})^{-1}$]. The conductivities of today's polyacetylenes approach that of metallic copper. In general, the method of electrolysis affects to a great degree the electrical and physical properties of the polymer [18]. Polypyrrole is the best known of the conductive polymers. It does not require added dopants to be conductive [19]. Its chemical structure is written as

Conductive-nonconductive switching

This polymer "switches" between a black conductive form and a yellow nonconductive form by changing the applied potential on the underlying electrode so as to reduce or oxidize the π-electron network of the polymer. This switching may occur once per second. Polypyrrole films can be very easily obtained electrooxidatively in acetonitrile or water at platinum electrodes and can be peeled off from the electrode as free-standing films for various studies [20]. An account of various applications of polypyrrole films was recently given by Moutet [21].

The general mechanism of anodic electropolymerization of five-membered rings such as those of pyrrole or thiophene is analogous to that of the couplings of aromatic compounds [22]:

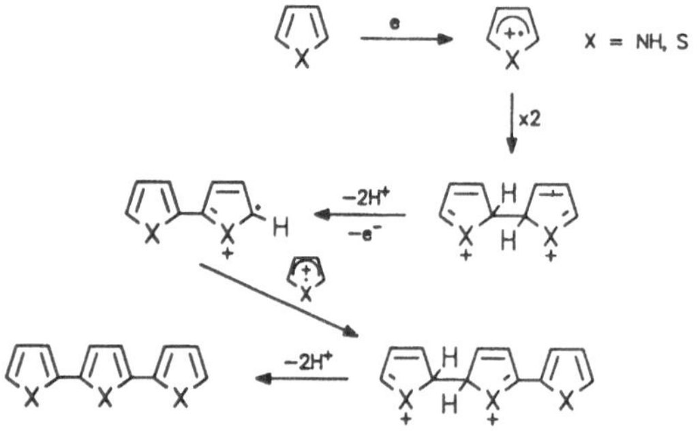

Polyaniline films

Polymers of pyrrole, aniline, phenols, thiophene, furan and selenophene have been studied and continue to be studied very actively, especially since these polymers can be very conveniently prepared by simple electrochemical methods [23]. Polyaniline films are prepared by anodic oxidation of the monomer in aqueous hydrochloric acid solution [24].

Photo-sensitivity

Electrochemical and photoelectrochemical synthesis of polypyrrole films containing TiO_2 particles exhibit photosensitivity. The electrosynthesis was carried out in ethanol under illumination of TiO_2 and in aqueous TiO_2 suspensions containing small amounts of electrolytes [25].

Methylthio-phene polymer

Electropolymerization can often be easily accomplished in undivided cells. Electropolymerization of 3-methylthiophene monomer in nitrobenzene/tetrabutylammonium hexafluorophosphate medium affords highly conducting films [26]. The films were $2000-4000\,\text{Å}$ thick and had conductivities of $895-1975\,\text{Scm}^{-1}$. Electrooxidative polymerization of aminopyridine was carried out in an undivided cell fitted with platinum electrodes, in $MECN/TetNBF_4$ solution [27]. The structure of the polymer was given as,

Aminopyridine polymer

Electronically conducting polymers were obtained by electrolysis of a mixture of 3-dodecylthiophene and 3-methylthiophene in nitrobenzene/Et_4NPF_6 medium [28]. An undivided cell fitted with an indium/tin oxide anode and a platinum cathode was used. The electrolysis was carried out at $5\,°C$ and with a current of $2\,mA/cm^2$. Some electropolymerizations may be carried out in undivided cells using only the monomers without any solvent. Murray and coworkers [29] polymerized aniline and pyrrole by electrolyzing the monomers under nitrogen without any dilution of the monomer, except addition of the electrolyte, Bu_4NClO_4 or Et_4NOTS.

Organic polymers formed on electrode surfaces as films or in solution are expected to find applications as "composite" electrode materials or as polymers per se [30].

5.5 Applications of Electrochemical Methodology to the Synthesis of Natural Products

The synthesis of natural products is one of the most challenging tasks in synthetic organic chemistry. An important common characteristic of biosynthesis and electroorganic synthesis is that they involve reactions occurring under *mild conditions*. Various bond formations and particularly synthesis of cyclic intermediates for total synthesis of natural products may often be best achieved by electrochemical methods [1–4].

It is interesting to note that the oldest and first electroorganic method, that is, the Kolbe synthesis, has also been one from among the first methods to be applied to the electrochemical synthesis of natural products. The total syntheses of pentacyclosqualene [5] and α-onocerine [6] were accomplished some years ago starting with the carboxylic acids (**1**) and (**2**). The desired products were synthesized by well-planned combinations of Kolbe dimerization and chemical steps – admittedly, steps that only experts may hope to carry out successfully!

Kolbe synthesis

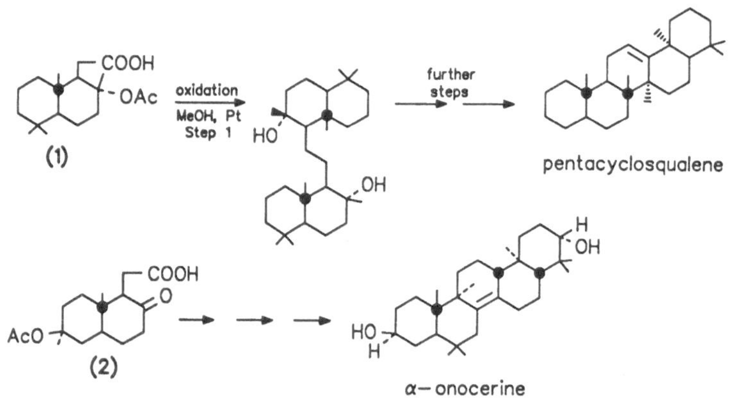

Pheromone synthesis   The Kolbe oxidation has been applied to the synthesis of some phero-mones. Mascalure, the pheromone of the housefly, has been made by coelectrolysis of oleic and heptanoic acids [2,7] in 80% yield.

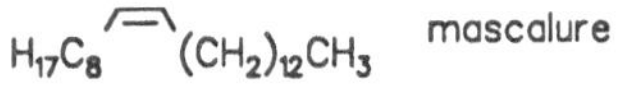

mascalure

Klunenburg and Schäfer synthesized dispalure (**6**), the pheromone of the gypsy moth by double Kolbe electrolysis of Z-4-octanediacid half-ester (**4**) which was made from the cyclooctadiene (**3**). The synthesis is summarized in Scheme 5.3

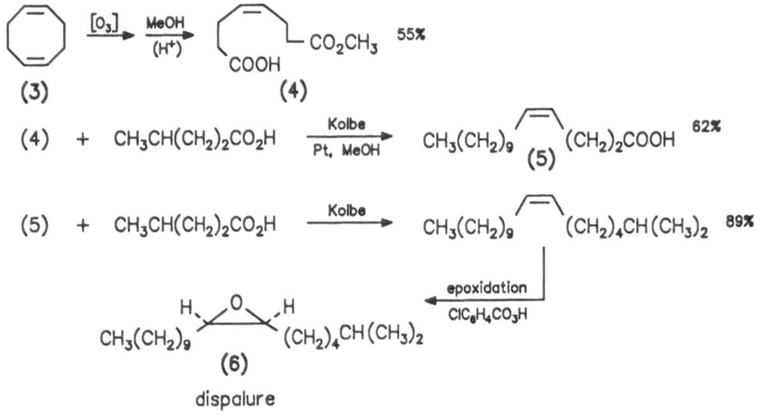

Scheme 5.3

The Kolbe electrolysis was conducted in methanol containing the mixed acids and some KOH to neutralize about 1% of the acids, in a cell fitted with platinum electrodes and under a current density of $240 \, \text{mA/cm}^2$. Distillation of solvent followed by epoxidation of the electrolysis product gave the dispalure product. The pheromone of the eastern spruce budworm E-11-tetradecenal (**3**) and also the tetradecenol intermediate (**4**) shown below were prepared by combining Kolbe-cross couplings with known chemical methods [9], Scheme 5.4

Kolbe cross couplings

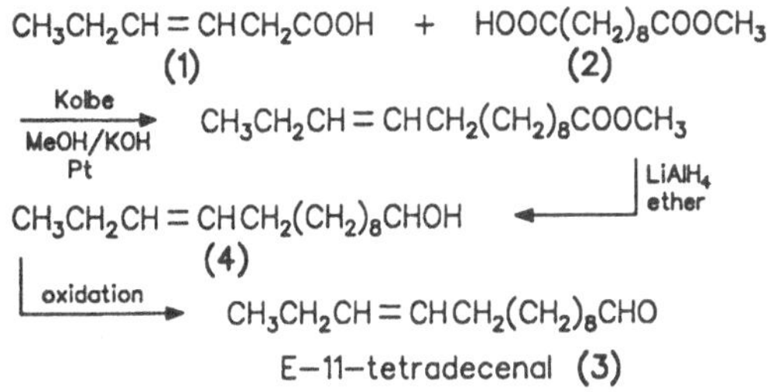
CH3CH2CH=CHCH2COOH + HOOC(CH2)8COOCH3
(1)
(2)
Kolbe
MeOH/KOH
Pt
CH3CH2CH=CHCH2(CH2)8COOCH3
LiAlH4
ether
CH3CH2CH=CHCH2(CH2)8CHOH
(4)
oxidation
CH3CH2CH=CHCH2(CH2)8CHO
E-11-tetradecenal (3)

Scheme 5.4

The Kolbe reaction was carried out in an undivided cell with two platinum ($7\,cm^2$) electrodes closely positioned to minimize the IR drop of the methanolic solution containing the substrates 1.4 g of (**1**) and 0.5 g of (**2**) and 0.8 g of KOH. The cell voltage was adjusted so as to allow a current of $0.2\,A/cm^2$ to pass through the solution. The electrolyzed solution was diluted with water neutralized and extracted with ether (57% yield). After purification by elution chromatography the product was chemically transformed by reduction and oxidation to the desired pheromone *E*-11-tetradecenal. The electrochemical oxidation of a series of isoquinoline alkaloids was studied in depth by Bobbitt et al. in relation to biosynthetic oxidations involving C—C and C—O—C dimer formations [9a]. It was concluded that the anodic oxidation was more successful in regard to specificity than the catalytic oxidation method over platinum catalyst. The oxidation of corypalline (**1**) was studied in 0.1 M HCl and 0.1 M $Na_2B_4O_7$ solutions under controlled potential at 5–10 °C and under a nitrogen atmosphere. Anodes of Pt, Hg, and graphite felt were used. The reaction and product yields are shown in Scheme 5.5

Isoquinoline alkaloids Biosynthetic oxidations

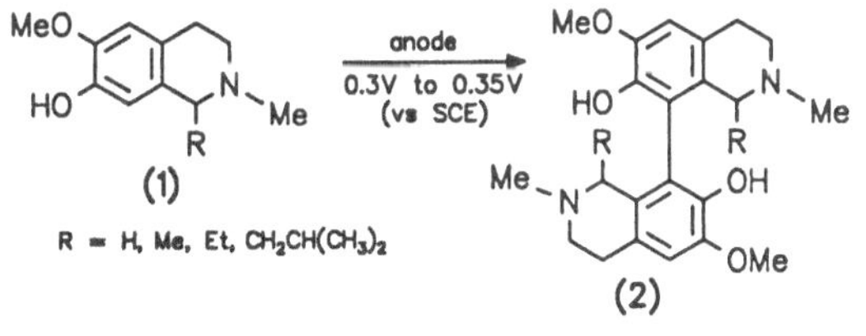
MeO
HO
N
Me
R
(1)
R = H, Me, Et, CH2CH(CH3)2
anode
0.3V to 0.35V
(vs SCE)
MeO
HO
N
Me
R
R
Me
N
OH
OMe
(2)

Medium	Electrode, %yield of (2)		
$H_2O_2/Na_2B_4O_7$	Pt, 50%	Hg, 70%	Graphite, 85%
H_2O/HCl	10–20%	—	35–40%

Scheme 5.5

5.6 Electroreductive Cyclizations for Natural Product Synthesis (Little-Baizer ERC Reaction)

A potentially most useful electroreductive cyclization reaction (ERC) has emerged from the recent studies of professors R. D. Little and the late M. M. Baizer at the University of California [10]. This novel electrochemical reaction can be of great synthetic value for the synthesis of natural products. It provides a new and efficient route [11] "for the construction of ring systems such as the five-membered ring subunit of sterpurene and the five- and six-membered rings found in the bicyclo 3,2,1 subunit of quadrone."

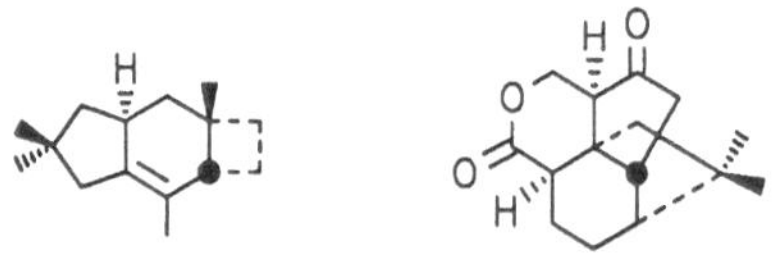

Construction of ring systems

1—sterpurene   quadrone

The general ERC reaction is synoptically expressed as follows:

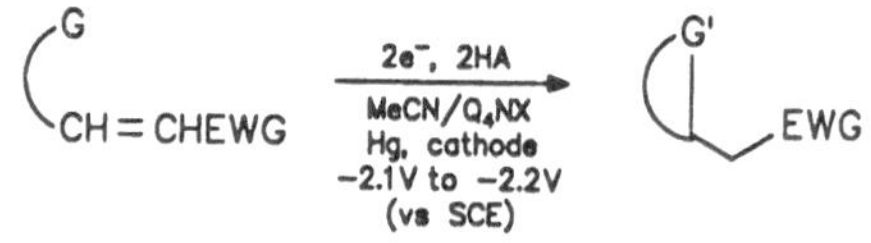

G = CH₂OMe, COR, C(=NR)R'
G' = CH₃, CH₂OH, CHOHR, CH(NHR)R'
EWG = electron withdrawing group, eg. CO₂CH₃
HA = proton source

The electroreduction proceeds by an one-electron transfer from the mercury cathode to the unsaturated unit, CH=CHEWG, of the substrate molecule, producing the anion radical. The anion radical undergoes closure onto G and protonates to give a neutral intermediate radical which electronates and protonates to afford finally the cyclized product. Suitable proton sources were found to be the carbon acids dimethyl and diethyl malonates. Thus reduction of the keto ester (1) to the cyclized product (2) occurs by an ECE process, Scheme 5.6

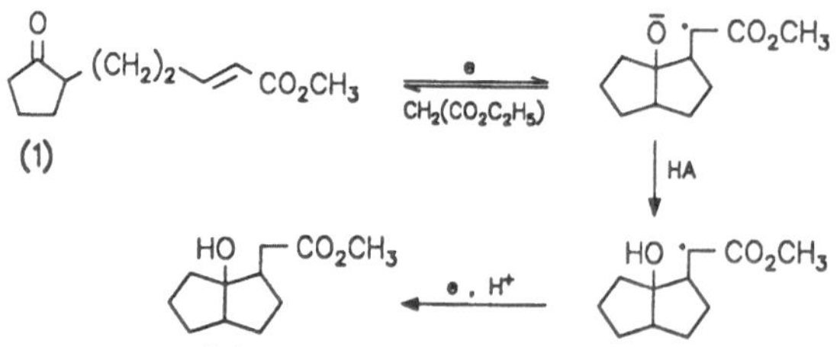

Scheme 5.6

The ERC reaction is carried out in an H-type cell with mercury pool as cathode at −2.2 V (vs SCE) and a platinum foil as anode. The solvent for both anolyte and catholyte is dry degassed acetonitrile containing Et$_4$NOTs as electrolyte. The catholyte contains the substrate and the proton source in the molar ratio 4:1, respectively. The electrolyzed solution (the catholyte) is treated with acetic acid and the products after workup with water are isolated by chromatography (silica gel, 15–40% ether in skellysolve F). Table 5.1 includes typical substrates and products obtained by the ERC method. It is seen that the ERC method gives products with varying stereoselectivities with the OH and the ester group, *trans* to each other about the new σ-bond. The *trans* forms predominate when lower temperatures are used in the electrolysis. The results in Table 5.1 refer to electrolysis in 10% aqueous acetonitrile with Et$_4$NOTs as electrolyte. The electrolysis can be conducted in CH$_3$CN/H$_2$O media with careful addition (dropwise of acetic acid to prevent pH changes higher than the starting pH value, to prevent saponification).

Table 5.1 Typical substrates and products concerning the ERC method. (Only *trans* isomers are shown)
J. Org. Chem. 53:2287, 1988. Copyright of the American Chemical Society, 1988

substrate	product	*trans : cis*	%yield
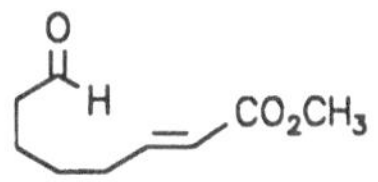	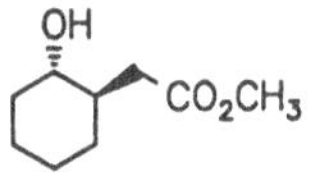	1.8 : 1	72
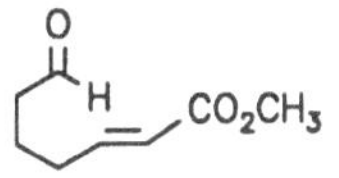	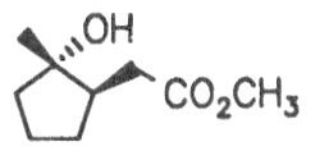	1.4 : 1	70
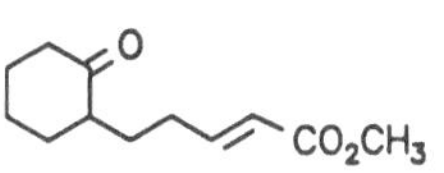	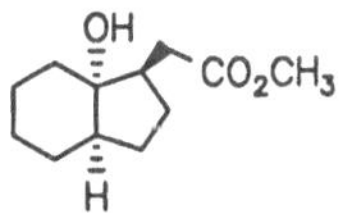	12.5 : 1	74
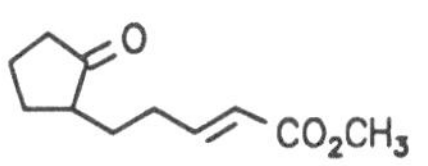	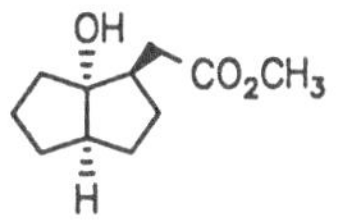	11.4 : 1	79

In the total synthesis of 1-sterpurene (**6**), Scheme 5.7, (which is a fungus metabolite agent of the silver leaf disease) the ERC method was used [12] for the stereoselective cyclization via conversion of the bisenoate (**1**) to the trans diester (**2**). The electrolysis was performed in MeCN/0.2 M Bu$_4$NBr solution containing CeCl$_3$ (1.3 equivalents), 2 equivalents of CH$_2$(CO$_2$Et)$_2$ and 0.033 M of the bisenoate. The cathode was a mercury pool at a

Synthesis of 1-sterpurene

potential of $-2.3\,\mathrm{V}$ (vs SCE). It was assumed that a complex of $CeCl_3$ with the substrate was formed which favored the desired stereoselectivity (14.8:1 *trans*:*cis*) for the desired diester intermediate products (2).

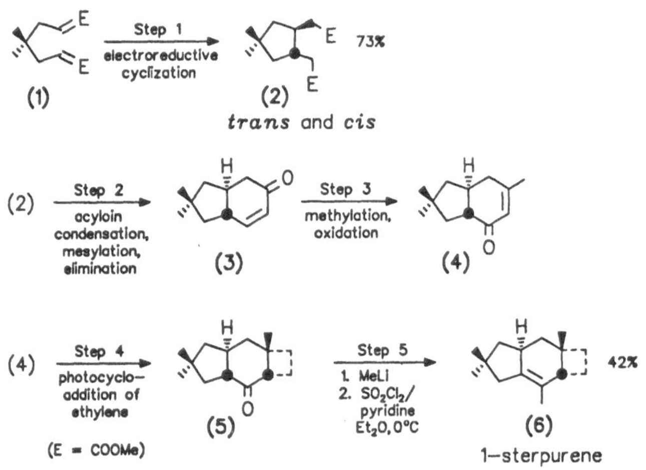

Scheme 5.7

Quadrone   The ERC method was used for the total synthesis of quadrone [13]. This synthesis is very complex to be described here. The reaction starts with an electroreductive, stereoselective cyclization step with the unsaturated ester (1) via which the first of the four rings of quadrone are produced. By a sequence of several, very sophisticated, chemical steps the required structure was achieved which led to the quadrone product, Scheme 5.8

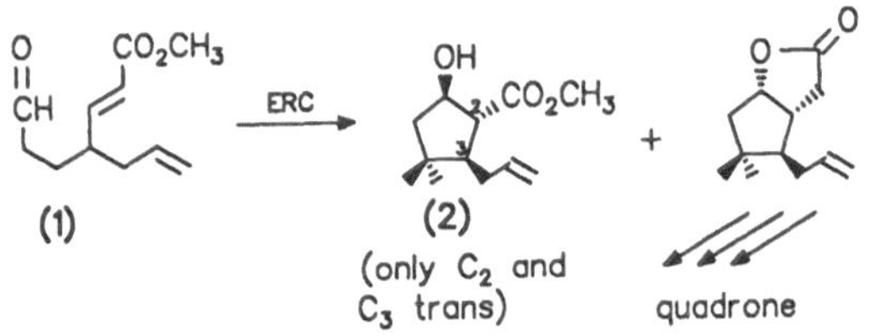

Scheme 5.8

5.7 Electrogenerative Systems

According to Langer [1] "electrogenerative processes are those in which favorable thermodynamics are exploited together with requisite kinetic and mechanistic factors to produce a desired chemical while generating by-product electricity." This very interesting electrochemically methodology has not been fully appreciated, although it has considerable practical potential [2].

The most fundamental requirement for the establishment of an electro-generative system is that the cell reaction be exothermic ($\Delta G° \ll 0$). The thermodynamic cell efficiency defined as $\Delta G°/\Delta H°$ increases as the entropy term $T\Delta S°$ decreases, since $\Delta G° = \Delta H° - T\Delta S°$.

The generated cell voltage, Eg, theoretically would be given as

$$E_g = E_{cell} - IR$$

where I and R denote current and resistance and $E°$ is the standard thermodynamic potential. The theoretical voltage output of an electro-generative cell can be estimated from the thermodynamic relation of the energy $\Delta G°$ of the cell reaction with the cell potential $E°$

$$\Delta G° = -nFE°$$

For example, in the air/ethanol electrogenerative cell in which ethanol is oxidized to acetaldehyde the thermodynamic parameters are [1]:

$$\Delta H° = -52 \text{ kcal/mol}$$
$$\Delta G° = -48 \text{ kcal/mol}$$
$$E° = 1.05 \text{ V}$$

The measured cell potential exhibited by this cell was found to be 0.64 V. Sometimes values of experimental cell potentials happen to be quite close to theoretical values even when the processes are irreversible. The electro-generative hydrogenations of benzene and ethylene are two such examples [1]:

$$C_6H_6 + 3H_2 \longrightarrow C_6H_{12} \quad \begin{array}{l} E° = 0.169 \text{ V} \\ E_{exp} = 0.142 \text{ V} \end{array}$$

$$C_2H_4 + H_2 \longrightarrow C_2H_6 \quad \begin{array}{l} E° = 0.522 \text{ V} \\ E_{exp} = 0.502 \text{ V} \end{array}$$

A simple, typical schematic construction of an electrogenerative cell is given below with an olefin, S, and Cl_2 as reactants, Figure 5.2.

Electrodes for electrogenerative cells

Electrogenerative cells are fitted with porous catalytic electrodes in order to minimize voltage losses caused by increasing overpotentials at higher current densities (overpotentials increase with current density). Catalytic electrodes can be made by bonding metal powders (e.g. platinum black) with teflon or polyethylene [3]. Recently, Stafford [4] reported the electro-generative oxidation of propylene on a pilot-plant scale in a cell with a gas-diffusion palladium electrode. The major products were acrolein, acrylic acid and CO_2. Langer [5] et al. developed an electrogenerative process whereby selective oxidation of alcohols to esters or carboxylic acids takes place. The electrocatalysts were transition metal surfaces modified by treatment with SO_2. Chemisorbed sulfur species covered the

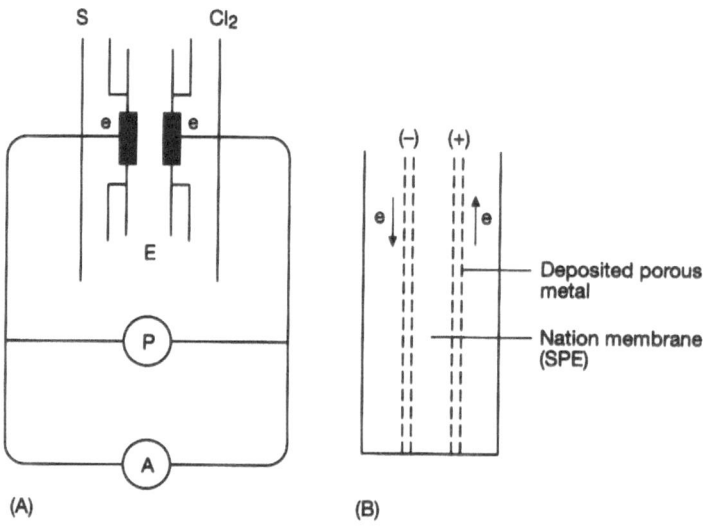

Fig. 5.2A,B. Schematic representation of electrogenerative cells. (Al Type A. e, electrodes E, electrolyte flowing between electrodes. P and A, potentiometer and ammeter. B Type B. SPE-type electrogenerative cell

electrocatalyst particle. Ethanol was thus oxidized to ethyl acetate with 43% conversion and $3.4\,A/cm^2$ byproduct electricity.

Zirconia electrochemical reactor Michaels and Vayenas [6] developed an electrogenerative process for the oxidative dehydrogenation of ethylbenzene to produce styrene, using a zirconia electrochemical reactor. The process was proposed as an alternative to the endothermic dehydrogenation of ethylbenzene (580 °C) and its equilibrium limited conversion (55–60%). The electrogenerative process allows quantitative conversion in a single pass with cogeneration of electricity. Also, using this type of zirconia solid electrolyte reactor, methanol was converted to formaldehyde with 85–92% selectivity [7]. The electrogenerative cell was described as follows:

$$CH_3OH,\ Ag/ZrO_2 \qquad 8\%\ Y_2O_3/Ag,\ air$$
$$(820 - 970\,K)$$

$$CH_3OH\ +\ 1/2O_2\ \longrightarrow\ H_2CO\ +\ H_2O$$

Pietsch and Langer studied the electrogenerative preparation of halohydrins and halogenated alkenes and indicated the possible integration of an ethylene-chlorine cell with a chlor-alkali facility [8]. The overall electrogenerative reaction was given as follows:

$$C_2H_4\ +\ 0.4\,NaCl\ +\ 1.2\,H_2O\ \longrightarrow\ 0.2\,C_2H_4Cl\ +\ 0.8\,C_2H_4O\ +\ 0.4\,NaOH\ +\ H_2$$

Recently Otsuka and coworkers [9] reported an electrogenerative system for the selective synthesis of acetaldehyde. The cell is described as:

$$C_2H_4 \; + \; H_2O, \;\; Pd/silica \; wool \; disc \; holding \; H_3PO_4 \; H_2O/Pd, \; O_2$$
$$Pt$$

In this fuel-type cell C_2H_4 is oxidized to acetaldehyde with greater than 97% selectivity. The pertinent electrode reactions are:

$$anode, \;\; C_2H_4 \; + \; H_2O \;\; \longrightarrow \;\; CH_3CHO \; + \; 2H^+ \; + \; 2e^-$$

$$cathode, \;\; 1/2O_2 \; + \; 2H^+ \; + \; 2e \;\; \longrightarrow \;\; H_2O$$

This system avoids the use of corrosive aqueous solutions ($PdCl_2$, $CuCl_2$, HCl) needed in the chemical method, and facilitates greatly the separation of reactants and products.

Electrogenerative systems using biomass-derived materials

Electrogenerative systems using biomass-derived materials are studied for the purpose of exploiting biomass wastes and perhaps producing useful products and coproduct electricity [10]. Oxidation of carbohydrates and alcohols of biomass origin take place spontaneously in aqueous NaOH solution at metal oxide electrodes (Ni, Ag). The electrogenerative cell is acidobasic. The anolyte is alkaline and the catholyte is aqueous sulfuric or hydrochloric acid in which a platinum or graphite is used as cathode. The catholyte contains oxidizing ions such as Fe^{3+}, NO_3^+, Ce^{4+}, etc. With glucose as the biomass material in the basic anolyte compartment cell voltages as high as 1.3 V and currents of about 50 mA/cm^2 are attained at 40°C. Gluconates and saccharates appear to be among the oxidation products. In the absence of oxidizing ions in the catholyte hydrogen is copiously evolved at the platinum cathode at 40–60°C. The electrogenerative cell is shown schematically in Figure 5.3

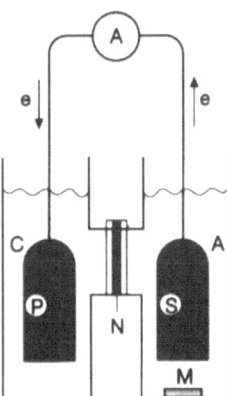

Fig. 5.3. Acidobasic electrogenerative cell utilizing biomass. *P*, Platinum cathode *S*, Silver anode, *N* Nation cation exchange membrane A, anolyte NaOH/H$_2$O/glucose C, cutholyte HCl/H$_2$O, for molecular hydrogen evolution

In this type of cell (silver electrode) propargyl alcohol in spontaneously oxidized to propiolic acid (50% yield) with byproduct electricity.

5.8 Electrogenerated Bases

Proton transfer reagents

Certain organic compounds may be converted electrochemically into strong bases in suitable organic solutions. These electrogenerated bases, EGB, can be used in situ as proton transfer reagents analogous to electron transfer mediators in indirect electrochemical reactions [1]. An EGB abstracts protons from carbon acids and thus may initiate a sequence of reactions. If an EGB is regenerated in situ it can act electrocatalytically. Substances capable of affording EGBs are called probases, PB. Radical anions and anions may act as EGBs and also as nucleophiles and as reducing agents depending on their structure. Steric hindrance at the basic

Probases center makes an EGB more effective as a base than as a nucleophile. In the presence of suitable electrophiles chemical reactions occur, as in the case with CO_2 whereby carboxylations are effected according to an overall scheme,

$$PB + e \longrightarrow EGB^- \quad \text{cathode}$$
$$EGB^- + RH \longrightarrow EGBH + R^-$$
$$R^- + CO_2 \longrightarrow RCO_2^-$$
$$EGBH - e \longrightarrow PB + H^+ \quad \text{anode}$$

Upon reduction of a PB the produced EGB may be in the form of an anion radical, an anion, or a dianion. Phenols, amides, and carboxylic acids afford EGBs by one-electron reduction:

$$RXH \overset{\bullet}{\longrightarrow} RX^- + 1/2H_2$$
$$X = N, O, C$$

Efficient probases

Some typical efficient probases are pyrrolidones, N-methylacetamide, L-tryptophan, methionine, azobenzenes, Ph_3CH and CCl_4 (giving CCl_3^-) and oxygen giving the superoxide ion. Utley et al. recently prepared organic bases from the PBs quinomethides (**1**) (**2**) and derivatives of fluoren-9-ylidenemethane (**3**) and demonstrated their use in Wittig-type reactions [2,3].

R = H   X = CN
R = Br   X = CN
R = H   X = CO₂Et

Cyclic voltammetry at a mercury electrode indicates that these substances are reversibly reduced in DMF solutions at the low potentials of -0.1 to -0.2 and -0.7 to -0.9 (vs Ag/AgI) giving anion radicals and dianions which are able to abstract protons from phosphonium salts added in the electrolysis medium. These salts are reduced at -1.4 to -1.45 V. Preparative electrolyses (Wittig-reaction route) were carried out in divided cells with a mercury pool cathode in DMF solvent containing the PB, the phosphonium salt and the carbonyl compound. Suitable electrolytes are Bu_4NBF_4, $LiClO_4$ or $MgClO_4$. The preparation of vitamin A acetate [3], Scheme 5.9, was effected (30–40% yields with $LiClO_4$ as electrolyte; no yield was possible with $Bu_4N^+BF_4$ in this case).

Wittig reaction route

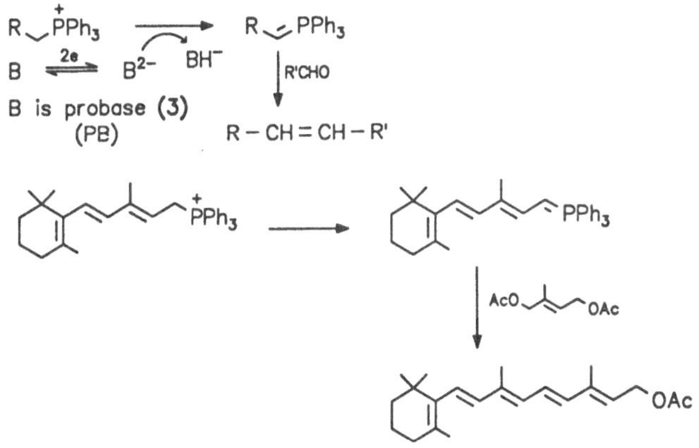

Scheme 5.9

Regioselective alkylation of the cyclic ketone shown below was effected [4] by electrolysis in the presence of the PB, Ph_3CH:

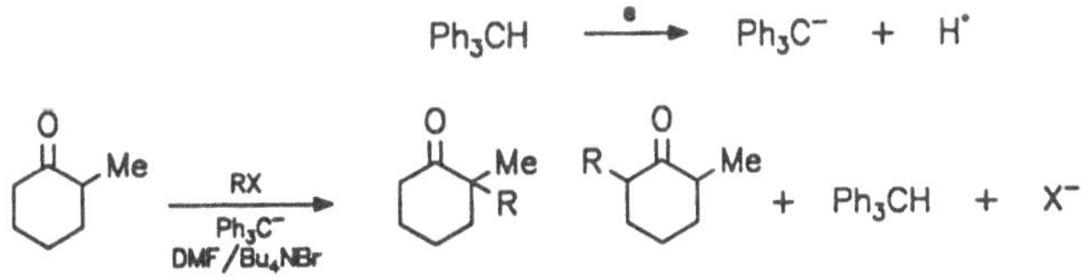

An interesting isomerization was reported by Kimura et al. involving an aromatic enol ether and an EGB as shown below [5]:

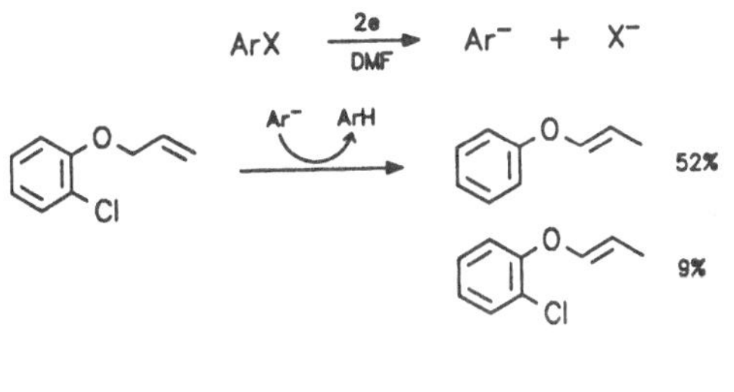

An efficient acylation reaction most probably promoted by an electrogenerated acid (EGA) catalyst was recently reported by Gatti [6]. Electrogenerated acids, like EGBs, may act as catalysts [7], in reactions involving electron-rich substrates as in the following examples [6]:

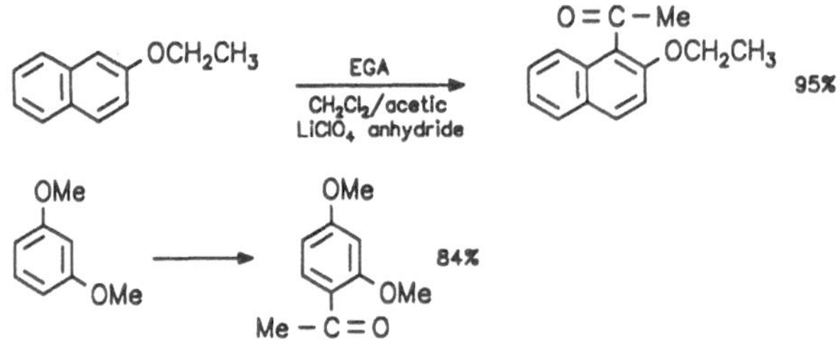

5.9 Electroreduction of Enones (α,β-Unsaturated Ketones)

The enone is a very useful functional group in organic synthesis, in general. Its electrochemistry has been widely studied and was reviewed by Little and Baizer [1]. Only a few typical reactions will be included here.

Hydrodimerizations

Reductive hydrodimerizations of enones take place at mercury electrodes in aqueous and in organic solutions. Intermolecular and intramolecular hydrodimerizations are possible [2,3]:

$$PhCH{=}CHCOCH_3 \; + \; H_2C{=}CHCN \text{ (excess)} \xrightarrow[\text{AcOH, } -1.4\,V\,(SCE)]{Et_4NOTs}$$

$$CH_3COCH_2CHPhCHPhCH_2COCH_3 \text{ (32\%)} \; + \; CH_3COCH_2CHPhCH_2CH_2CN \text{ (10\%)}$$

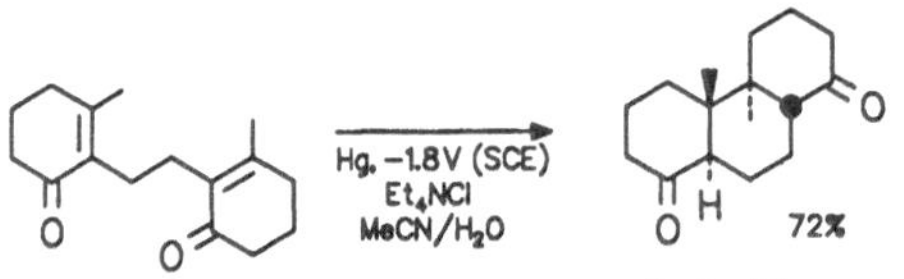

(ref. 3)

The anion radical of an enone can act as an EGB as in the following example [4]:

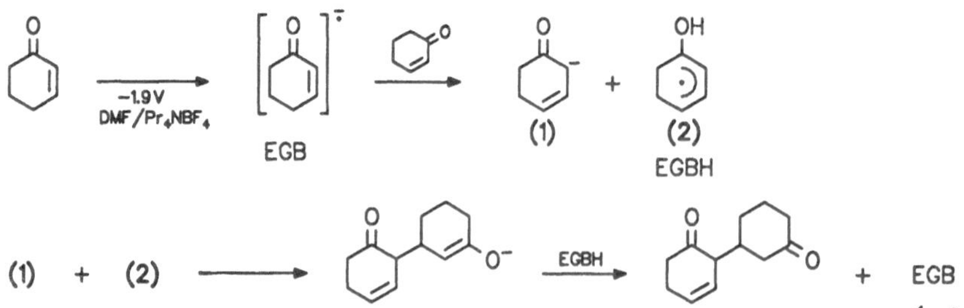

(ref. 4)

Pinacolizations and saturation of C=C bonds are also possible reductive reactions of enones. Electroreductive cyclizations of certain enone esters occur readily in dry DMF/Bu$_4$NBF$_4$ in divided cells with mercury cathodes [5]. With tosylateal a leaving group the following cyclization occurs:

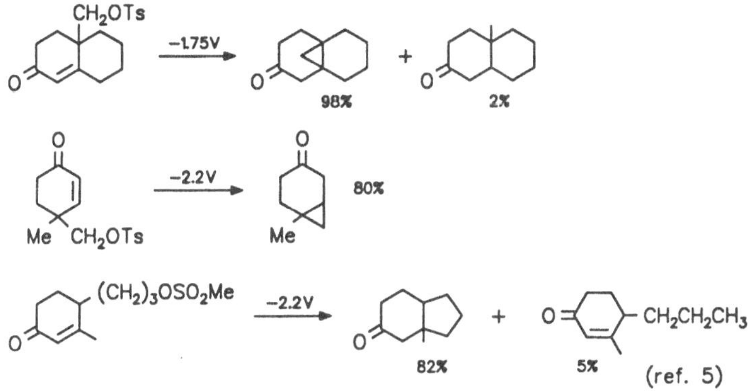

Pinacolizations and saturation of C=C bonds Cyclizations

Electroreduction of nonconjugated enones makes possible the construction of five- and six-membered rings [6].

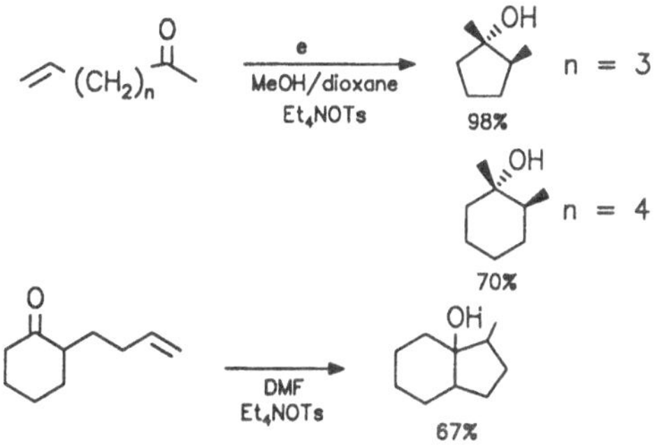

Ring construction

5.10 Oxygenations. The Superoxide Ion

Molecular oxygen is easily electroreduced in both aqueous and organic media. In aqueous acidic media oxygen is reduced to H$_2$O and in basic media to OH$^-$ ion. The oxygen molecule is assumed to accept an electron from the cathode to become a superoxide ion usually written as O$_2^-$. In aqueous media this ion is rapidly decomposed as indicated below:

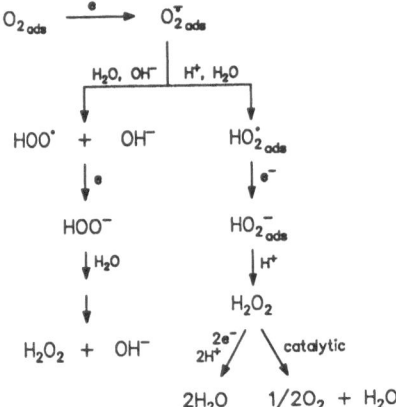

Various functions of the superoxide ion

In dry organic solvents molecular oxygen is reduced at the cathode at $-0.8\,V$ vs SCE to give the $O_2^{\bar{\cdot}}$ ion. In these media the superoxide ion is fairly soluble and quite stable and can function in several useful ways for synthetic purposes [1–5]: as an electron transfer agent, a radical, a nucleophile and an efficient electrogenerated base (EGB). As an EGB superoxide ion is able to abstract protons from carbon acids of low pKa. As a result sequential reactions are possible as indicated below in the presence of both $O_2^{\bar{\cdot}}$ and O_2:

$$O_2 + e \longrightarrow O_2^{\bar{\cdot}}$$
$$RH + O_2^{\bar{\cdot}} \longrightarrow R^- + HO^{\cdot}$$
$$R^- + O_2 \longrightarrow \text{oxygenated product}$$

RH = carbon acid

Carbon acids suitable for such reactions are represented by the general formula, CA, below [5]:

$$H-\overset{\overset{\displaystyle X}{|}}{\underset{\underset{\displaystyle Y}{|}}{C}}-Z$$

CA

X = alkyl, aryl, ArO

Y = H, alkyl, aryl, Z

Z = electron withdrawing group

Oxygenation of N-N-diethylamide

Thus electrolysis in the presence of both $O_2^{\bar{\cdot}}$ and O_2 brings about the oxygenation of esters, nitriles, N-N-dialkylamides, sulfones and nitro compounds. The esters MeCH$_2$COOEt, in Ph$_2$CHCOOMe for example were transformed to the corresponding hydroxy analogs MeCH(OH)COOEt (15%) Ph$_2$C(OH)COOMe (40%). Similarly oxidative decyanations of nitriles such as Ph$_2$CHCN and Ph$_2$CH(CH$_3$)CN Ph$_2$CH(CH$_3$)CN gave the ketonic products Ph$_2$CO (95%) and PhCOCH$_3^-$ (72%). Typically, such electrolytic oxygenations are performed in divided, H-type cells, with mercury pool cathodes ($-1.0\,V$ vs SCE) and platinum anodes in dry DMF/Bu$_4$NBr media containing 5–10 mmol of substrate per 100 ml solvent. A stream of dry oxygen is passed through the catholyte so that both $O_2^{\bar{\cdot}}$ and

O_2 are present for the oxygenation reaction. The products are isolated by distilling off the solvent in vacuo, repeated extractions with ether and washing with minimum amounts of water [5]. The following possible reaction sequence was suggested for the oxygenation of *N-N*-diethylamide:

Ph$_2$CH(CH$_3$)-CN

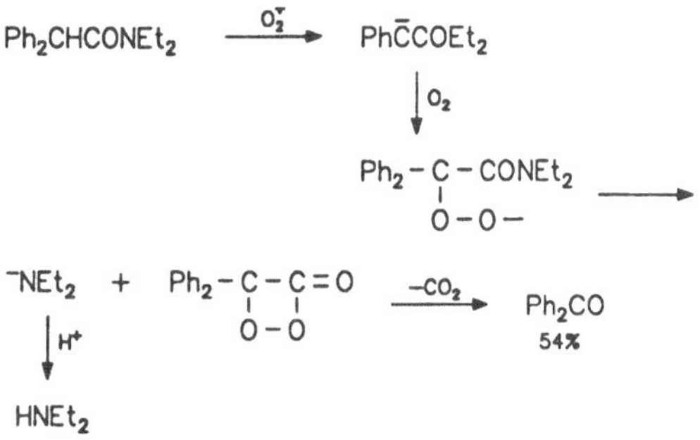

Cleavages via in situ generated superoxide ion occur as exemplified in the oxygenation of chalcones in the presence of oxygen [6], Scheme 5.10:

Oxygenation of chalcones

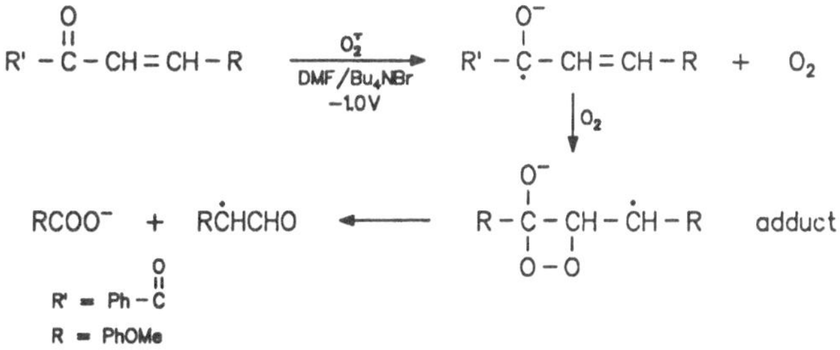

Scheme 5.10

Nitroalkanes are converted to ketones by electrolysis in an H-type cell in MeCN under a stream of oxygen [6a]:

Ketone conversion

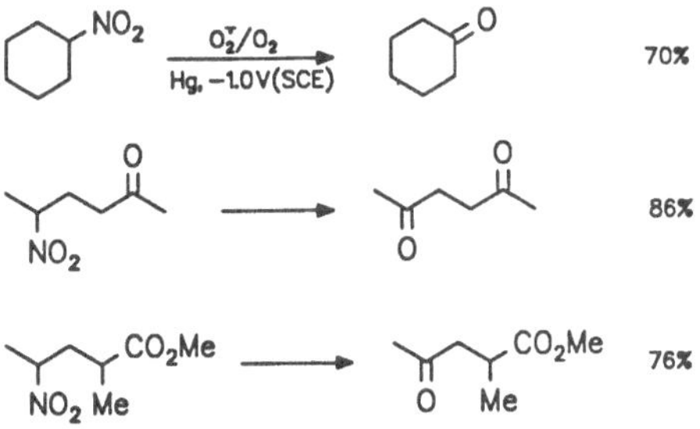

In these reactions electrogenerated superoxide ion abstracts a proton from the nitroalkane to give the nitronate ion which is then oxidized by molecular oxygen to give the ketone. The superoxide ion reacts with organohalides by an S$_N$2 process [6b]:

$$RX + O_2^{\bar{\cdot}} \longrightarrow RO_2^{\cdot} + X^-$$

$$RO_2^{\cdot} + O_2^{\bar{\cdot}} \longrightarrow RO_2^- + O_2$$

$$RO_2^- + RX \longrightarrow ROOR + X^-$$

Also with esters,

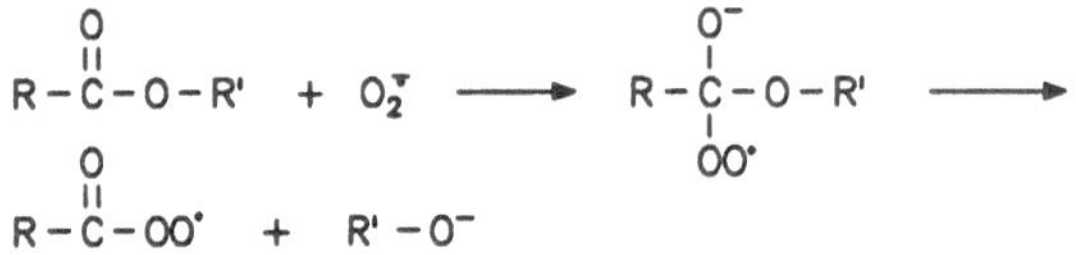

(**Note**: Some organic carbanions or dicarbanions may absorb oxygen. Reactions carried out in the presence of oxygen may yield oxygenated side products.)

Tosylhydrazones are fragmented by superoxide ion to give alkenes. Secondary alcohols react with superoxide to give ketones [7]. Nicotine is converted to cotinine [8],

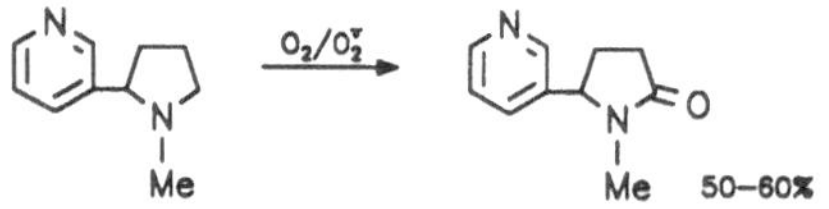

Diphenylmethane, cyclohexane and *ortho*- and *para*-cresols give benzophenone, cyclohexanone and benzylic acids, respectively, in 55–68% yields [9]. Catalytically induced oxygenations are effected in the presence of cuprous ions and air (oxygen) with benzene as substrate, apparently via the Cu(I)O$_2$ adduct [9a]:

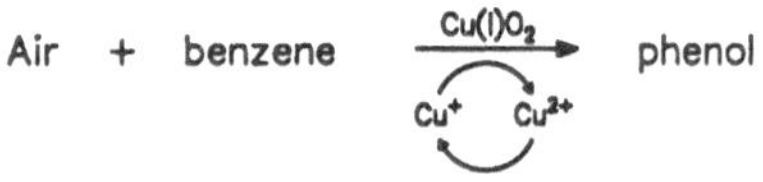

Epoxidation of enones   In situ epoxidation of enones in the presence of oxygen and a carbon acid gave high yields of products [10]:

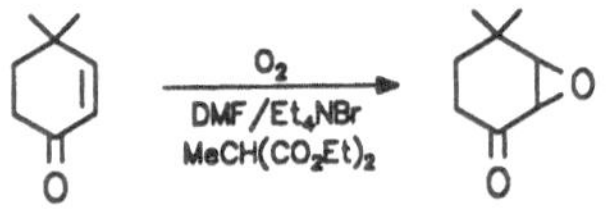

Dehydrogenation of dihydrophenzine   Rapid dehydrogenation of dihydrophenazine occurs by a concerted transfer of an N-H proton and a N-H hydrogen atom [11]:

$$O_2 \xrightarrow{e} O_2^{\bar{\cdot}}$$

$$PhNHNHPh + O_2^{\bar{\cdot}} \xrightarrow[DMF]{k} [PhNNPh]^{\bar{\cdot}} + H_2O_2$$

$$(k > 100\ M^{-1}sec^{-1})$$

An interesting oxygenation process was reported by Yoshida et al. [12], in which oxygen, olefins and 1,3-diketones react to give [2+2+2] cyclo-addition products by anodic electroinitiation. The organic substrate is deelectronated and the resulting radicals react with molecular oxygen by a radical chain mechanism as is depicted below with cyclopentadione and styrene as substrates:

Cycloaddition products by anodic electro-initiation

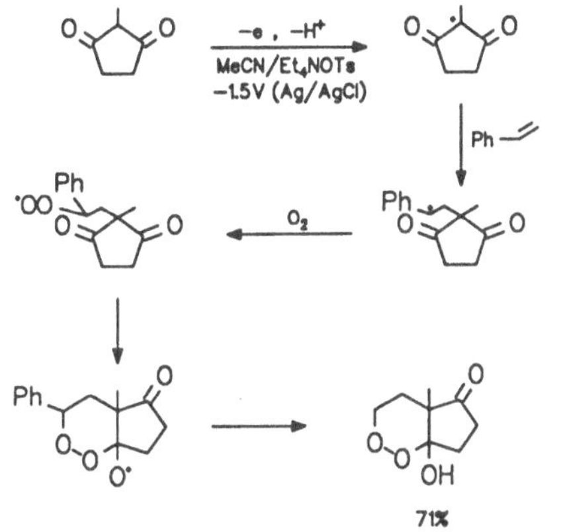

Bard and Mayeda [13] obtained singlet oxygen by electrolytic oxidation of ferrocene and reduction of oxygen and the subsequent or parallel homogeneous electron transfer reaction,

Singlet oxygen

$$\text{Ferrocene}^{+\cdot} \;+\; O_2^{\tau} \longrightarrow \text{Ferrocene}^{\cdot} \;+\; {}^1O_2$$

Singlet Oxygen enters in chemical reactions such as,

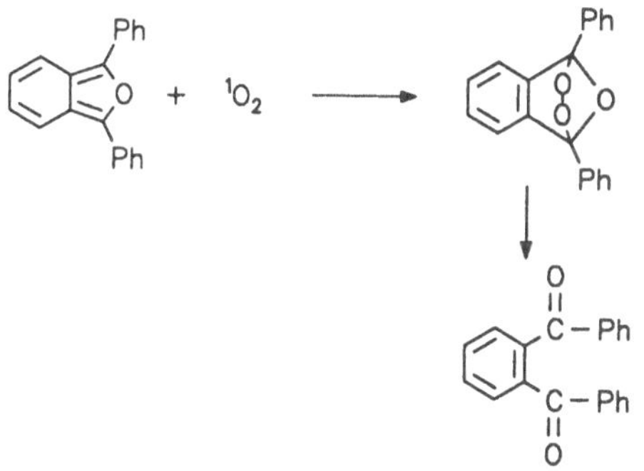

References

Section 5.1 Electrolysis in Two-phase Systems

1. Feess H, Wendt H (1980) J Chem Techn Biotechn 30:297
2. Pletcher D, Camlen PH, Healy KP (1982) J Appl Electrochem 12:693
3. Eberson L, Helgèe B (1978) Acta Chem Scand B32:313
4. Georges J, Desmettre S (1986) Electrochim Acta 31:1519

5. Do JS, Chou T-C (1989) J Appl Electrochem 19:922
6. McIntire GL, Blount NH (1979) J Am Chem Soc 101:7720
7. McIntire GL (1981) J Electrochem Soc 128:427
8. Laurent E, Raunier G, Tomalla M (1982) Nouv J Chim 6:515
9. Girault HH (1987) Electrochim Acta 32:383
10. Forsyth SR, Pletcher D, Healy KP (1987) J Appl Electrochem 17:905
11. Fleischmann M, Tennakoun CLK, Baufield HA, William RJ (1983) J Appl Electrochem 13:593
12. Vettrorazzi N, Fernandez H, Silber JJ, Sereno L (1990) Electrochim Acta 35:1081
13. Bhakte RC (1987) J Macrom Sci-Chem A24:993
14. Knarr RT, Velasco M, Lynn S, Tobias CW (1992) J Electrochem Soc 139:948

Section 5.2 Solid Polymer Electrolyte-Electrolysis

1. Nakgima H et al. (1987) Electrochimica Acta 32:791
2. Rubinstein I, Bard AJ (1981) J Am Chem Soc 103:5007
3. Srinivasann S, Lu WP (1979) J Appl Electrochem 9:269
4. Verbrugge MW, Hills RE (1990) J Electrochem Soc 137:893
5. Yeo RS (1979) J Electrochem Soc 103:533
6. Ogumi Z et al. (1983) Electrochim Acta 28:1687
7. Ogumi Z, Inaba M, Ohashi S, Ushida M, Takehara ZI (1987) Electrochim Acta 33:365
8. Ukihashi H et al. (1986) Synthesis 513
9. Tada H, Kawahara H (1988) J Electrochem Soc 135:1728
10. Kita H, Nagagima H (1986) Electrochim Acta 31:193
11. Fabiunke R et al. (1989) Dechema-monog 112:299–315
12. Enea O (1988) J Electrochem Soc 135:1601
13. Bidan G, Ehui B (1989) J Chem Soc Chem Commun 1568

Section 5.3 Reactions via Electrode Dissolution

1. Silvestri G, Gambino S, Filardo C (1991) Acta Chem Scand 45:487, review
2. Danley DE (1979) Kirk-Othmer encyclopedia of chemistry and technology, vol 8, 3rd edn. Wiley, p 762
3. Silvestri G, Gambino S, Filardo G (1987) Electrochim Acta 32:965
4. Gambino S, Silvestri G (1973) Tetrahedron Lett 32:3125
5. Folest JC, Guibe S, Nedelec JY, Perichon J (1990) J Chem Res 258
6. Lehmkul H, et al. (1975) Justus Liebigs Ann 672
7. Urabe N, Hory F, Kenkyu Hokoko (1973) 2381
8. Pierre G, Limosin D (1988) Makromol Chem 189:1484
9. Silvestri G, Gambino S, Filardo G (1989) J Appl Electrochem 19:946
10. Said FF, Tuck DG (1982) Inorg Chem Acta 59 1; Chada RK, Kumar R, Tuck DG (1987) Can J Chem 65:1336
11. Sausa A et al. (1990) J Chem Soc Dalton Trans 2101
12. Chaussard J, Folest JC, Nedelec JY, Perichon J, Sibille S, Troupel M (1990) Synthesis 5 369

Section 5.4 Electropolymerizations. Conductive Polymers

1. Roberts R et al. (1982) Industrial applications of electroorganic synthesis. Ann Arbor Science Publishers, Ann Arbor
2. Kanazawa KK, Diaz AF, Street GB (1980) J Electrochem Soc 127:83C
3. Rubinstein I (1983) J Polymer Sci 21 3035; Yamamoto K et al. (1988) Polymer Bull 19:533

4. Soubiran P, Aeiych S, Aaron JJ, Delamar M, Lacaze PC (1988) J Electrochem Soc 251:89
5. Leclerc M, Gway J, DaoLe H (1988) J Electroanal Chem 251:21
5a. Pelter A, Maud JM, Jenkins I, Sadeka C, Coles G (1989) Tetrahedron Lett 30:3461
6. Zecchin S, Tomat R, Sciavon G, Cotti G (1988) Synth Metals 25:393
7. Yamamoto K, Asada T, Nishide H, Tsuchida E (1990) Bull Chem Soc (Japan) 63:1216
8. Tashiro K, Matsumura K, Kobayashi M (1990) J Phys Chem 94 3197
9. Little WA (1964) Phys Rev A134:1416
10. Chiang CK et al. (1978) J Chem Phys 69:5098
11. Seymour RB (ed) (1981) Conductive polymers, Plenum, New York; (1987) Polymer 28:533
12. Metzer RM et al. (1987) Synthetic Metals 18:797
13. Reynolds JR (1988) Chemtech (July) 440
14. Makaruk L, Pron A (1990) Intern Polym Sci Techn 17 T81
15. Genies EM, Tsintavis C (1988) J Electroanal Chem 200 127
16. Frommer JF, Chance RR (1990) In: Grayson M, Kroschwitz J (eds) Encyclopedia of polymer science and engineering, vol 5, 2nd edn. Wiley, New York, p 461
17. Ferraris JP, Skiles GD (1987) Polymer 28:582
18. Marcos ML et al. (1987) Electrochim Acta 32:1453
19. Andrieux CP et al. (1990) J Am Chem Soc 112:2439
20. Kyriacou D, unpublished work
21. Deronjier A, Moutet J-C (1989) Acc Chem Res 22:249
22. Ponkali J (1992) Chem Rev 92:711–738
23. Diaz AF, Kanazawa KK, Gardini GP (1979) J Chem Soc Chem Commun 635
24. Gottsfield S, Redondo A (1987) J Electrochem Soc 134:271
25. Kawan K, Mihara N, Kawabata S, Yoneyama H (1990) J Electrochem Soc 137:1793
26. Poncali J, Yassar A, Garnier F (1988) J Chem Soc Chem Commun 581
27. Hayat V et al. (1986) Polymer Commun 27:362
28. Sato M, Shimitzu T, Yamauchi A (1990) Makromol Chem 191:313
29. McCarley RL, Morita M, Wilbourn KO, Murray RW (1988) J Electroanal Chem 245:321
30. Marks J (1985) Science 227:881

Section 5.5 and 5.6 Applications of Electrochemical Methodology to the Synthesis of Natural Products

1. Nelson RF (1975) In: Weinberg NL (ed) Techniques of electroorganic synthesis. Wiley-Interscience, New York, p 269
2. Scheafer HJ (1978) Electrosynthesis of natural products. IUPAC Int Symposium, British Library London
3. Weedon BCL (1963) Adv Org Chem 1:1
4. Little RD, Fox DP, Moens L, Wolin R, Baizer MM (1987) In: Torii S (ed) Recent advances in electroorganic synthesis. Kadansha, Tokyo, Japan
5. Corey EJ, Sauers RR (1959) J Am Chem Soc 81:1739
6. Storc G, Meisels A, Davis JE (1963) J Am Chem Soc 85:3419
7. Grible JW, Sanstead JK (1973) J Chem Soc Chem Commun 735; Seidel W, Schäfer HJ, ref. 2
8. Klünenburg H, Schäfer HJ (1978) Angew Chem, Int Ed Eng 17:47
9. Singh KN, Singh M, Misra RA (1991) Ind J Chem 30B:867
9a. Bobbitt JM, Yagi H, Shibuay S, Stork JT (1971) J Org Chem 36:3006
10. Little RD, Fox DP, van Hijfte L, Denneker R, Sowell G, Wolin RL, Moens L, Baizer MM (1988) J Org Chem 53:2287

11. Fox DP, Little RD, Baizer MM (1985) J Org Chem 50:2202; Nujent ST, Baizer MM, Little RD (1982) Tetrahedron Lett 23:1339; Amptch MA, Little RD, private communication; Bode HE, Sowell CG, Little RD (1990) Tetrahedron Lett 31:2225
12. Moens L, Baizer MM, Little RD (1986) J Org Chem 51:4497
13. Sowell CG, Wolin RL, Little RD (1990) Tetrahedron Lett 31:485

Section 5.7 Electrogenerative Systems

1a. Langer SH, Card JC, Forel MJ (1986) Pure Appl Chem 58:895
1b. Sakellaropoulos GP, Langer SH (1976) J Catalysis 44:25; Langer SH (1975) J Electrochem Soc 122:242
 2. Langer SH, Sakellaropoulos GP (1979) Ind Eng Chem Res Dev 18:507
 3. Langer SH, Collec-Rios JA (1985) Chemtech April, p 226; also ref. 1b
 4. Stafford GP (1987) Electrochim Acta 32:1137
 5. Langer SH, Foral MG, Card JC (1988) U.S. Patent, 4793905
 6. Michael J, Vayenas C (1984) International Society of Electrochemistry Meeting, extended abstracts, Berkeley, California, p 763
 7. Neophytides S, Vayenas CG (1990) J Electrochem Soc 137:839
 8. Pietsch SJ, Langer SH (1979) AIChE Symposium Series, 1857551
 9. Otsuka K, Shimitzu Y, Yamanaka I (1990) J Electrochem Soc 137:2076
10. Kyriacou D (unpublished work)

Section 5.8 Electrogenerated Bases

1. Baizer MM (1984) Tetrahedron, 40:935; Hallcher RD, White DA, Baizer MM (1979) J Electrochem Soc 126:404; Saveant JM, Bin SK (1977) J Org Chem 42:1249; Komori T, et al. (1987) Bull Chem Soc, Japan 60:3365
2. Goulard MDF, Ling-Chung SK, Utley JHP (1987) Tetrahedron Lett 6083
3. Mheta RR, Pardini VL, Utley JHP (1982) J Chem Soc Perkin Trans 1:2921
4. Fuchigani T, Suzuki K, Nonaka T (1990) Electrochim Acta 35:239
5. Kimura M, Mihahara H, Moritani H, Sawaki Y (1990) J Org Chem 55:3897
6. Gatti N (1990) Tetrahedron Lett 31:3933
7. Uneama K (1987) Topics in Curr Chem 142:167

Section 5.9 Electroreduction of Enones (α,β-unsaturated ketones)

1. Little RD, Baizer MM (1989) In: Patai S (ed) The chemistry of enones, Wiley, New York, chap. 14
2. Baizer MM, Andersen JD (1965) J Org Chem 30:3138
3. Mandell L, Daley RF, Day J (1976) J Org Chem 41:4087
4. Baizer MM (1984) Tetrahedron 40:1935
5. Gassman PG, Rasmy OM, Burdock TO, Saito K (1981) J Org Chem 40:5455
6. Shono T, Nishiguchi I, Ohmitzu H, Mitani M (1978) J Am Chem Soc 100:545

Section 5.10 Oxygenations, Superoxide Ion

 1. Allen PM, Hess U, Foote GS, Baizer MM (1982) Synthet Commun 12:1236
 2. Lee-Ruff E (1979) Chem Soc Rev 12:105
 3. Wilkshire J, Sawer DT (1979) Acc Chem Res 12:105
 4. Nanni EJ, Sawer DT (1980) J Am Chem Soc 102, 7591
 5. Little RD et al. (1983) Acta Chem Scand B37:509
 6. Singh M, Singh KN, Mistra RA (1991) Bull Chem Soc Japan 64:2599
6a. Monte WT, Baizer MM, Little RD (1983) J Org Chem 48:803
6b. Bere AL, Berger Y (1966) Bull Soc Chim 2263

7. Singh M, Mistra RA (1989) Synthesis 5:403
8. (1983) Chem Abst 99, p 657 (710556)
9. Zutshi K, Rastogi R, Dixit G (1984) International Society of Electrochemistry, extended abstracts Berkeley, California, p 760
9a. Kinoshita T, Horada J, Ito S, Sasaki K (1983) Angew Chem 22:502
10. Mitchio S, Baizer MM (1983) J Org Chem 48:9931
11. Raez OA, Vuldez CM, Fector J, Sawer DT (1988) J Org Chem 53:2166
12. Yoshida J, Sakaguchi K, Isoe S, Hirotsu K (1987) Tetrahedron Lett 28:667
13. Bard AJ, Mayeda EA (1973) J Am Chem Soc 95:6223

Subject Index

Active Electrode Area   13
adipic acid   49
adiponitrile   111
adsorption at electrode surfaces   15, 16, 26
acetals   29, 50, 95
acetamidations   86
acetamido compounds   44, 50, 51
acenaphthene   97
acetoxylations
   mechanism   83
   intramolecular   95
   of alkyl aromatics   85
   of piperidines   84, 85
   trifluoroacetoxylations   86
acetone   114
acetophenones   112, 113
acetylenes   151, 183, 201
acidobasic cell   211
acylations   136, 155, 214
acylprolines, acylpipecolic acids   67
acylated quinone imine ketals   58, 59
acrylic acid   115, 209
acrolein   209
alcohols
   oxidation products   28, 29
   at oxide covered electrodes   30, 31, 32
   at adatom electrodes   36
   aliphatic acids   120
alkaloids   73, 205
alkanes, oxidation   20, 21
alkanoates   49
   coupling   49
alkenes
   cyclopropanations   125
alkoxylations   92, 93
alkyl halides, iodides   63
amides   57, 58
amines, oxidation   32, 33, 55−57
amino acids   182
amidines   64
aminopyrimidines   69
aminotriazines   69
aminopyrazolines   70
aminophthalimides   73, 131, 132

aniline   134
anion radicals   110, 111, 150, 214
anodic reaction
   fundamental nature   18
   direct, indirect   4
   intramolecular cyclizations   24, 25
anthracene   100, 172, 187
aspartic acid   182
azobenzene, redox catalyst   188

Baizer process   111
basicity   26
benzene   128, 173
benzophenone   155
benzoquinone   173, 181
benzyl alcohol   120, 192
benzyl benzoate   175
benzylic esters   149
benzyne   128
biomass, anodic degradation   41
bismesityl   23
Birch reductions   110
butadiene   90
Butler-Volmer rate equation   10

camphor   45
carbenes
   dichloro compounds   123
   cyclopropanations with   124, 125
   couplings   126
carbon acids   26
carbon disulfide   75
carbon-silicon bonds   158
carbonyls, couplings   113
carboxylic acids
   Kolbe synthesis   47−49, 53
   mixed couplings   49
cathodic reactions   2, 3, 5
   couplings   6, 115
cation radicals   26, 55, 57, 60
catechol   38
cell potential   9
   and free energy   9
cells with consumable anodes   125

cinnolines   134
citronellol   99, 178
composite electrodes   134
consumable anodes   120
coumestan compounds   37, 38
couplings, in general   5
cresols   39
current
    types of   2, 8
    exchange   8, 9, 11
    limiting   13
    efficiency   14
cyanation
    mechanism   79
    typical reactions   27,80 – 82
cyanide ion   79
cyanoporphyrins   67
cyclohexane, oxidation   21
cyclopropanes   94

deelectronations   18
decarboxylations   50
dehalogenations   127
    couplings   128
    of dimethyl maleate with alkyl dihalides   126
dehydrodimerizations of hydrocarbons   19
deoxygenations   149
N,N-dialkylamides   58
dibenzyl ether   64
diglyme as solvent   186
diffusion coefficient   133
dihydrobenzoin   117
dihydroxcoumestan   38
dihydroxypyrimidines   73
diketones   49
dimethylformamide   58, 96
dimethylphenol   39
dimethylpyrrolidinium ion   188
diphenoquinone mediator   41
dissolvable anode methodology   196
    organometallics
doping of electronically conductive polymers
    201
double layer, electrical   14, 212
double mediator systems   171
1,3-dithiane   76
dithiolanes   77
durene   23

electroactivity   6
Electrodes
    (anode, cathode, working, counter)   2
electrode potential   8, 9
    physical meaning of   9

Electrolysis
    basic apparatus   2
    solvents   3
    in two-phase solvent systems   192
    conditions   3, 12, 13
    basic physical-chemical processes   3
electrocarboxylations   151
electrocatalysis   31, 113, 114, 132, 133, 170, 188, 195
electron transfers
    possible reaction barriers   6, 7
    inner, outer, spheres   168
electrohydrogenation   114
electrophydrogenolysis   122, 128, 129
electrogenerative systems
    Langer's definition   208
    cell construction, schematic   210, 211
    using biomass   211
electrogenerated bases (EGB)   212
electroorganic reaction, generalized   4
electropolymerizations   198
electrophiles   25, 151
electrophore   113, 114
electroreduction, direct, C–H bond formations   111
equilibrium potential   8
ethylcyanide   33
ethers, aromatic, alkoxy   60, 61
    bis-enol   62
    mixed   100
esters   63, 121, 175
enamino ketones, preparation   140

Faraday's Laws   14
Fenton's reagent   174
flavonoids   179
five-member ring, polymerizations   202
Friedel-Crafts catalyst, with aluminum anodes   196
films, electrochromic   199
fluorine   158
fluorosulfuric acid, carboxylic acid oxidations in   52

gas diffusion electrodes   151
glucose, oxidation   28, 176
glycerol, oxidation   31
graft-copolymerization   174
graphite intercalation compounds as amalgams   185
Grignard pair   110

half-cell, anodic, cathodic reactions   3
halide ions, mediators   174
halides, general cathodic reduction mechanism   122
halodiphenylphosphines   156
halofunctionalizations   88
heteroaromatics   68
heteroazolium salts   142

heterocyclics 65, 68, 145
heterocyclizations 195
hydrocarbons
    highly strained 124
    main anodic oxidations 10
    in FSO₃H 19, 20
    in SO₂ 22
    in organic solvents 22, 23
hydrazine derivatives 46
hydrogenations, catalytic 112, 113
hydrodimers, formations 112, 116, 117
hydrogen peroxide 193
hydroxy groups
    direct anodic oxidation 28, 29
    in aprotic media 29
    at oxide covered electrodes 30, 31, 33
    oxidation mechanism at NiOOH electrodes 30, 31
hydroquinones 100
hydroxy radicals, adsorbed 36
p-hydroxybenzaldehyde 117
hydroxycoumarin 38
hydroxylation of styrenes 91
α-hydroxylimine 118
5-hydroxytryptamine (5 H-T) 72
hypervalent iodanyl radicals 64

indirect processes 172
    electron transfer mechanisms 168
    paired 172
    epoxidations 175
indandiones 116
indigo 154
indoles 80
immobilized oxidant 177
intramolecular cyclizations 24, 25, 114
iodides, mediators 176
iodinations 90
iodobenzene, oxidation 65
iodonium ion 63, 90
isopropyl alcohol 170

Kolbe synthesis, in SPE cells 47
    decarboxylative possible routes 48
    aldehydes from 50
    Kolbe radicals 49
Ketones
    reduction 112, 113
    cyanoketones, cyclizations 108
    splitting of C-H bonds 118, 158
    α-cleavages 43
    α,α'-dihaloketones, cyclization 124

lactones 45, 115
lawronitrile 155

leucoindigo 154
lithiated ketals 95
Little-Baizer ERC reaction 206
lignin-organosol, anodic oxidation 42
living polymer 194
locations of reacting species in double layer 16
Lontrel 129

magnesium anode 124
maleic acid 182
mesitylene 23
mesylates 148
metal complexes, novel 121, 159, 184
metallic ions, mediators 182
metal/SPE interfaces, electrocatalysts 151
metal carboxylates and alkoxides 120
methane from CO₂ 151
methoxylations
    N-methylamides 58
    methylacrylate 52
    methylcyclohexanones 45
    1-methylpyrrole 81
    methoxylation vs cyanation 82
methylene groups, chlorination 89
monosaccharides, direct oxidation 43
Mn³⁺/Mn⁺ couple, mediator 172

natural product synthesis 203
    Kolbe oxidative steps in 204, 205
    pheromones 204
    1-sterpurene 208
    quadrone 208
    corypalline 205
Nerst's equation 7
nitrate ion, mediator 178
nitration of aromatics 178
nitrite anion 137
nitroalkanes 217
nitroanions 138
nitriles 55, 56, 154, 175
nitronium ion 101
nitrosophenol 182
m-nitrotoluene 182
nucleophiles 25
nucleophilicity 26

octaethylporphyrins 62
n-octanol 195
n-octane, oxidations in MeCN/NO₃⁻ 22
olefins activated, reduction 111
    from dinitro's 137
organic metals 201
organohalides, iodides 63
organoelemental bonds 197

organic cation radicals   178
oxadiazolines   46, 47
oxalate esters   149
oxidations, paired   173
oxide covered electrodes   30, 31, 54, 55
    for carboxylic acids   32, 33
    oxidation of amines, aromatics and steroids   32, 33
oxolanes   49
1,3-oxazolidines   95
oximes   154
oxyen
    reduction   215, 216
    superoxide ion   216
    electrolysis in oxygen-superoxide solutions   217, 218
    singlet oxygen   219
oxyselenylation   177

palladium-carbon catalyst   85
paired processes   172
periodate-iodate catalyst   177
perfluorocyclopentene-butene polymers   133
phase transfer catalysts   192
phenols   36 – 38, 148
    anodic couplings   39, 41
    hydroxylation   91
phenothiazene   68
phenoxonium ion   39
phenylazolidine-3-ones   71
phenylvinylsulfone   147
phosphines   157
    complexes   185
phosphorus compounds   166
photosynthesis   153, 154
potassium iodide catalyst   176
polymer cells   151, 209
polymer films on electrodes   181
    electronic conductors   201, 203
polymerizations   199, 200
polypyrrole   202
pyrazines   139
pyrazoles   70
pyridinations   100
pyrimidines   137
pyrrolidines   72

quinoline   138
quinomethides   212
quinones   38, 42
quinone imineketals   54, 58
    monoketals   65

Raney nickel electrodes   133

redox mediators   167, 171
reversible electrodic reactions   8, 11, 13
Ritter-type electrochemical reactions   86
rose-oxide   100
ruthenates   170
ruthenium electrode   151
rhodium complexes   188

saccarin   171
sacrificial anodes   124, 126
sebacic acid   49
selenium electrode   132
selenophene   202
silanes   197
sililating agents   197
silver cathode   202
solar cells   159
solid polymer electrolyte (SPE)   194, 195
steroids   32, 117
Steckhan's organic mediators   179
styrene polymer   193
sulfide salts   74
sulfenyl, sulfonyl compounds   147
sulfones   147
sulfoxides   147
sulfonium salts   77, 78
sulfur, elemental   145
Swenton's EEC_rC_p mechanism   59, 60

underpotential deposition (UPD)   36
Urethanes   58, 95
unsaturated carboxylic acids   120
unsaturated ketones (enones)   214, 215

vicinal nitro compounds   137
vitamin B_{12} 183, 188
4-vinylpyridine   200
vitamin A acetate   213
voltage distribution in cells   9
voltammetric curves   13
    reaction reversibility   181
    alkane oxidation   21
    propargyl alcohol   34

Wittig-type reactions with EBG   213

Yoshida beads-oxidants   177

zinc electrode   126
zinc chloride   90
    octaethylporphyrins   67
    oxalate   151
p-zylene   82, 85

Springer-Verlag
and the Environment

We at Springer-Verlag firmly believe that an international science publisher has a special obligation to the environment, and our corporate policies consistently reflect this conviction.

We also expect our business partners – paper mills, printers, packaging manufacturers, etc. – to commit themselves to using environmentally friendly materials and production processes.

The paper in this book is made from low- or no-chlorine pulp and is acid free, in conformance with international standards for paper permanency.